Quiver Representations and Quiver Varieties

GRADUATE STUDIES
IN MATHEMATICS **174**

Quiver Representations and Quiver Varieties

Alexander Kirillov Jr.

American Mathematical Society
Providence, Rhode Island

EDITORIAL COMMITTEE

Dan Abramovich
Daniel S. Freed (Chair)
Gigliola Staffilani
Jeff A. Viaclovsky

2010 *Mathematics Subject Classification.* Primary 16G20; Secondary 14C05, 14D21, 16G60, 16G70, 17B10, 17B22, 17B67.

For additional information and updates on this book, visit
www.ams.org/bookpages/gsm-174

Library of Congress Cataloging-in-Publication Data

Names: Kirillov, Alexander A., 1967-
Title: Quiver representations and quiver varieties / Alexander Kirillov, Jr.
Description: Providence, Rhode Island : American Mathematical Society, [2016] | Series: Graduate studies in mathematics ; volume 174 | Includes bibliographical references and index.
Identifiers: LCCN 2016018803 | ISBN 9781470423070 (alk. paper)
Subjects: LCSH: Directed graphs. | Representations of graphs. | Graph theory. | AMS: Associative rings and algebras – Representation theory of rings and algebras – Representations of quivers and partially ordered sets. msc | Algebraic geometry – Cycles and subschemes – Parametrization (Chow and Hilbert schemes). msc | Algebraic geometry – Families, fibrations – Applications of vector bundles and moduli spaces in mathematical physics (twistor theory, instantons, quantum field theory). msc | Associative rings and algebras – Representation theory of rings and algebras – Representation type (finite, tame, wild, etc.). msc | Associative rings and algebras – Representation theory of rings and algebras – Auslander-Reiten sequences (almost split sequences) and Auslander-Reiten quivers. msc | Nonassociative rings and algebras – Lie algebras and Lie superalgebras – Representations, algebraic theory (weights). msc | Nonassociative rings and algebras – Lie algebras and Lie superalgebras – Root systems. msc | Nonassociative rings and algebras – Lie algebras and Lie superalgebras – Kac-Moody (super)algebras; extended affine Lie algebras; toroidal Lie algebras. msc
Classification: LCC QA166.15 .K75 2016 | DDC 512/.46–dc23 LC record available at https://lccn.loc.gov/2016018803

Copying and reprinting. Individual readers of this publication, and nonprofit libraries acting for them, are permitted to make fair use of the material, such as to copy select pages for use in teaching or research. Permission is granted to quote brief passages from this publication in reviews, provided the customary acknowledgment of the source is given.

Republication, systematic copying, or multiple reproduction of any material in this publication is permitted only under license from the American Mathematical Society. Permissions to reuse portions of AMS publication content are handled by Copyright Clearance Center's RightsLink® service. For more information, please visit: http://www.ams.org/rightslink.

Send requests for translation rights and licensed reprints to reprint-permission@ams.org.

Excluded from these provisions is material for which the author holds copyright. In such cases, requests for permission to reuse or reprint material should be addressed directly to the author(s). Copyright ownership is indicated on the copyright page, or on the lower right-hand corner of the first page of each article within proceedings volumes.

© 2016 by the author. All rights reserved.
Printed in the United States of America.

∞ The paper used in this book is acid-free and falls within the guidelines
established to ensure permanence and durability.
Visit the AMS home page at http://www.ams.org/

10 9 8 7 6 5 4 3 2 1   21 20 19 18 17 16

To my children: Vanya, Elena, and Andrew

Contents

Preface	xi

Part 1. Dynkin Quivers

Chapter 1.	Basic Theory	3
§1.1.	Basic definitions	3
§1.2.	Path algebra; simple and indecomposable representations	7
§1.3.	K-group and dimension	11
§1.4.	Projective modules and the standard resolution	11
§1.5.	Euler form	15
§1.6.	Dynkin and Euclidean graphs	16
§1.7.	Root lattice and Weyl group	20
Chapter 2.	Geometry of Orbits	23
§2.1.	Representation space	23
§2.2.	Properties of orbits	24
§2.3.	Closed orbits	26
Chapter 3.	Gabriel's Theorem	31
§3.1.	Quivers of finite type	31
§3.2.	Reflection functors	32
§3.3.	Dynkin quivers	38
§3.4.	Coxeter element	41
§3.5.	Longest element and ordering of positive roots	43

Chapter 4. Hall Algebras		47
§4.1.	Definition of Hall algebra	47
§4.2.	Serre relations and Ringel's theorem	52
§4.3.	PBW basis	56
§4.4.	Hall algebra of constructible functions	61
§4.5.	Finite fields vs. complex numbers	66
Chapter 5. Double Quivers		69
§5.1.	The double quiver	69
§5.2.	Preprojective algebra	70
§5.3.	Varieties $\Lambda(\mathbf{v})$	72
§5.4.	Composition algebra of the double quiver	75

Part 2. Quivers of Infinite Type

Chapter 6. Coxeter Functor and Preprojective Representations		83
§6.1.	Coxeter functor	84
§6.2.	Preprojective and preinjective representations	86
§6.3.	Auslander–Reiten quiver: Combinatorics	88
§6.4.	Auslander–Reiten quiver: Representation theory	92
§6.5.	Preprojective algebra and Auslander–Reiten quiver	96
Chapter 7. Tame and Wild Quivers		103
§7.1.	Tame-wild dichotomy	103
§7.2.	Representations of the cyclic quiver	105
§7.3.	Affine root systems	106
§7.4.	Affine Coxeter element	107
§7.5.	Preprojective, preinjective, and regular representations	112
§7.6.	Category of regular representations	113
§7.7.	Representations of the Kronecker quiver	118
§7.8.	Classification of regular representations	121
§7.9.	Euclidean quivers are tame	126
§7.10.	Non-Euclidean quivers are wild	127
§7.11.	Kac's theorem	129

Chapter 8.	McKay Correspondence and Representations of Euclidean Quivers	133
§8.1.	Finite subgroups in SU(2) and regular polyhedra	133
§8.2.	ADE classification of finite subgroups	135
§8.3.	McKay correspondence	141
§8.4.	Geometric construction of representations of Euclidean quivers	146

Part 3. Quiver Varieties

Chapter 9.	Hamiltonian Reduction and Geometric Invariant Theory	159
§9.1.	Quotient spaces in differential geometry	159
§9.2.	Overview of geometric invariant theory	160
§9.3.	Relative invariants	163
§9.4.	Regular points and resolution of singularities	168
§9.5.	Basic definitions of symplectic geometry	171
§9.6.	Hamiltonian actions and moment map	174
§9.7.	Hamiltonian reduction	177
§9.8.	Symplectic resolution of singularities and Springer resolution	180
§9.9.	Kähler quotients	182
§9.10.	Hyperkähler quotients	186

Chapter 10.	Quiver Varieties	191
§10.1.	GIT quotients for quiver representations	191
§10.2.	GIT moduli spaces for double quivers	195
§10.3.	Framed representations	200
§10.4.	Framed representations of double quivers	204
§10.5.	Stability conditions	206
§10.6.	Quiver varieties as symplectic resolutions	210
§10.7.	Example: Type A quivers and flag varieties	212
§10.8.	Hyperkähler construction of quiver varieties	216
§10.9.	$\mathbb{C}^\times$ action and exceptional fiber	219

Chapter 11.	Jordan Quiver and Hilbert Schemes	225
§11.1.	Hilbert schemes	225
§11.2.	Quiver varieties for the Jordan quiver	227
§11.3.	Moduli space of torsion free sheaves	230
§11.4.	Anti-self-dual connections	235
§11.5.	Instantons on $\mathbb{R}^4$ and ADHM construction	238
Chapter 12.	Kleinian Singularities and Geometric McKay Correspondence	241
§12.1.	Kleinian singularities	241
§12.2.	Resolution of Kleinian singularities via Hilbert schemes	243
§12.3.	Quiver varieties as resolutions of Kleinian singularities	245
§12.4.	Exceptional fiber and geometric McKay correspondence	248
§12.5.	Instantons on ALE spaces	253
Chapter 13.	Geometric Realization of Kac–Moody Lie Algebras	259
§13.1.	Borel–Moore homology	259
§13.2.	Convolution algebras	261
§13.3.	Steinberg varieties	264
§13.4.	Geometric realization of Kac–Moody Lie algebras	266
Appendix A.	Kac–Moody Algebras and Weyl Groups	273
§A.1.	Cartan matrices and root lattices	273
§A.2.	Weight lattice	274
§A.3.	Bilinear form and classification of Cartan matrices	275
§A.4.	Weyl group	276
§A.5.	Kac–Moody algebra	277
§A.6.	Root system	278
§A.7.	Reduced expressions	280
§A.8.	Universal enveloping algebra	281
§A.9.	Representations of Kac–Moody algebras	282
Bibliography		285
Index		293

Preface

This book is an introduction to the theory of quiver representations and quiver varieties. It is based on a course given by the author at Stony Brook University. It begins with basic definitions and ends with Nakajima's work on quiver varieties and the geometric realization of Kac–Moody Lie algebras.

The book aims to be a readable introduction rather than a monograph. Thus, while the first chapters of the book are mostly self-contained, in the second half of the book some of the more technical proofs are omitted; we only give the statements and some ideas of the proofs, referring the reader to the original papers for details.

We tried to make this exposition accessible to graduate students, requiring only a basic knowledge of algebraic geometry, differential geometry, and the theory of Lie groups and Lie algebras. Some sections use the language of derived categories; however, we tried to reduce their use to a minimum.

The material presented in the book is taken from a number of papers and books (some small parts are new). We provide references to the original works; however, we made no attempt to discuss the history of the work. In many cases the references given are the most convenient or easy to read sources, rather than the papers in which the result was first introduced. In particular, we heavily used Crawley-Boevey's lectures [**CB1992**], Ginzburg's notes [**Gin2012**], and Nakajima's book [**Nak1999**].

Acknowledgments. The author would like to thank Pavel Etingof, Victor Ginzburg, Radu Laza, Hiraku Nakajima, Olivier Schiffmann, Jason Starr, and Jaimie Thind for many discussions and explanations. Without them, this book would never have been written.

In addition, I would also like to thank Ljudmila Kamenova and the anonymous reviewers for their comments on the preliminary version of this book and my son Andrew Kirillov for his help with proofreading.

Part 1

Dynkin Quivers

Chapter 1

Basic Theory

1.1. Basic definitions

Definition 1.1. A *quiver* $\vec{Q}$ is a directed graph; formally, it can be described by a set of vertices I, a set of edges Ω, and two maps $s, t \colon \Omega \to I$ which assign to every edge its source and target respectively. (Other references sometimes use h (for "head") and t (for "tail"), or "in" and "out", instead of source and target.)

We will commonly write $h \colon i \to j$ to indicate that edge h has source i and target j.

Throughout this book, we will always assume that the sets of edges and vertices are finite. Unless otherwise stated, we will also assume that $\vec{Q}$ is connected.

If we forget the directions of edges in $\vec{Q}$, we get a graph Q. Thus, one can also think of $\vec{Q}$ as a graph Q along with an orientation, i.e. choosing, for each edge of Q, which of the two endpoints is the source and which is the target.

Throughout the book, we will use the notation **k** for the ground field, which can be any field; unless specified otherwise, all vector spaces and linear maps are considered over the field **k**.

Definition 1.2. A representation of a quiver $\vec{Q}$ is the following collection of data:

- For every vertex $i \in I$, a vector space V_i over **k**.
- For every edge $h \in \Omega$, $h \colon i \to j$, a linear operator $x_h \colon V_i \to V_j$.

A morphism of representations $f\colon V \to W$ is a collection of linear operators $f_i\colon V_i \to W_i$ which commute with the operators x_h: if $h\colon i \to j$, then $f_j x_h = x_h f_i$. It is clear that morphisms $V \to W$ form a vector space, which we will denote by $\mathrm{Hom}_{\vec{Q}}(V,W)$, or just $\mathrm{Hom}(V,W)$ when there is no ambiguity.

As usual, we will also use the notation $\mathrm{End}_{\vec{Q}}(V) = \mathrm{Hom}_{\vec{Q}}(V,V)$ for the algebra of endomorphisms of a representation V and $\mathrm{Aut}_{\vec{Q}}(V) = \{f \in \mathrm{End}_{\vec{Q}}(V) \mid f \text{ is invertible}\}$ for the group of automorphisms of V.

Throughout the book, unless stated otherwise, we will only be considering finite-dimensional representations, i.e. those where each space V_i is finite-dimensional. We will denote the category of finite-dimensional representations of quiver $\vec{Q}$ by $\mathrm{Rep}(\vec{Q})$. Our main goal will be obtaining a complete classification of representations.

Example 1.3. Let

(1.1) $$\vec{Q} = \bullet \circlearrowright$$

(this quiver is called the Jordan quiver). Then a representation of this quiver is a pair (V,x), where V is a **k**-vector space and $x\colon V \to V$ is a linear map. Thus, classifying representations of $\vec{Q}$ is equivalent to classifying linear operators up to a change of basis, or matrices up to conjugacy. This is a classical problem of linear algebra; over an algebraically closed field, the classification is given by Jordan canonical form. Note that the answer depends on the ground field **k**: if **k** is not algebraically closed, the answer is different.

Example 1.4. Let $\vec{Q} = \underset{1}{\bullet} \longrightarrow \underset{2}{\bullet}$. Then

$$\mathrm{Rep}(\vec{Q}) = \{(V_1, V_2, x) \mid V_1, V_2 \text{ -- vector spaces,}$$
$$x\colon V_1 \to V_2 \text{ -- a linear operator}\}.$$

In this case, it is known that by a change of basis in V_1, V_2, any such operator can be brought to the form

$$x = \begin{pmatrix} I_{r \times r} & 0 \\ 0 & 0 \end{pmatrix},$$

where $I_{r \times r}$ is the $r \times r$ unit matrix. Note that in this case, the classification does not depend on the field **k**.

Example 1.5. Let $\vec{Q} = \underset{1}{\bullet} \rightrightarrows \underset{0}{\bullet}$ (this quiver is called the Kronecker quiver). Classifying representations of this quiver is equivalent to classifying pairs of linear operators $x, y\colon V_1 \to V_2$. This is a significantly more

1.1. Basic definitions

difficult problem than classifying a single linear operator. In this case a complete classification is known but is rather complicated; we will discuss it in Section 7.7. For now, let us just consider classification of representations with $\dim V_1 = \dim V_2 = 1$. In this case, choosing a basis in V_1, V_2, we can treat x, y as numbers; change of basis acts on x, y by $x \mapsto \lambda x$, $y \mapsto \lambda y$. Therefore, isomorphism classes of representations are in bijection with $\mathbf{k}^2/\mathbf{k}^\times = \{0\} \cup \mathbb{P}^1(\mathbf{k})$. In particular, if the field $\mathbf{k}$ is infinite, then there are infinitely many isomorphism classes.

Operations with representations. The category $\mathrm{Rep}(\vec{Q})$ is endowed with the following operations, similar to those in the category of group representations:

- **Direct sums:** If $V, W \in \mathrm{Rep}(\vec{Q})$, we define their direct sum $V \oplus W \in \mathrm{Rep}(\vec{Q})$ by $(V \oplus W)_i = V_i \oplus W_i$, with the obvious definition of operators x_h.

- **Subrepresentations and quotients:** A *subrepresentation* $V \subset W$ is a collection of vector subspaces $V_i \subset W_i$ such that $x_h V \subset V$: for any edge $h\colon i \to j$, we have $x_h(V_i) \subset V_j$. In this situation, we can also define the quotient representation W/V by $(W/V)_i = W_i/V_i$, with the obvious definition of x_h.

- **Kernel and image:** For any morphism of representations $f\colon V \to W$, we define representations $\mathrm{Ker}\, f$ by

$$(\mathrm{Ker}\, f)_i = \mathrm{Ker}(f_i \colon V_i \to W_i),$$

and $\mathrm{Im}\, f$ by

$$(\mathrm{Im}\, f)_i = \mathrm{Im}(f_i \colon V_i \to W_i).$$

It is easy to check that images and kernels defined above satisfy the usual properties such as $\mathrm{Im}\, f \simeq V/(\mathrm{Ker}\, f)$ (in other words, $\mathrm{Rep}(\vec{Q})$ is an *abelian* category over $\mathbf{k}$).

Using these notions, we can rewrite the results of Example 1.4 by saying that any representation is isomorphic to a direct sum of the following representations:

$$\underset{1}{\overset{\mathbf{k}}{\bullet}} \xrightarrow{1} \underset{2}{\overset{\mathbf{k}}{\bullet}} \qquad \underset{1}{\overset{0}{\bullet}} \longrightarrow \underset{2}{\overset{\mathbf{k}}{\bullet}} \qquad \underset{1}{\overset{\mathbf{k}}{\bullet}} \longrightarrow \underset{2}{\overset{0}{\bullet}}$$

Subspace problem. Consider the quiver

$$\vec{Q} = \bullet \longrightarrow \bullet \longleftarrow \bullet$$

and let V be a representation of $\vec{Q}$:

$$\overset{V_1}{\bullet} \overset{x_1}{\longrightarrow} \overset{V_0}{\bullet} \overset{x_2}{\longleftarrow} \overset{V_2}{\bullet}$$

Then we can write $V_1 = \operatorname{Ker}(x_1) \oplus V_1'$, $V_2 = \operatorname{Ker}(x_2) \oplus V_2'$; thus, any representation V is isomorphic to a direct sum $n_1 S(1) \oplus n_2 S(2) \oplus V'$, where

$$S(1) = \overset{\mathbf{k}}{\bullet} \overset{0}{\longrightarrow} \overset{0}{\bullet} \overset{0}{\longleftarrow} \overset{}{\bullet}, \qquad S(2) = \overset{0}{\bullet} \overset{0}{\longrightarrow} \overset{0}{\bullet} \overset{}{\longleftarrow} \overset{\mathbf{k}}{\bullet}$$

and V' is a representation with x_1, x_2 injective. Therefore, classifying representations of $\vec{Q}$ easily reduces to classification of triples (V_0, V_1, V_2), where V_0 is a vector space and V_1, V_2 are subspaces of V_0. Standard results of linear algebra show that in this case one can introduce a basis $v_j, j \in J$, in V_0 and two subsets $J_1, J_2 \subset J$ so that

- $v_j, j \in J_1$, is a basis in V_1,
- $v_j, j \in J_2$, is a basis in V_2,
- $v_j, j \in J_1 \cap J_2$, is a basis in $V_1 \cap V_2$.

In terms of quiver representations, this can be rewritten as follows: any representation of the quiver $\vec{Q}$ is isomorphic to a direct sum of the following representations:

(1.2)

$$\begin{array}{ccc}
\overset{\mathbf{k}}{\bullet} \overset{0}{\to} \overset{0}{\bullet} \overset{0}{\leftarrow} \overset{}{\bullet} & \overset{0}{\bullet} \overset{0}{\to} \overset{0}{\bullet} \overset{}{\leftarrow} \overset{\mathbf{k}}{\bullet} & \overset{0}{\bullet} \overset{}{\to} \overset{\mathbf{k}}{\bullet} \overset{}{\leftarrow} \overset{0}{\bullet} \\
\overset{\mathbf{k}}{\bullet} \overset{}{\to} \overset{\mathbf{k}}{\bullet} \overset{0}{\leftarrow} \overset{}{\bullet} & \overset{0}{\bullet} \overset{}{\to} \overset{\mathbf{k}}{\bullet} \overset{}{\leftarrow} \overset{\mathbf{k}}{\bullet} & \overset{\mathbf{k}}{\bullet} \overset{}{\to} \overset{\mathbf{k}}{\bullet} \overset{}{\leftarrow} \overset{\mathbf{k}}{\bullet}
\end{array}$$

(arrows $\mathbf{k} \to \mathbf{k}$ are identities; arrows $\mathbf{k} \to 0$ and $0 \to \mathbf{k}$ are obviously zero).

In a similar way, one easily sees that the problem of classification of representations of the quiver

is equivalent to the problem of classifying triples of subspaces in a vector space. (An answer to this problem will be given later; see Example 1.13.)

1.2. Path algebra; simple and indecomposable representations

Let $\vec{Q}$ be a quiver. A *path* of length l in $\vec{Q}$ is a sequence of edges such that the source of each next edge coincides with the target of the previous one:
$$p = (h_l, \ldots, h_1), \qquad s(h_{i+1}) = t(h_i).$$
We define the source and target of a path in the obvious way: $s(h_l, \ldots, h_1) = s(h_1)$, $t(h_l, \ldots, h_1) = t(h_l)$.

We define multiplication of paths by concatenation:
(1.3)
$$(h_l, \ldots, h_1)(h'_m, \ldots, h'_1) = \begin{cases} (h_l, \ldots, h_1, h'_m, \ldots, h'_1), & s(h_1) = t(h'_m), \\ 0, & \text{otherwise.} \end{cases}$$

Thus, a path of length l can be written as a product of l edges. It is also convenient to extend this definition allowing paths of length zero; formally, we introduce elements e_i with $s(e_i) = t(e_i) = i$ and extend multiplication by
(1.4)
$$e_i p = \begin{cases} p, & t(p) = i, \\ 0, & \text{otherwise,} \end{cases} \qquad p e_i = \begin{cases} p, & s(p) = i, \\ 0, & \text{otherwise.} \end{cases}$$

Note that, in particular, this implies
$$e_i e_j = \delta_{ij} e_i.$$

Definition 1.6. The path algebra $\mathbf{k}\vec{Q}$ of quiver $\vec{Q}$ is the algebra with the basis given by paths in $\vec{Q}$ (including paths of length zero) and multiplication defined by (1.3), (1.4).

The following properties of the path algebra are immediate from the definition.

(1) $\mathbf{k}\vec{Q}$ is an associative algebra with unit $1 = \sum e_i$.
(2) $\mathbf{k}\vec{Q}$ is naturally $\mathbb{Z}_+$-graded by path length, and $(\mathbf{k}\vec{Q})_0 = \bigoplus \mathbf{k} e_i$ is semisimple.
(3) $\mathbf{k}\vec{Q}$ is finite-dimensional iff the quiver $\vec{Q}$ contains no oriented cycles.
(4) Elements e_i are indecomposable projectors. (Recall that a projector e is decomposable if it is possible to write it as a sum of nonzero orthogonal projectors: $e = e' + e''$, with $(e')^2 = e'$, $(e'')^2 = e''$, and $e' e'' = e'' e' = 0$.)

Theorem 1.7. *The category of representations of $\vec{Q}$ (not necessarily finite-dimensional) is equivalent to the category $\mathbf{k}\vec{Q}$-mod of left $\mathbf{k}\vec{Q}$-modules.*

Proof. Let $V = (V_i, x_h)$ be a representation of $\vec{Q}$. Define, for each path p, operators $x_p \colon V_{s(p)} \to V_{t(p)}$ by $x_{e_i} = \mathrm{id}_{V_i}$, $x_{h_l \ldots h_1} = x_{h_l} \cdots x_{h_1}$. It is easy to see that so-defined operators, extended by zero, define an action of the path algebra on the space $\tilde{V} = \bigoplus_{i \in I} V_i$.

Conversely, given a $\mathbf{k}\vec{Q}$-module M, define $M_i = e_i M$; then for every edge $h \colon i \to j$ in $\vec{Q}$, we have $h(M_i) \subset M_j$; thus, we get a representation of $\vec{Q}$.

Therefore, we have defined functors $\mathrm{Rep}(\vec{Q}) \to \mathbf{k}\vec{Q}\text{-mod}$, $\mathbf{k}\vec{Q}\text{-mod} \to \mathrm{Rep}(\vec{Q})$; one easily sees that these two functors are inverse to each other and thus give an equivalence $\mathrm{Rep}(\vec{Q}) \simeq \mathbf{k}\vec{Q}\text{-mod}$. $\square$

Therefore, all the usual notions and results of the theory of modules over associative algebras can be translated to representations of quivers.

Definition 1.8. A representation $V \in \mathrm{Rep}(\vec{Q})$ is called

- *simple* (or irreducible) if it contains no nontrivial subrepresentations,
- *semisimple* if it is isomorphic to a direct sum of simple representations,
- *indecomposable* if it cannot be written as a direct sum of nonzero subrepresentations.

Example 1.9. For the quiver $\vec{Q} = \underset{1}{\bullet} \longrightarrow \underset{2}{\bullet}$, representations $\underset{1}{\overset{\mathbf{k}}{\bullet}} \longrightarrow \underset{2}{\overset{0}{\bullet}}$ and $\underset{1}{\overset{0}{\bullet}} \longrightarrow \underset{2}{\overset{\mathbf{k}}{\bullet}}$ are simple, and representation $\underset{1}{\overset{\mathbf{k}}{\bullet}} \overset{1}{\longrightarrow} \underset{2}{\overset{\mathbf{k}}{\bullet}}$ is indecomposable but not semisimple.

Our first goal is classification of simple representations. This is easy, at least when $\vec{Q}$ has no oriented cycles. Namely, for every $i \in I$ define representation $S(i)$ by

$$(1.5) \qquad S(i)_j = \begin{cases} \mathbf{k}, & i = j, \\ 0, & i \neq j, \end{cases}$$

and all $x_h = 0$. It is obvious that each $S(i)$ is simple and that they are pairwise nonisomorphic.

Theorem 1.10. *Let $\vec{Q}$ be a quiver without oriented cycles. Then representations $S(i)$, $i \in I$, form a full list of simple representations of $\vec{Q}$.*

Proof. Assume that V is a simple representation. Consider the set $I' = \{i \mid V_i \neq 0\}$. Since $\vec{Q}$ contains no oriented cycles, there must exist $i \in I'$ such that there are no edges $i \to j, j \in I'$. Therefore, V contains a subrepresentation V' given by $V'_i = V_i$, $V'_j = 0$ for $j \ne i$. Since V is simple, this implies $V = V'$, and $\dim V_i = 1$. □

However, as Example 1.9 shows, not every representation is semisimple (in fact, most representations are not); thus, knowing simple representations is not enough to classify all representations. On the other hand, the following theorem shows that it would be enough to know all indecomposable representations.

Theorem 1.11. *Any finite-dimensional representation of a quiver $\vec{Q}$ can be written as a direct sum of indecomposable representations, and such decomposition is unique up to reordering.*

Indeed, this is just a special case of the well-known Krull–Schmidt theorem for artinian modules (see, for example, [**ARS1997**, Theorem 2.2]).

Therefore, our main goal will be classification of indecomposable representations of $\vec{Q}$. We will denote

(1.6)
$$\operatorname{Ind}(\vec{Q}) = \{\text{isomorphism classes of nonzero} \\ \text{indecomposable representations of } \vec{Q}\}.$$

Example 1.12. For the quiver $\vec{Q} = \underset{1}{\bullet} \longrightarrow \underset{2}{\bullet}$, results of Section 1.1 show that indecomposable representations of $\vec{Q}$ are

$$S(1) = \underset{1}{\overset{\mathbf{k}}{\bullet}} \xrightarrow{0} \underset{2}{\bullet}, \qquad S(2) = \underset{1}{\overset{0}{\bullet}} \xrightarrow{\mathbf{k}} \underset{2}{\bullet}, \qquad I_{12} = \underset{1}{\overset{\mathbf{k}}{\bullet}} \xrightarrow{\mathbf{k}} \underset{2}{\bullet}.$$

Similarly, for the quiver $\vec{Q} = \bullet \longrightarrow \bullet \longleftarrow \bullet$ there are six indecomposable representations, shown in (1.2).

Example 1.13. Let

(1.7)
$$\vec{Q} = \begin{array}{c} \bullet \\ \searrow \\ \bullet \longleftarrow \bullet \\ \nearrow \\ \bullet \end{array}$$

As was mentioned at the end of Section 1.1, classifying representations of this quiver is essentially equivalent to classifying triples of subspaces in a vector space. Consider representations shown in Figure 1.1, where for the first three rows the arrows are defined in the obvious way (all arrows between one-dimensional spaces are isomorphisms) and for the last representation, arrows are given by injections whose images are in general position (i.e. any two of the three one-dimensional subspaces are linearly independent). Then it is easy to see that each of them is indecomposable; later we will show that this is a full list of indecomposable representations, thus giving a complete answer to the problem of classifying triples of subspaces in a vector space (see Example 3.22).

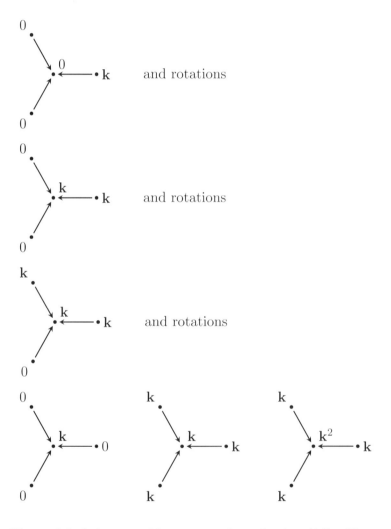

Figure 1.1. Indecomposable representations of quiver (1.7). "Rotations" mean all representations obtained from a given one by rotating the quiver.

1.3. K-group and dimension

Recall that for any abelian category $\mathcal{C}$, one can define the Grothendieck group $K(\mathcal{C})$ (also called the K-group) as the abelian group generated by symbols $[M]$, $M \in \mathcal{C}$, with relations $[A] - [B] + [C] = 0$ for every short exact sequence $0 \to A \to B \to C \to 0$. In particular, this implies that $[A] = [B]$ if A, B are isomorphic and that $[A \oplus B] = [A] + [B]$.

In this section we discuss the Grothendieck group of the category of representations of a quiver $\vec{Q}$; we will denote it simply $K(\vec{Q})$.

Example 1.14. Let $\vec{Q} = \bullet$, so that $\text{Rep}(\vec{Q}) = \mathcal{V}ec$ is the category of finite-dimensional vector spaces. Since every vector space is isomorphic to a direct sum of copies of $\mathbf{k}$, we see that in this case $K(\vec{Q}) = \mathbb{Z}$ is the free group generated by the element $[\mathbf{k}]$. One can also describe the isomorphism $K(\vec{Q}) \simeq \mathbb{Z}$ by $[M] \mapsto \dim M$.

The following theorem is a natural generalization of this result. For a representation V of $\vec{Q}$, define its *graded dimension* $\mathbf{dim}\, V \in \mathbb{Z}^I$ by

$$(1.8) \qquad (\mathbf{dim}\, V)_i = \dim(V_i).$$

Theorem 1.15. *Let $\vec{Q}$ be a quiver without oriented cycles. Then the map*

$$\mathbf{dim} \colon K(\vec{Q}) \to \mathbb{Z}^I$$
$$[V] \mapsto \mathbf{dim}\, V$$

is an isomorphism.

Proof. Since every representation has a composition series with simple factors, by Theorem 1.10 this implies that $[V] = \sum n_i [S(i)]$. Thus, classes $[S(i)]$ generate $K(\vec{Q})$.

Next, note that $\mathbf{dim}$ is well defined on $K(\vec{Q})$; since $\mathbf{dim}\, S(i) = e_i$ are independent in $\mathbb{Z}^I$, this implies that $[S(i)]$ are independent in $K(\vec{Q})$ and thus are free generators of the abelian group $K(\vec{Q})$. $\square$

1.4. Projective modules and the standard resolution

Recall that a module P over an associative algebra A is projective iff the functor $\text{Hom}(P, -)$ is exact; in particular, a direct summand of a free module is projective. Thus, for the path algebra $A = \mathbf{k}\vec{Q}$, the modules

$$(1.9) \qquad \begin{aligned} P(i) &= Ae_i \qquad (i \in I) \\ &= \{\text{linear combinations of paths starting at } i\} \end{aligned}$$

are projective, as $A = \bigoplus_i P(i)$. (Note that $P(i)$ can be infinite-dimensional.)

Theorem 1.16. *For any representation V of $\vec{Q}$, we have $\operatorname{Hom}_{\vec{Q}}(P(i), V) = V_i$.*

Proof. For any $v \in V_i$, define the corresponding homomorphism $P(i) \to V$ by $p \mapsto x_p(v)$ if p is a path starting at i. One easily sees that this is an isomorphism of vector spaces. $\square$

Example 1.17. Let $\vec{Q} = \underset{1}{\bullet} \longrightarrow \underset{2}{\bullet}$. Then

$$P(1) = \underset{1}{\overset{\mathbf{k}}{\bullet}} \overset{1}{\longrightarrow} \underset{2}{\overset{\mathbf{k}}{\bullet}},$$

$$P(2) = \underset{1}{\overset{0}{\bullet}} \longrightarrow \underset{2}{\overset{\mathbf{k}}{\bullet}} = S(2).$$

It is easy to see that if $\vec{Q}$ has no oriented cycles, then $P(i)$ is finite-dimensional and in the Grothendieck group we have $[P(i)] = \sum_j p_{ji}[S(j)]$, where $p_{ji} = $ (number of paths $i \to j$); in particular, it is possible to order vertices so that the matrix p_{ji} is upper triangular with ones on the diagonal and thus invertible. Therefore, classes $[P(i)]$ also form a basis of $K(\vec{Q})$.

Theorem 1.18. *Assume that $\vec{Q}$ has no oriented cycles. Then the modules $\{P(i), i \in I\}$ form the full set of nonzero indecomposable projective objects in $\operatorname{Rep}(\vec{Q})$.*

Proof. Indecomposability of $P(i)$ follows from $\operatorname{Hom}(P(i), P(i)) = \mathbf{k}$. To show that it is a full set of indecomposable projectives, assume that P is a projective module. Let $P' = \bigoplus n_i P(i)$, where $n_i = \dim \operatorname{Hom}(P, S(i))$. Then it follows from Theorem 1.16 that $\operatorname{Hom}(P, S(i)) \simeq \operatorname{Hom}(P', S(i))$, which (using projectivity of P) easily implies that for any V, one has an isomorphism $\operatorname{Hom}(P, V) \simeq \operatorname{Hom}(P', V)$. Thus, $P \simeq P'$. $\square$

We will use representations $P(i)$ to construct a projective resolution for any representation V of $\vec{Q}$.

First, we review some general theory. Let A be an associative algebra with unit; then for any module M we have

$$M \simeq A \otimes_A M = A \otimes_{\mathbf{k}} M / I,$$

where the subspace I is generated by elements $ab \otimes m - a \otimes bm$, $a, b \in A$, $m \in M$. Moreover, it suffices to take elements b from some set of generators of A: if $L \subset A$ is a subspace such that elements $l \in L$ generate A, then I is spanned by $al \otimes m - a \otimes lm$, $a \in A$, $l \in L$, $m \in M$. Thus, we have an exact sequence of A-modules (all tensor products are over $\mathbf{k}$):

(1.10)
$$A \otimes L \otimes M \xrightarrow{d_1} A \otimes M \xrightarrow{d_0} M \to 0,$$
$$d_1 \colon a \otimes l \otimes m \mapsto al \otimes m - a \otimes lm,$$
$$d_0 \colon a \otimes m \mapsto am.$$

This is a beginning of a free resolution of M; the next term would come from relations among generators $l \in L$, and so on.

This has a modification: if A_0 is a subalgebra, and L is a subspace such that $A_0 L \subset L$, $LA_0 \subset L$, and A_0, L generate A, then we have an exact sequence

(1.11)
$$A \otimes_{A_0} L \otimes_{A_0} M \xrightarrow{d_1} A \otimes_{A_0} M \xrightarrow{d_0} M \to 0$$

with the morphisms defined in the same way as before. The proof of this modification is left as an easy exercise to the reader.

We can now apply this in the case $A = \mathbf{k}\vec{Q}$.

Theorem 1.19. *Let $\vec{Q}$ be a quiver with set of vertices I and set of edges Ω. For any $\mathbf{k}\vec{Q}$-module V, we have the following short exact sequence of $\mathbf{k}\vec{Q}$-modules:*

(1.12)
$$0 \to \bigoplus_{h \in \Omega} P(t(h)) \otimes \mathbf{k}h \otimes V_{s(h)} \xrightarrow{d_1} \bigoplus_{i \in I} P(i) \otimes V_i \xrightarrow{d_0} V \to 0,$$

where $P(i)$ is the projective module defined by (1.9), and the differentials are defined by

$$d_1(p \otimes h \otimes v) = ph \otimes v - p \otimes x_h(v),$$
$$d_0(p \otimes v) = x_p(v).$$

This resolution will be called the standard resolution *of V.*

Proof. Exactness at V, $\bigoplus P(i) \otimes V_i$ follows from (1.11) if we let $A_0 = $ paths of length zero $= \bigoplus \mathbf{k}e_i$, and $L = \bigoplus_{h \in \Omega} \mathbf{k}h$, which gives $A \otimes_{A_0} V = \bigoplus Ae_i \otimes e_i V = \bigoplus P(i) \otimes V_i$.

To prove that d_1 is injective, assume

(1.13)
$$d_1\left(\sum_n p_n \otimes h_n \otimes v_n\right) = \sum p_n h_n \otimes v_n - p_n \otimes x_{h_n}(v_n) = 0.$$

Let $l = $ maximal length of the paths p_n appearing in this sum. Then taking the terms of length $l + 1$ in (1.13) we get

$$\sum p_k h_k \otimes v_k = 0,$$

where the sum is taken over all k such that $l(p_k) = l$.

On the other hand, for different h_k, the paths of the form $p_k h_k$ are linearly independent, which leads to a contradiction. $\square$

In particular, this shows that the category of (not necessarily finite-dimensional) representations of $\vec{Q}$ has enough projectives, so we can define Ext functors in the usual way. We will denote Ext functors in this category by $\operatorname{Ext}^i_{\vec{Q}}(V,W)$.

Corollary 1.20. *For any $V, W \in \operatorname{Rep}(\vec{Q})$, we have $\operatorname{Ext}^i_{\vec{Q}}(V,W) = 0$ for any $i > 1$.*

Indeed, Ext^i can be computed using a projective resolution of V, and every representation V has a projective resolution of the form $0 \to P_1 \to P_0 \to V \to 0$.

Example 1.21. Let $V = S(i)$ be a simple module; then the standard resolution (1.12) becomes
$$0 \to \bigoplus_{h\colon i \to j} P(j) \to P(i) \to S(i) \to 0.$$

Using this and Theorem 1.16, we can easily compute $\operatorname{Ext}^n(S(i), W)$. In particular,
$$\dim \operatorname{Hom}(S(i), S(j)) = \delta_{ij},$$
$$\dim \operatorname{Ext}^1(S(i), S(j)) = \text{number of edges } i \to j \text{ in } \vec{Q}.$$

Categories satisfying $\operatorname{Ext}^i(V, W) = 0$ for any $i > 1$ are called *hereditary*. They also admit another characterization given below.

Corollary 1.22. *If $P \in \operatorname{Rep}(\vec{Q})$ is projective, then any subrepresentation of P is also projective.*

Proof. Let $V \subset P$; denote $W = P/V$, so we have a short exact sequence
$$0 \to V \to P \to W \to 0.$$
Then, for any X, we have the corresponding long exact sequence of Ext functors:
$$\cdots \to \operatorname{Ext}^1(W, X) \to \operatorname{Ext}^1(P, X) \to \operatorname{Ext}^1(V, X) \to \operatorname{Ext}^2(W, X) \to \cdots.$$
Since $\operatorname{Ext}^1(P, X) = 0$ (P is projective) and $\operatorname{Ext}^2(W, X) = 0$ by the previous corollary, we have $\operatorname{Ext}^1(V, X) = 0$, which implies that V is projective. $\square$

Most of the statements above can be repeated for injective modules. Namely, we have the following proposition, the proof of which is left to the reader.

Theorem 1.23. *Let $\vec{Q}$ be a quiver. Define, for any $i \in I$,*

(1.14) $$Q(i) = (e_i A)^*,$$

where $A = \mathbf{k}\vec{Q}$ is the path algebra and $*$ is the graded dual: $(e_i A)^* = \bigoplus_l (e_i A^l)^*$, where A^l is the span of paths of length l. (Note that $Q(i)$ has a natural structure of a left A-module.) Then we have the following statements:

(1) *For $V \in \operatorname{Rep} \vec{Q}$, we have natural isomorphisms $\operatorname{Hom}_{\vec{Q}}(V, Q(i)) \simeq V_i^*$.*

(2) *If $\vec{Q}$ has no oriented cycles, then modules $Q(i)$, $i \in I$, form a full set of indecomposable injective representations of $\vec{Q}$.*

(3) *Any representation V of $\vec{Q}$ has an injective resolution of the form*
$$0 \to V \to I_0 \to I_1 \to 0.$$

1.5. Euler form

For any two representations V, W of quiver $\vec{Q}$, define $\langle V, W \rangle \in \mathbb{Z}$ by

(1.15) $$\langle V, W \rangle = \sum (-1)^i \dim \operatorname{Ext}^i(V, W) = \dim \operatorname{Hom}(V, W) - \dim \operatorname{Ext}^1(V, W).$$

(Since $\operatorname{Ext}^i = 0$ for $i > 1$, there are no higher terms.)

Example 1.24. For $V = P(i)$, $\langle P(i), W \rangle = \dim W_i$ (see Theorem 1.16). In particular,
$$\langle P(i), S(j) \rangle = \delta_{ij}.$$
Similarly, if $W = Q(i)$ is the indecomposable injective, then
$$\langle V, Q(i) \rangle = \dim V_i.$$

Theorem 1.25. *The number $\langle V, W \rangle$ only depends on $\mathbf{dim}\, V$, $\mathbf{dim}\, W$ and thus defines a bilinear form on $\mathbb{Z}^I$, called the* Euler *form. Moreover, for $\mathbf{v}, \mathbf{w} \in \mathbb{Z}^I$, we have*

(1.16) $$\langle \mathbf{v}, \mathbf{w} \rangle = \sum_{i \in I} \mathbf{v}_i \mathbf{w}_i - \sum_{h \in \Omega} \mathbf{v}_{s(h)} \mathbf{w}_{t(h)}.$$

Proof. It follows from the long exact sequence of Ext functors that if we have a short exact sequence $0 \to A \to B \to C \to 0$, then $\langle A, W \rangle - \langle B, W \rangle + \langle C, W \rangle = 0$. Thus, we can use the projective resolution $0 \to P_1 \to P_0 \to V \to 0$ of V constructed in Theorem 1.19 to compute $\langle V, W \rangle$. Since $\langle P(i), W \rangle = \dim W_i$, this gives
$$\langle V, W \rangle = \langle P_0, W \rangle - \langle P_1, W \rangle$$
$$= \sum_i \dim V_i \dim W_i - \sum_h \dim V_{s(h)} \dim W_{t(h)}. \qquad \square$$

Example 1.26. $\langle S(i), S(j)\rangle = \delta_{ij} - $ (number of edges $i \to j$).

Note that $\langle\,,\,\rangle$ is not symmetric. We will frequently use the symmetrized Euler form

$$(1.17) \qquad (\mathbf{v}, \mathbf{w}) = \langle \mathbf{v}, \mathbf{w}\rangle + \langle \mathbf{w}, \mathbf{v}\rangle = \sum_i \mathbf{v}_i \mathbf{w}_i(2 - 2n_{ii}) - \sum_{i \neq j} n_{ij} \mathbf{v}_i \mathbf{w}_j,$$

where n_{ij} is the number of (unoriented) edges between i and j in the graph Q. We will also use the associated quadratic form, called the *Tits form*:

$$(1.18) \qquad q_{\vec{Q}}(\mathbf{v}) = \tfrac{1}{2}(\mathbf{v}, \mathbf{v}) = \langle \mathbf{v}, \mathbf{v}\rangle.$$

Note that the symmetrized Euler form and the Tits form are independent of orientation: they only depend on the underlying graph Q. In fact, they can be defined for any graph Q.

1.6. Dynkin and Euclidean graphs

Let $\vec{Q}$ be a quiver and let Q be the underlying graph. Throughout this section, we assume that Q is connected.

Definition 1.27. Let Q be a connected graph. Q is called *Dynkin* if the associated Tits form q_Q defined by (1.18) is positive definite. Q is called *Euclidean* if the Tits form is positive semidefinite.

Theorem 1.28. *Let Q be a connected graph.*

(1) *Q is Dynkin iff it is one of the graphs shown in Figure* 1.2.

(2) *Q is Euclidean iff it is one of the graphs shown in Figure* 1.3. *In this case, the radical of the symmetrized Euler form is one-dimensional:*

$$\{x \in \mathbb{Z}^I \mid (x, y) = 0 \text{ for all } y\} = \mathbb{Z}\delta,$$

where $\delta = \sum \delta_i \alpha_i$ can be chosen so that $\delta_i > 0$, $\min \delta_i = 1$ (this uniquely determines δ).

Proof. We will give an outline of the proof, skipping some details. First, it can be shown by explicit computation that each of the graphs shown in Figure 1.3 is indeed Euclidean, with the radical spanned by the element δ. Since each graph Q in Figure 1.2 is a full subgraph (i.e. obtained by taking some subset of vertices and all edges between these vertices) in a graph $\hat{Q}$ of Figure 1.3, the Tits form q_Q is the restriction of the Tits form $q_{\hat{Q}}$; this easily implies that each graph of Figure 1.2 is indeed Dynkin. This proves one direction.

1.6. Dynkin and Euclidean graphs

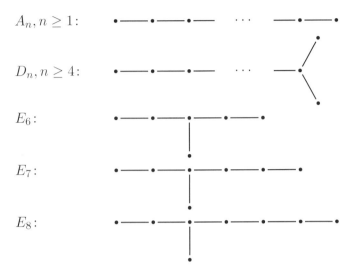

Figure 1.2. Dynkin graphs. In all cases, the subscript n is equal to the number of vertices.

Now, let Q be a connected graph. Clearly there can be three possibilities:

(1) **Q is one of the graphs of Figure 1.3.** In this case, Q is Euclidean.

(2) **Q contains one of the graphs Q' of Figure 1.3 as a proper subgraph (not necessarily full).** In this case, we claim that the Tits form of Q is indefinite.

Indeed, if Q' contains all vertices of Q (but not all the edges), then let $\delta = \sum \delta_i \alpha_i$ be the element defined by Figure 1.3. Since all $\delta_i > 0$, it is easy to see that $(\delta, \delta)_Q < (\delta, \delta)_{Q'} = 0$.

If Q contains vertices which are not in Q', let i_0 be a vertex which is not in Q' and which is connected to at least one vertex of Q'. Let $\delta = \delta_{Q'}$ be defined by Figure 1.3. We claim that $(\alpha_0, \delta) \le -1$: indeed, $(\alpha_0, \delta) = -\sum_i \delta_i n_{0i}$; since i_0 is connected to at least one vertex j of Q', we have $n_{j_0}\delta_j \ge 1$. In addition, the same reasoning as in the previous paragraph shows that $(\delta, \delta) \le 0$.

Thus, if we let $x = \alpha_0 + m\delta$, then

$$(x,x) = (\alpha_0, \alpha_0) + 2m(\alpha_0, \delta) + m^2(\delta, \delta) \le (\alpha_0, \alpha_0) - 2m < 0$$

for large enough m.

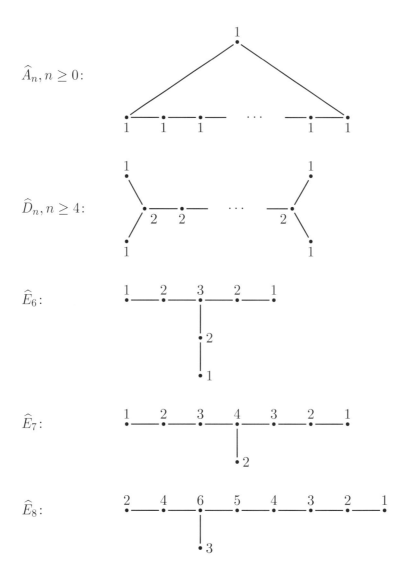

Figure 1.3. Euclidean graphs. In all cases, the subscript n is equal to (the number of vertices) -1. The labels δ_i written next to the vertices are the coordinates of the element $\delta = \sum \delta_i \alpha_i$, which generates the radical of the Tits form.

(3) **Q does not contain any of the graphs of Figure 1.3 as a subgraph.**

In this case, Q contains no cycles (otherwise it would contain $\widehat{A}_n$ as a subgraph) so it is a tree. Next, it has no vertices of valency more than 3 (otherwise it would contain $\widehat{D}_4$) and has at most one vertex of valency 3 (otherwise it would contain $\widehat{D}_n$). Thus, it

1.6. Dynkin and Euclidean graphs

is either a graph of type A_n or a "star" graph, consisting of three branches meeting at a point v. Denote the lengths of these branches by l_1, l_2, l_3 (the length is the number of vertices in a branch, including the central vertex v).

Since Q does not contain $\widehat{E}_6$, $\min l_i = 2$. Since it does not contain $\widehat{E}_7, \widehat{E}_8$, it is easy to see that either we must have $l_1 = l_2 = 2$, l_3 — arbitrary (in which case Q is a graph of type D_n) or we must have $l_1 = 2$, $l_2 = 3$, $l_3 = 3, 4,$ or 5, which gives $Q = E_6, E_7, E_8$ respectively.

Thus, in this case Q is one of the graphs of Figure 1.2. □

Remark 1.29. Dynkin graphs shown in Figure 1.2 are exactly the same as simply-laced Dynkin diagrams in the theory of semisimple Lie algebras and root systems (see, e.g., [**Ser2001**]).

Similarly, Euclidean graphs (except for $\widehat{A}_0$) are exactly the Dynkin diagrams of the (untwisted) affine simply-laced Lie algebras (see [**Kac1990**]). Instead of notation $\widehat{X}_n$, some references use notation $\widetilde{X}_n$ or $X_n^{(1)}$.

For future use, we also mention the following result.

Theorem 1.30. *For $p, q, r \geq 1$, let $\Gamma(p, q, r)$ be the "star" graph consisting of three branches of lengths p, q, r meeting at a central vertex i_0 (branch length is the number of vertices in a branch, including the central vertex; if one of p, q, r is 1, then we only have 2 branches). Then $\Gamma(p, q, r)$ is Dynkin if and only if the following inequality holds:*

$$\tag{1.19} \frac{1}{p} + \frac{1}{q} + \frac{1}{r} > 1.$$

Moreover, every Dynkin graph is of the form $\Gamma(p, q, r)$.

Proof. Elementary computations show that possible triples (p, q, r) satisfying (1.19) are (up to permutation)

$$(p, q, 1), \quad p, q \geq 1,$$
$$(p, 2, 2), \quad p \geq 2,$$
$$(p, 3, 2), \quad p = 3, 4, 5.$$

This matches the classification of Dynkin graphs given in Theorem 1.28: $(p, q, 1)$ corresponds to A_n, $n = p + q - 1$; $(p, 2, 2)$ corresponds to D_{p+2}; and $(p, 3, 2)$, for $p = 3, 4, 5$, corresponds to E_6, E_7, E_8 respectively. □

Exercise 1.31. Show that one direction of Theorem 1.30 (Dynkin implies (1.19)) can be deduced without using the classification result. Namely, let $\mathbf{v} \in \mathbb{R}^I$ be defined by $\mathbf{v}_0 = 1$, where 0 is the central vertex, and $\mathbf{v}_i = (l-k)/l$

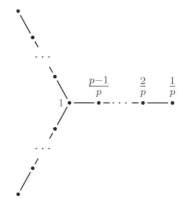

Figure 1.4

if i is a vertex on branch of length l, at distance k from the center (see Figure 1.4).

Prove that then $(\mathbf{v}, \alpha_i) = 0$ for any $i \neq 0$ and

$$(\mathbf{v}, \mathbf{v}) = \left(\frac{1}{p} + \frac{1}{q} + \frac{1}{r}\right) - 1.$$

1.7. Root lattice and Weyl group

The Dynkin and Euclidean graphs which appeared in the last section also appear in the classification of finite-dimensional and affine Lie algebras. We will be frequently using some terminology and notation from the theory of Lie algebras and root systems. Detailed information can be found in Appendix A; here, we just give the list of notation.

Let Q be a graph with the set of vertices I; we will assume that Q is connected and has no edge loops.

We define the corresponding Cartan matrix

(1.20) $$C_Q = 2I - A,$$

where A is the adjacency matrix of Q:

$$A_{ij} = \text{number of edges between vertices } i, j \in I.$$

This matrix can be used to define a Kac–Moody algebra $\mathfrak{g}_Q$ (see details in Appendix A). In particular, this gives rise to the root lattice and the Weyl group.

Namely, the root lattice L is defined by

(1.21) $$L = \bigoplus_{i \in I} \mathbb{Z}\alpha_i \simeq \mathbb{Z}^I.$$

1.7. Root lattice and Weyl group

The generators α_i will be called *simple roots*. If $\vec{Q}$ is a quiver with the underlying graph Q, then we have defined, for every $V \in \operatorname{Rep} \vec{Q}$, its dimension $\mathbf{dim}(V) \in L$ so that

(1.22) $$\alpha_i = \mathbf{dim}\, S(i).$$

For an element $\mathbf{v} = \sum \mathbf{v}_i \alpha_i \in L$, we will write

(1.23) $$\mathbf{v} \geqslant 0 \text{ iff } \mathbf{v}_i \geq 0 \text{ for all } i.$$

We will also write $\mathbf{v} \geqslant \mathbf{w}$ if $\mathbf{v} - \mathbf{w} \leqslant 0$. We will use the notation

(1.24) $$L_+ = \mathbb{Z}_+^I = \{\mathbf{v} \in L \mid \mathbf{v} \geqslant 0\}.$$

The Cartan matrix defines an integral symmetric bilinear form on L, which we call the symmetrized Euler form:

(1.25) $$(\alpha_i, \alpha_j) = C_{ij} = 2\delta_{ij} - A_{ij}.$$

It can also be interpreted in terms of the representation theory of quivers (see Section 1.5).

For every $i \in I$, we define the corresponding *simple reflection*

(1.26) $$s_i \colon L \to L \colon \alpha \mapsto \alpha - (\alpha, \alpha_i)\alpha_i.$$

Definition 1.32. The Weyl group W of Q is the subgroup in $\operatorname{GL}(L \otimes_{\mathbb{Z}} \mathbb{R})$ generated by simple reflections s_i, $i \in I$.

It is easy to show that the action of W preserves the bilinear form $(\,,\,)$ on L.

Finally, the Lie algebra $\mathfrak{g}$ also defines a *root system* $R \subset L$. A general definition of it can be found in Section A.6. However, in the special case of Dynkin and Euclidean graphs, the root system admits a very easy description:

$$R = \{\alpha \in L - \{0\} \mid (\alpha, \alpha) \leq 2\}.$$

It can be shown (but is not obvious) that the root system has a decomposition

(1.27) $$R = R_+ \sqcup R_-, \quad R_+ = R \cap L_+, \quad L_- = -L_+.$$

Roots $\alpha \in R_+$ will be called positive roots, and elements of R_-, negative.

A root α is called *real* if $\alpha = w\alpha_i$ for some $w \in W$, $i \in I$; all other roots are called *imaginary*, so that

$$R = R^{re} \sqcup R^{im}.$$

Theorem 1.33.

(1) *If Q is Dynkin, then there are no imaginary roots, and*
$$R = R^{re} = \{\alpha \in L \mid (\alpha, \alpha) = 2\}.$$

(2) *If Q is a Euclidean graph, then*
$$R^{im} = \{n\delta, n \in \mathbb{Z} - \{0\}\},$$
$$R^{re} = \{\alpha \in L \mid (\alpha, \alpha) = 2\},$$

where δ is as in Theorem 1.28.

Chapter 2

Geometry of Orbits

In this chapter we present a geometric approach to quiver representations, identifying the set of isomorphism classes of representations with orbits of certain group actions. We assume that the reader is familiar with basic notions of algebraic geometry, such as affine algebraic varieties, Zariski topology, etc.

Throughout this chapter, $\mathbf{k}$ is an algebraically closed field; unless specified otherwise, the words open, closed, etc. refer to *Zariski topology*. Notation $\overline{X}$ stands for the closure of X in Zariski topology. We also fix a quiver $\vec{Q}$ with set of vertices I and set of oriented edges Ω.

2.1. Representation space

Let V be a finite-dimensional representation of the quiver $\vec{Q}$; denote $\mathbf{v} = \dim V$ (see (1.8)). By choosing a basis in each V_i, we can identify $V_i \simeq \mathbf{k}^{\mathbf{v}_i}$; thus, the structure of a representation of $\vec{Q}$ is described by a collection of matrices $x_h \in \mathrm{Hom}_{\mathbf{k}}(\mathbf{k}^{\mathbf{v}_i}, \mathbf{k}^{\mathbf{v}_j}) = \mathrm{Mat}_{\mathbf{v}_j \times \mathbf{v}_i}(\mathbf{k})$, for each edge $h\colon i \to j$, or equivalently, by a vector x in the space

$$(2.1) \qquad R(\vec{Q}, \mathbf{v}) = \bigoplus_{h \in \Omega} \mathrm{Hom}_{\mathbf{k}}(\mathbf{k}^{\mathbf{v}_{s(h)}}, \mathbf{k}^{\mathbf{v}_{t(h)}}).$$

Conversely, every $x \in R(\vec{Q}, \mathbf{v})$ defines a representation

$$V^x = (\{\mathbf{k}^{\mathbf{v}_i}\}, \{x_h\}).$$

The space $R(\vec{Q}, \mathbf{v})$ will be called the *representation space*. We will usually omit $\vec{Q}$ in the notation, writing just $R(\mathbf{v})$. Also, we will occasionally use the notation
$$R(V) = \bigoplus_{h \in \Omega} \mathrm{Hom}_{\mathbf{k}}(V_{s(h)}, V_{t(h)})$$
for an I-graded vector space $V = \bigoplus V_i$. Obviously, $R(V) \simeq R(\dim V)$.

Consider now the group

(2.2) $$\mathrm{GL}(\mathbf{v}) = \prod_{i \in I} \mathrm{GL}(\mathbf{v}_i, \mathbf{k})$$

which acts in $R(\mathbf{v})$ by conjugation.

Theorem 2.1. *Two elements $x, x' \in R(\mathbf{v})$ define isomorphic representations of $\vec{Q}$ iff they are in the same $\mathrm{GL}(\mathbf{v})$ orbit.*

The proof of this theorem is left to the reader as an easy exercise.

In other words, we have a bijection
$$\begin{pmatrix} \text{isomorphism classes of representations} \\ \text{of graded dimension } \mathbf{v} \end{pmatrix} \leftrightarrow (\mathrm{GL}(\mathbf{v})\text{-orbits in } R(\mathbf{v})).$$

We will denote the orbit of $x \in R(\mathbf{v})$ by $\mathbb{O}_x$; abusing the language, we will also use notation $\mathbb{O}_V$ for the orbit corresponding to a representation V.

Example 2.2. Let $\vec{Q} = \underset{1}{\bullet} \longrightarrow \underset{2}{\bullet}$ and $\mathbf{v} = (1,1)$. Then $R(\mathbf{v}) = \mathbf{k}$ and $\mathrm{GL}(\mathbf{v}) = \mathbf{k}^\times \times \mathbf{k}^\times$, acting on R by $(\lambda, \mu)(x) = \lambda x \mu^{-1}$. In this case, there are exactly two orbits, $\{0\}$ and $\mathbf{k}^\times$, corresponding to representations $\underset{1}{\overset{\mathbf{k}}{\bullet}} \overset{0}{\longrightarrow} \underset{2}{\overset{\mathbf{k}}{\bullet}}$ and $\underset{1}{\overset{\mathbf{k}}{\bullet}} \overset{1}{\longrightarrow} \underset{2}{\overset{\mathbf{k}}{\bullet}}$ respectively.

2.2. Properties of orbits

The group $\mathrm{GL}(\mathbf{v})$ is a linear algebraic group acting on a finite-dimensional vector space. General theory of algebraic groups (which is mostly parallel to the theory of Lie groups) gives certain properties of orbits, which we list here without a proof; proofs can be found, for example, in [**Spr1998**] or in [**OV1990**, Chapter 3].

(1) Each orbit is a nonsingular algebraic variety.
(2) Closure of an orbit (in Zariski and for $\mathbf{k} = \mathbb{C}$, also in the analytic topology) is a union of orbits. Moreover, $\overline{\mathbb{O}} - \mathbb{O}$ is a union of orbits of smaller dimensions.

2.2. Properties of orbits

(3) For any $x \in R(\mathbf{v})$, the stabilizer subgroup

$$G_x = \{g \in \mathrm{GL}(\mathbf{v}) \mid g(x) = x\}$$

is a closed algebraic subgroup in $\mathrm{GL}(\mathbf{v})$, and we have natural isomorphisms

(2.3)
$$\begin{aligned}\mathbb{O}_x &= \mathrm{GL}(\mathbf{v})/G_x,\\ T_x\mathbb{O}_x &= T_1\mathrm{GL}(\mathbf{v})/T_1(G_x),\end{aligned}$$

where T_1 stands for the tangent space at identity. In particular,

(2.4)
$$\dim \mathbb{O}_x = \dim \mathrm{GL}(\mathbf{v}) - \dim G_x.$$

(4) There exists at most one orbit of dimension equal to $\dim R(\mathbf{v})$; if it exists, it is open and dense in $R(\mathbf{v})$ (in Zariski topology and, for $\mathbf{k} = \mathbb{C}$, also in the analytic topology).

In addition, we have the following result, which is specific to quiver representations.

Theorem 2.3. *For every $x \in R(\mathbf{v})$, the stabilizer G_x is connected in Zariski topology; for $\mathbf{k} = \mathbb{C}$, it is also connected in the analytic topology.*

Proof. By definition,

$$G_x = \{g \in \mathrm{GL}(\mathbf{v}) \mid g_j x_h g_i^{-1} = x_h \text{ for any edge } h\colon i \to j\} = \mathrm{Aut}_{\vec{Q}}(V^x),$$

where V^x is the representation of $\vec{Q}$ corresponding to $x \in R(\mathbf{v})$. But $\mathrm{Aut}_{\vec{Q}}(V)$ is the subset in $\mathrm{End}_{\vec{Q}}(V)$ consisting of operators satisfying $\det A \neq 0$. Thus, $\mathrm{Aut}_{\vec{Q}}(V)$ can be written as $L - X$, where L is a vector space and X is an algebraic subvariety of codimension 1. This immediately implies that it is Zariski connected (any open subset in $L - X$ is also open in L, and thus any two nonempty open subsets intersect).

For $\mathbf{k} = \mathbb{C}$, connectedness in the analytic topology follows from the observation that if $m, m' \in L - X$, then the complex line through m, m' intersects X at finitely many points and thus m, m' can be connected by a path in $L - X$. $\square$

Note that for $\mathbf{k} = \mathbb{R}$, the stabilizers can be disconnected in the analytic topology.

Finally, we notice that the tangent and normal spaces to the orbit can be described in terms of quiver representations.

Theorem 2.4. *Let $x \in R(\mathbf{v})$ and let $V^x \in \text{Rep}(\vec{Q})$ be the corresponding representation. Then*

(1) $T_x \mathbb{O}_x \simeq \text{End}(\mathbf{v})/\text{End}_{\vec{Q}}(V^x)$, *where* $\text{End}(\mathbf{v}) = \bigoplus_{i \in I} \text{End}(\mathbf{k}^{\mathbf{v}_i})$.

(2) *Let* $N_x \mathbb{O}_x = T_x R(\mathbf{v})/T_x \mathbb{O}_x$ *be the normal space at x to the orbit. Then* $N_x \mathbb{O}_x = \text{Ext}^1_{\vec{Q}}(V^x, V^x)$.

Proof. First, note that $T_1(\text{GL}(\mathbf{v})) = \text{End}(\mathbf{v})$. Next, since $G_x = \text{Aut}_{\vec{Q}}(V^x)$ (see the proof of Theorem 2.3), we have $T_1 G_x = \text{End}_{\vec{Q}}(V^x)$, which, together with (2.3), proves the first statement of the theorem.

To prove the second statement, we use the following lemma.

Lemma 2.5. *For any $V, W \in \text{Rep}(\vec{Q})$, we have an exact sequence*

$$0 \to \text{Hom}_{\vec{Q}}(V, W) \to \bigoplus_i \text{Hom}_{\mathbf{k}}(V_i, W_i)$$

$$\to \bigoplus_h \text{Hom}_{\mathbf{k}}(V_{s(h)}, W_{t(h)}) \to \text{Ext}^1_{\vec{Q}}(V, W) \to 0.$$

Indeed, this lemma is immediately obtained by applying the functor $\text{Hom}_{\vec{Q}}(-, W)$ to the standard resolution (1.12) and using $\text{Hom}(P(i), W) = W_i$, $\text{Ext}^1(P(i), W) = 0$.

In particular, applying this lemma for $V = W$, we get an exact sequence

$$0 \to \text{End}_{\vec{Q}}(V^x) \to \text{End}(\mathbf{v}) \to R(\mathbf{v}) \to \text{Ext}^1_{\vec{Q}}(V, V) \to 0.$$

Since by the first part of the theorem we have $\text{End}(\mathbf{v})/\text{End}_{\vec{Q}}(V^x) = T_x \mathbb{O}_x$, this exact sequence can be rewritten as the short exact sequence

$$0 \to T_x \mathbb{O}_x \to R(\mathbf{v}) \to \text{Ext}^1_{\vec{Q}}(V^x, V^x) \to 0,$$

thus proving the second statement of the theorem. $\square$

Corollary 2.6. *The orbit $\mathbb{O}_V$ is open iff $\text{Ext}^1_{\vec{Q}}(V, V) = 0$.*

2.3. Closed orbits

In this section, we discuss orbits corresponding to semisimple representations of $\vec{Q}$. We show that these are exactly the same as closed orbits.

Theorem 2.7. *Let $0 \to M_1 \to M \to M_2 \to 0$ be a short exact sequence of representations of $\vec{Q}$. Then*

$$\overline{\mathbb{O}}_M \supset \mathbb{O}_{M_1 \oplus M_2}.$$

2.3. Closed orbits

Proof. Choose an identification $M \simeq M_1 \oplus M_2$ as a vector space. Then each operator x_h, $h \in \Omega$, will be written in a block triangular form:
$$x_h = \begin{pmatrix} x_1 & x_{12} \\ 0 & x_2 \end{pmatrix},$$
where x_1, x_2 are the restrictions of x_h to M_1, M_2 respectively, and x_{12} is some linear operator $M_2 \to M_1$.

Now consider the one-parameter subgroup $g(t), t \in \mathbf{k}^\times$, in $\mathrm{GL}(\mathbf{v})$ defined by
$$g(t) = \begin{pmatrix} t & 0 \\ 0 & 1 \end{pmatrix}.$$
Then
$$g(t) x_h g(t)^{-1} = \begin{pmatrix} x_1 & t x_{12} \\ 0 & x_2 \end{pmatrix}.$$
Taking the limit as $t \to 0$, we see that the closure of the orbit $\mathbb{O}_x$ contains $\bar{x}_h = \begin{pmatrix} x_1 & 0 \\ 0 & x_2 \end{pmatrix}$, which corresponds to representation $M_1 \oplus M_2$. $\square$

It turns out that this in fact gives a complete answer to the question of inclusion for orbit closures. Namely, let
$$(2.5) \qquad V = V^{\geq 0} \supset V^{\geq 1} \supset \cdots \supset V^{\geq l} = \{0\}$$
be a decreasing filtration of V by subrepresentations. For any such filtration, define the associated graded representation by
$$\operatorname{gr} V = \bigoplus V^{\geq n}/V^{\geq n+1}.$$

Theorem 2.8. *Let V, V' be representations with $\dim V = \dim V'$. Then:*

(1) *If $V' \simeq \operatorname{gr} V$ for some filtration of the form (2.5), then $\mathbb{O}_{V'} \subset \overline{\mathbb{O}}_V$.*

(2) *If $\mathbb{O}_{V'}$ is closed, then the converse statement also holds: if $\mathbb{O}_{V'} \subset \overline{\mathbb{O}}_V$, then $V' \simeq \operatorname{gr} V$ for some filtration of the form (2.5).*

Proof. It easily follows from Theorem 2.7 that for each filtration of the form (2.5), we have $\mathbb{O}_{\operatorname{gr} V} \subset \overline{\mathbb{O}}_V$ (use induction in length of filtration).

To prove the converse statement, we will use some methods of the Geometric Invariant Theory; we will return to a detailed discussion of these methods later (see Chapter 9). For now, we only need the following result, the proof of which can be found in [**Kem1978**, Theorem 1.4].

Lemma 2.9. *Let $x \in R(\mathbf{v})$ and let $Y \subset R(\mathbf{v})$ be a closed $\mathrm{GL}(\mathbf{v})$-invariant set such that $Y \cap \overline{\mathbb{O}}_x \neq \varnothing$. Then there exists a one-parameter subgroup $\lambda \colon \mathbf{k}^\times \to \mathrm{GL}(\mathbf{v})$ such that $\lim_{t \to 0} \lambda(t).x \in Y$.*

Thus, assume that $\mathbb{O}_{V'} \subset \overline{\mathbb{O}}_{V^x}$. By the lemma, there exists a one-parameter subgroup $\lambda(t)$ such that the limit $\bar{x} = \lim_{t\to 0} \lambda(t).x$ exists and $V' \simeq V^{\bar{x}}$. This subgroup defines an algebraic action of $\mathbf{k}^\times$ on V; by a well-know result from the theory of algebraic groups (see [**Spr1998**]), we can decompose each V_i into eigenspaces:
$$V_i = \bigoplus_{n \in \mathbb{Z}} V_i^n, \qquad \lambda(t)|_{V_i^n} = t^n.$$
Thus, for each edge $h\colon i \to j$, we have $x_h = \bigoplus x_h^{mn}$, $x_h^{mn}\colon V_i^n \to V_j^m$. Obviously, we have $\lambda(t)x = \sum t^{m-n} x_h^{mn}$, so the limit $\bar{x} = \lim_{t \to 0} \lambda(t).x$ exists iff $x_h^{mn} = 0$ for all $m < n$. This implies that each of the spaces
$$V^{\geq k} = \bigoplus_{m \geq k} V^m$$
is x-stable, so we get a decreasing filtration of quiver representations
$$\cdots \supset V^{\geq -1} \supset V^{\geq 0} \supset V^{\geq 1} \supset \cdots$$
with $V^{\geq n} = 0$ for $n \gg 0$, and $V^{\geq n} = V$ for $n \ll 0$. The corresponding graded representation is given by $\operatorname{gr} V = \bigoplus V^{\geq n}/V^{\geq n+1} = \bigoplus V^n$ with the operators assigned to edges being $\bigoplus x_h^{nn} = \lim_{t\to 0} \lambda(t).x_h = \bar{x}_h$; thus, $\operatorname{gr} V \simeq V^{\bar{x}} \simeq V'$. $\square$

In particular, we can define the *semisimplification* of V to be the semisimple representation
$$(2.6) \qquad V^{ss} = \bigoplus W^n/W^{n+1},$$
where $V = W^0 \supset W^1 \supset \cdots \supset W^n = \{0\}$ is a composition series, i.e. a filtration such that all quotients W^n/W^{n+1} are simple. By the Jordan–Hölder theorem, V^{ss} does not depend on the choice of the composition series.

We can now formulate the main result of this section.

Theorem 2.10.

(1) *An orbit $\mathbb{O}_V$ is closed if and only if V is semisimple.*

(2) *Closure of every orbit $\mathbb{O}_V$ contains a unique closed orbit, namely $\mathbb{O}_{V^{ss}}$.*

Proof. By Theorem 2.8, $\mathbb{O}_V$ is closed if and only if for any filtration, we have $V \simeq \operatorname{gr} V$; it is easy to see that this happens iff V is semisimple. The second part immediately follows from Theorem 2.8 and the Jordan–Hölder theorem. $\square$

Corollary 2.11. *If $\vec{Q}$ has no oriented cycles, then $R(\mathbf{v})$ contains a unique closed orbit, $\{0\}$, which lies in the closure of any other orbit.*

2.3. Closed orbits

Indeed, by Theorem 1.10 in this case the only simple representations are $S(i)$, and thus for any semisimple representation one has $x_h = 0$.

Example 2.12. Consider the quiver ↻. In this case, $R(\mathbf{v}) = \mathrm{Mat}_n(\mathbf{k})$ is the space of $n \times n$ matrices, with the group $\mathrm{GL}(n)$ acting by conjugation. Thus, the orbits are exactly the conjugacy classes of matrices. By the results above, closed orbits correspond to diagonalizable matrices, and for every matrix A, the unique closed orbit contained in the closure $\overline{\mathbb{O}}_A$ is the orbit of the diagonalizable matrix A^{ss} with the same eigenvalues as A.

Finally, let us also study the orbits of maximal dimension (this, in particular, includes open orbits if they exist). It turns out that these orbits are "as indecomposable as possible". More precisely, we have the following theorem.

Theorem 2.13. *Let $\mathbb{O} \subset R(\mathbf{v})$ be an orbit of maximal possible dimension. Then the corresponding representation V has decomposition*

$$V \simeq \bigoplus I_k,$$

where I_k are indecomposable representations such that $\mathrm{Ext}^1(I_k, I_l) = 0$ for $k \neq l$.

Proof. Let $V = \bigoplus I_k$ be a decomposition of V as a direct sum of indecomposable representations. Assume that for some $k \neq l$, we have $\mathrm{Ext}^1(I_k, I_l) \neq 0$; without loss of generality, we may assume that $\mathrm{Ext}^1(I_2, I_1) \neq 0$. Then there exists a nontrivial extension $0 \to I_1 \to J \to I_2 \to 0$. For such an extension, $J \not\simeq I_1 \oplus I_2$: indeed, by applying the functor $\mathrm{Hom}(I_2, -)$ to this short exact sequence, we see that

$$\dim \mathrm{Hom}(I_2, J) = \dim \mathrm{Hom}(I_2, I_1) + \dim W < \dim \mathrm{Hom}(I_2, I_1 \oplus I_2),$$

where $W = \mathrm{Ker}(\mathrm{Hom}(I_2, I_2) \to \mathrm{Ext}^1(I_2, I_1))$.

By Theorem 2.8, in this case $\mathbb{O} \subset \overline{\mathbb{O}_{V'}}$, where $V' = J \oplus I_3 \oplus \cdots \not\simeq V$. But this contradicts the fact that the dimension of $\mathbb{O}$ is the maximal possible. $\square$

Chapter 3

Gabriel's Theorem

In this chapter, we will study representations of Dynkin quivers. The main result of this section is the famous theorem due to Gabriel: a quiver has a finite number of indecomposable representations if and only if it is Dynkin. Our proof is based on the use of so-called *reflection functors*, introduced by Bernstein, Gelfand, and Ponomarev.

3.1. Quivers of finite type

Let **k** be an algebraically closed field.

Definition 3.1. A quiver $\vec{Q}$ is of *finite type* if for any $\mathbf{v} \in \mathbb{Z}_+^I$, the number of isomorphism classes of indecomposable representations of dimension $\mathbf{v}$ is finite.

Example 3.2. It follows from results of Section 1.1 that quiver $\underset{1}{\bullet} \longrightarrow \underset{2}{\bullet}$ is of finite type.

The Jordan quiver $\bullet \circlearrowleft$ is not of finite type (over an algebraically closed field, for every d there is exactly one one-parameter family of indecomposable representations, namely J_λ, $\lambda \in \mathbf{k}$,—the Jordan block of size d with eigenvalue λ). The Kronecker quiver $\underset{1}{\bullet} \rightrightarrows \underset{0}{\bullet}$ is also not of finite type: in Example 1.5, we constructed an infinite number of representations of dimension $(1,1)$.

We will discuss the representation theory of the Kronecker quiver later, in Chapter 7.

The main result of this chapter is the following theorem, due to Gabriel [**Gab1972**].

Theorem 3.3. *A connected quiver $\vec{Q}$ is of finite type if and only if the underlying graph Q is Dynkin (see Definition 1.27). Such a quiver will be called a* Dynkin *quiver.*

Recall that in Section 1.6, we gave a complete list of all Dynkin graphs.

Representation theory of quivers which are not of finite type will be considered in Part 2.

We will now prove one direction (finite type implies Dynkin); the proof of the converse statement will be given in Section 3.3, where we also give detailed information about representations of such quivers.

Assume that $\vec{Q}$ is of finite type. Since every representation is a direct sum of indecomposables, this implies that for any $\mathbf{v} \in \mathbb{Z}_+^I$, there is a finite number of isomorphism classes of representations of dimension $\mathbf{v}$. By Theorem 2.1, this means that the action of the group $\mathrm{GL}(\mathbf{v})$ on the representation space $R(\vec{Q}, \mathbf{v})$ has only finitely many orbits; thus, there must exist an orbit $\mathbb{O}_x$ with $\dim \mathbb{O}_x = \dim R(\mathbf{v})$. Since $\dim \mathbb{O}_x = \dim \mathrm{GL}(\mathbf{v}) - \dim G_x$, where G_x is the stabilizer of x in $\mathrm{GL}(\mathbf{v})$, we get $\dim G_x = \dim \mathrm{GL}(\mathbf{v}) - \dim \mathbb{O}_x = \dim \mathrm{GL}(\mathbf{v}) - \dim R(\mathbf{v})$. On the other hand, since the subgroup of scalar matrices $\mathbf{k}^\times$ acts trivially in $R(\mathbf{v})$, $\dim G_x \geq 1$, so we have

$$\dim \mathrm{GL}(\mathbf{v}) - \dim R(\mathbf{v}) \geq 1,$$

$$\sum \mathbf{v}_i^2 - \sum_h \mathbf{v}_{s(h)} \mathbf{v}_{t(h)} \geq 1,$$

$$\langle \mathbf{v}, \mathbf{v} \rangle \geq 1,$$

where $\langle , \rangle$ is the Euler form (1.16). Thus, the Tits form $q(\mathbf{v}) = \langle \mathbf{v}, \mathbf{v} \rangle$ is positive for any nonzero $\mathbf{v} \in \mathbb{Z}_+^I$. It is easy to deduce from this that $q(\mathbf{v})$ is a positive definite quadratic form on $\mathbb{R}^I$. Thus, we have proved one direction of Gabriel's theorem: if a graph is of finite type, it must be a Dynkin graph.

3.2. Reflection functors

In this section we will introduce a new tool which will be used to complete the proof of Gabriel's theorem — reflection functors. Throughout this section, $\vec{Q}$ is an arbitrary quiver.

Definition 3.4. A vertex $i \in I$ is called a *sink* for $\vec{Q}$ if there are no edges $h \colon i \to j$.

Similarly, a vertex $i \in I$ is called a *source* for $\vec{Q}$ if there are no edges $h \colon j \to i$.

3.2. Reflection functors

Definition 3.5. Let $i \in I$ be a sink (respectively, a source) for $\vec{Q}$. We define the new quiver $\vec{Q}' = s_i^+(\vec{Q})$ (respectively, $\vec{Q}' = s_i^-(\vec{Q})$) by reversing the orientation of all edges incident to i:

$$\begin{array}{c}\text{(diagram)} \underset{s_i^-}{\overset{s_i^+}{\rightleftarrows}} \text{(diagram)}\end{array}$$

Note that if i is a sink for $\vec{Q}$, then i is a source for $\vec{Q}'$, and $s_i^- s_i^+(\vec{Q}) = \vec{Q}$.

Lemma 3.6. *If Q is a tree and Ω, Ω' are two orientations of Q, then Ω' can be obtained from Ω by a sequence of operations $s_i^\pm$.*

Proof. The proof goes by induction in the number of vertices. Let $k \in I$ be a vertex which is connected to exactly one other vertex; since Q is a tree, such a vertex exists. Let Q' be the graph obtained from Q by removing vertex k and the incident edge h; by the induction assumption, one can find a sequence of reflections $s_{i_1}^\pm, \ldots, s_{i_l}^\pm$ such that $s_{i_l}^\pm \ldots s_{i_1}^\pm \Omega$ coincides with Ω' on Q'. Multiplying if necessary by $s_k^\pm$ (depending on whether k is a source or sink), we see that Ω' can be obtained from Ω by a sequence of operations $s_i^\pm$. □

We will give a more precise statement later (see Corollary 6.18).

We now study the relation between representations of $\vec{Q}$ and $\vec{Q}'$.

Definition 3.7.

(1) Let $i \in I$ be a sink for $\vec{Q}$; denote $\vec{Q}' = s_i^+(\vec{Q})$. Define the functor $\Phi_i^+ \colon \text{Rep}(\vec{Q}) \to \text{Rep}(\vec{Q}')$ by

$$(\Phi_i^+(V))_j = V_j, \qquad j \neq i,$$
$$(\Phi_i^+(V))_i = \text{Ker}\left(\left(\bigoplus V_k\right) \to V_i\right),$$

where the sum is taken over all edges $h \colon k \to i$. We define on $\Phi_i^+(V)$ the structure of representation of $\vec{Q}'$ by defining for every edge $h \colon i \to k$ in $\vec{Q}'$ the corresponding operator x_h as the composition of embedding and projection

$$\text{Ker}\left(\left(\bigoplus V_k\right) \to V_i\right) \to \left(\bigoplus V_k\right) \to V_k.$$

Operators x_h for edges not incident to i are defined in the obvious way. The action of Φ_i^+ on morphisms is also defined in the obvious way.

(2) Let $i \in I$ be a source for $\vec{Q}$; denote $\vec{Q}' = s_i^-(\vec{Q})$. Define the functor $\Phi_i^- \colon \operatorname{Rep}(\vec{Q}) \to \operatorname{Rep}(\vec{Q}')$ by

$$(\Phi_i^-(V))_j = V_j, \quad j \neq i,$$
$$(\Phi_i^-(V))_i = \operatorname{Coker}\left(V_i \to \left(\bigoplus V_k\right)\right),$$

where the sum is taken over all edges $h \colon i \to k$. We define on $\Phi_i^-(V)$ the structure of representation of $\vec{Q}'$ by defining for every edge $h \colon k \to i$ in $\vec{Q}'$ the corresponding operator x_h as the composition of embedding and projection

$$V_k \to \left(\bigoplus V_k\right) \to \operatorname{Coker}\left(V_i \to \left(\bigoplus V_k\right)\right).$$

Operators x_h for edges not incident to i are defined in the obvious way. The action of Φ_i^- on morphisms is also defined in the obvious way.

The functors $\Phi_i^{\pm}$ were introduced by Bernstein, Gelfand, and Ponomarev [**BGP1973**] and are usually called *reflection functors* or *BGP reflection functors*.

Example 3.8. Let $\vec{Q} = \underset{1}{\bullet} \longrightarrow \underset{2}{\bullet}$. Then

$$\Phi_1^-\left(\underset{1}{\overset{\mathbf{k}}{\bullet}} \longrightarrow \underset{2}{\overset{0}{\bullet}}\right) = \underset{1}{\overset{0}{\bullet}} \longleftarrow \underset{2}{\overset{0}{\bullet}},$$

$$\Phi_2^+\left(\underset{1}{\overset{\mathbf{k}}{\bullet}} \longrightarrow \underset{2}{\overset{0}{\bullet}}\right) = \underset{1}{\overset{\mathbf{k}}{\bullet}} \overset{1}{\longleftarrow} \underset{2}{\overset{\mathbf{k}}{\bullet}},$$

$$\Phi_2^-\left(\underset{1}{\overset{\mathbf{k}}{\bullet}} \overset{1}{\longleftarrow} \underset{2}{\overset{\mathbf{k}}{\bullet}}\right) = \underset{1}{\overset{\mathbf{k}}{\bullet}} \longrightarrow \underset{2}{\overset{0}{\bullet}}.$$

Example 3.9. If i is a sink, then $\Phi_i^+(S(i)) = 0$. Similarly, if i is a source, then $\Phi_i^-(S(i)) = 0$.

The following theorems summarize some properties of these functors. We use notation $R^n \Phi$ for the derived functors of a left exact functor Φ, and $L^n \Phi$ for the derived functors of a right exact functor Φ.

Theorem 3.10. *The functor Φ_i^+ is left exact. Moreover, $R^n \Phi_i^+(V) = 0$ for all $n > 1$, and $R^1 \Phi_i^+(V) = 0$ iff V satisfies the following condition:*

$$(3.1) \qquad \left(\bigoplus_{h \colon k \to i} V_k\right) \to V_i \qquad \text{is surjective.}$$

We denote by $\operatorname{Rep}^{i,-}$ the full subcategory of $\operatorname{Rep}(\vec{Q})$ consisting of representations satisfying condition (3.1).

3.2. Reflection functors

Similarly, the functor Φ_i^- is right exact. Moreover, $L^n\Phi_i^-(V) = 0$ for all $n > 1$, and $L^1\Phi_i^-(V) = 0$ iff V satisfies the following condition:

$$(3.2) \qquad V_i \to \left(\bigoplus_{h:\, i \to k} V_k \right) \quad \text{is injective.}$$

We denote by $\operatorname{Rep}^{i,+}$ the full subcategory of $\operatorname{Rep}(\vec{Q})$ consisting of representations satisfying condition (3.2).

Proof. The most natural way to prove this theorem is as follows. Let $\operatorname{Com}(\vec{Q})$ be the category of complexes of representations of $\vec{Q}$; as is well known, this is an abelian category (see [**GM2003**]). Its objects can be described as collections of complexes $V_i^\bullet$ for each vertex $i \in I$, together with morphisms of complexes $x_h \colon V_i^\bullet \to V_j^\bullet$ for each $h \colon i \to j$.

Define the functor $R\Phi_i^+ \colon \operatorname{Rep}(\vec{Q}) \to \operatorname{Com}(\vec{Q}')$ by $R\Phi_i^+(V)_j = V_j$ for $j \ne i$ (considered as a complex concentrated in degree zero), and

$$R\Phi_i^+(V)_i = \left(\bigoplus V_k \to V_i \right)$$

(as a complex in degrees 0 and 1); the maps x_h are defined in the obvious way. Then one easily sees that $R\Phi_i^+$ is exact: it sends every short exact sequence of representations of $\vec{Q}$ to a short exact sequence of complexes, and $H^0(R\Phi_i^+) = \Phi_i^+$, $H^n(R\Phi_i^+) = 0$ for $n \ne 0, 1$, and $H^1(R\Phi_i^+)$ is the representation which is zero at vertices $j \ne i$, and at vertex i it has the vector space $\operatorname{Coker}(\bigoplus V_k \to V_i)$. Thus, $H^n(R\Phi_i^+(V)) = 0$ for $n > 0$ iff the map $\bigoplus V_k \to V_i$ is surjective. The statement about $L\Phi_i^-$ is proved in a similar way. $\square$

Theorem 3.11. *Let i be a sink for $\vec{Q}$, and let $\vec{Q}' = s_i^+(\vec{Q})$. Let $\operatorname{Rep}^{i,\pm}$ be as defined in Theorem 3.10.*

(1) *Functors Φ_i^+, Φ_i^- are adjoint to each other: for any $A \in \operatorname{Rep}(\vec{Q}')$, $B \in \operatorname{Rep}(\vec{Q})$, we have a natural isomorphism*

$$\operatorname{Hom}_{\vec{Q}}(\Phi_i^-(A), B) \simeq \operatorname{Hom}_{\vec{Q}'}(A, \Phi_i^+(B)).$$

(2) *For any $V \in \operatorname{Rep}(\vec{Q})$, $\Phi_i^+(V) \in \operatorname{Rep}^{i,+}(\vec{Q}')$; similarly, for any $V' \in \operatorname{Rep}(\vec{Q}')$, $\Phi_i^-(V') \in \operatorname{Rep}^{i,-}(\vec{Q})$.*

(3) *The restrictions*

$$\Phi_i^+ \colon \operatorname{Rep}^{i,-}(\vec{Q}) \to \operatorname{Rep}^{i,+}(\vec{Q}'),$$
$$\Phi_i^- \colon \operatorname{Rep}^{i,+}(\vec{Q}') \to \operatorname{Rep}^{i,-}(\vec{Q})$$

are inverse to each other and thus give equivalence of categories $\operatorname{Rep}^{i,-}(\vec{Q}) \simeq \operatorname{Rep}^{i,+}(\vec{Q}')$.

(4) If $V \in \text{Rep}^{i,-}(\vec{Q})$, then $\dim(\Phi_i^+(V)) = s_i(\dim V)$, where $s_i = s_{\alpha_i} \colon \mathbb{Z}^I \to \mathbb{Z}^I$ is the simple reflection (1.26).

Similarly, if $V' \in \text{Rep}^{i,+}(\vec{Q}')$, then $\dim(\Phi_i^-(V')) = s_i(\dim V')$.

Proof. The first part is given by explicit computation: it follows from the definition of Φ_i^- that a morphism $f \colon \Phi_i^-(A) \to B$ is completely determined by $\{f_j\}_{j \neq i}$, which have to satisfy the additional condition that the composition
$$A_i \to \bigoplus A_k \xrightarrow{f} \bigoplus B_k \to B_i$$
is zero (the sum is over all edges $h \colon k \to i$ in Q, respectively $i \to k$ in $\vec{Q}'$). Similarly, the right-hand side can also be identified with the same space.

The second part is immediate from the definitions. The isomorphism $\Phi_i^- \Phi_i^+(V) \simeq V$ follows from
$$\text{Coker}\Big(\text{Ker}(W \to V_i) \to W\Big) = W/\text{Ker}(W \to V_i) \simeq \text{Im}(W \to V_i) = V_i$$
if $W \to V_i$ is surjective (here we denoted $W = \bigoplus_{h\colon k \to i} V_k$). Isomorphism $\Phi_i^+ \Phi_i^-(V') \simeq V'$ is proved similarly.

Proof of the last part is given by an easy explicit computation. $\square$

Note that the proof also implies that for any $V \in \text{Rep}(\vec{Q})$ (not necessarily from $\text{Rep}^{i,-}(\vec{Q})$), we have a short exact sequence
$$0 \to \Phi_i^- \Phi_i^+(V) \to V \to \tilde{V} \to 0$$
where $\tilde{V}$ is concentrated at vertex i: $\tilde{V} \simeq nS(i)$.

For readers familiar with the theory of derived categories, we note that Theorem 3.10 and Theorem 3.11 can be reformulated in a more elegant way using the language of derived categories, as outlined in [**GM2003**, IV.4, Exercise 6]. Namely, denote by $D^b(\vec{Q}) = D^b(\text{Rep}(\vec{Q}))$ the derived category of representations of $\vec{Q}$; then the functors $\Phi_i^\pm$ can be easily extended to exact functors $R\Phi_i^+$, $L\Phi_i^-$ between $D^b(\vec{Q})$, $D^b(\vec{Q}')$.

Theorem 3.12. *In the assumptions of Theorem 3.11, the functors*
$$R\Phi_i^+ \colon D^b(\vec{Q}) \to D^b(\vec{Q}'),$$
$$L\Phi_i^- \colon D^b(\vec{Q}') \to D^b(\vec{Q})$$
are inverse to each other and thus give an equivalence of derived categories $D^b(\vec{Q}) \simeq D^b(\vec{Q}')$. Moreover, consider $\text{Rep} V$ as a subcategory in $D^b(\vec{Q})$ in the usual way; then, for $V \in \text{Rep}(\vec{Q})$, $R\Phi_i^+(V) \in \text{Rep}(\vec{Q}')$ iff $V \in \text{Rep}^{i,-}(\vec{Q})$, and similarly for $L\Phi_i^-$.

We skip the proof of this theorem.

3.2. Reflection functors

Corollary 3.13. *If i is a sink for $\vec{Q}$, $\vec{Q}' = s_i^+ \vec{Q}$, and $V \in \mathrm{Rep}^{i,-}(\vec{Q})$, $W \in \mathrm{Rep}(\vec{Q})$, then*
$$\mathrm{Hom}_{\vec{Q}}(V, W) = \mathrm{Hom}_{\vec{Q}'}(\Phi_i^+ V, \Phi_i^+ W).$$

Similarly, if i is a source for $\vec{Q}$, $\vec{Q}' = s_i^- \vec{Q}$, and $V \in \mathrm{Rep}(\vec{Q})$, $W \in \mathrm{Rep}^{i,+}(\vec{Q})$, then
$$\mathrm{Hom}_{\vec{Q}}(V, W) = \mathrm{Hom}_{\vec{Q}'}(\Phi_i^- V, \Phi_i^- W).$$

Proof. This immediately follows from Theorem 3.11: if $V \in \mathrm{Rep}^{i,-}(\vec{Q})$, then $V = \Phi_i^-(X)$ for some $X \in \mathrm{Rep}^{i,+}(\vec{Q}')$, so
$$\mathrm{Hom}_{\vec{Q}}(V, W) = \mathrm{Hom}_{\vec{Q}}(\Phi_i^-(X), W) = \mathrm{Hom}_{\vec{Q}'}(X, \Phi_i^+(W))$$
$$= \mathrm{Hom}_{\vec{Q}'}(\Phi_i^+ V, \Phi_i^+ W).$$

The second part is proved in the same way. $\square$

A similar statement for Ext functors is proved below.

Theorem 3.14. *Let $V \in \mathrm{Rep}(\vec{Q})$ be a nonzero indecomposable representation and let $i \in I$ be a sink (respectively, a source) for $\vec{Q}$. Then:*

(1) *If $V \simeq S(i)$, then $\Phi_i^\pm(V) = 0$.*

(2) *If $V \not\simeq S(i)$, then $V \in \mathrm{Rep}^{i,\mp}(\vec{Q})$, and $V' = \Phi_i^\pm(V)$ is a nonzero indecomposable representation of $\vec{Q}' = s_i^\pm(\vec{Q})$. In this case, $V \simeq \Phi^\mp(V')$.*

Proof. Part (1) is obvious; for part (2), we consider the case when i is a source. Then, note that if $f \colon V_i \to \bigoplus(V_k)$ is not injective, then we can write $V_i = \mathrm{Ker}(f) \oplus V_i''$, which gives the direct sum decomposition $V = V' \oplus V''$, where $V_i' = \mathrm{Ker}\, f$, $V_j' = 0$ for $j \neq i$, and $V_j'' = V_j$ for $j \neq i$. This contradicts the assumption that V is indecomposable. $\square$

Recall that for an element $\mathbf{v} \in \mathbb{Z}^I$, we write $\mathbf{v} \geqslant 0$ if $\mathbf{v}_i \geq 0$ for all i (see (1.23)).

Corollary 3.15. *In the assumptions of Theorem 3.14, $\Phi_i^\pm(V)$ is a nonzero indecomposable iff $s_i(\mathbf{dim}\, V) \geqslant 0$.*

Corollary 3.16. *If i is a sink for $\vec{Q}$, $\vec{Q}' = s_i^+ \vec{Q}$, and $V \in \mathrm{Rep}(\vec{Q})$, $W \in \mathrm{Rep}^{i,-}(\vec{Q})$, then*
$$\mathrm{Ext}^1_{\vec{Q}}(V, W) = \mathrm{Ext}^1_{\vec{Q}'}(\Phi_i^+ V, \Phi_i^+ W).$$

Similarly, if i is a source for $\vec{Q}$, $\vec{Q}' = s_i^- \vec{Q}$, and $V \in \mathrm{Rep}^{i,+} \vec{Q}$, $W \in \mathrm{Rep}(\vec{Q})$, then
$$\mathrm{Ext}^1_{\vec{Q}}(V, W) = \mathrm{Ext}^1_{\vec{Q}'}(\Phi_i^- V, \Phi_i^- W).$$

Proof. For the first statement, it suffices to consider the case when V is indecomposable. If $V = S(i) = P(i)$, then both sides are zero. Thus, we can assume that $V \not\simeq S(i)$, in which case, by Theorem 3.14, $V \in \text{Rep}^{i,-}(\vec{Q})$.

Let

(3.3) $$0 \to W \to A \to V \to 0$$

be an extension representing some class in $\text{Ext}^1(V, W)$. Note that since $R^1\Phi_i^+(V) = R^1\Phi_i^+(W) = 0$, a long exact sequence of derived functors implies that $R^1\Phi_i^+(A) = 0$, so $A \in \text{Rep}^{i,-}(\vec{Q})$.

Since $W \in \text{Rep}^{i,-}(\vec{Q})$, we have $R^1\Phi_i^+(W) = 0$, so the sequence

$$0 \to \Phi_i^+(W) \to \Phi_i^+(A) \to \Phi_i^+(V) \to 0$$

is also exact; this gives a map $\text{Ext}^1_{\vec{Q}}(V, W) \to \text{Ext}^1_{\vec{Q}'}(\Phi_i^+ V, \Phi_i^+ W)$.

A similar argument shows that Φ_i^- gives a map $\text{Ext}^1_{\vec{Q}'}(\Phi_i^+ V, \Phi_i^+ W) \to \text{Ext}^1_{\vec{Q}}(V, W)$. The fact that these two maps are inverse to each other is obvious. $\square$

Combining Corollary 3.13 and Corollary 3.16, we get the following result.

Theorem 3.17. *If i is a source for $\vec{Q}$, $\vec{Q}' = s_i^- \vec{Q}$ and $I_1, I_2 \in \text{Rep}(\vec{Q})$ are nonzero indecomposable representations such that $\Phi_i^-(I_1)$, $\Phi_i^-(I_2)$ are nonzero, then*

$$\text{Hom}_{\vec{Q}}(I_1, I_2) = \text{Hom}_{\vec{Q}'}(\Phi_i^- I_1, \Phi_i^- I_2),$$
$$\text{Ext}^1_{\vec{Q}}(I_1, I_2) = \text{Ext}^1_{\vec{Q}'}(\Phi_i^- I_1, \Phi_i^- I_2).$$

3.3. Dynkin quivers

In this section, we assume that $\vec{Q}$ is a Dynkin quiver. In particular, this implies that Q is a tree.

We will be heavily using results and terminology from the theory of root systems; see Appendix A for an overview and references. In particular, we will be using the following notation, introduced in Section 1.7:

$L = \bigoplus \mathbb{Z}\alpha_i$ — the root lattice.

$(\,,\,)$ the symmetrized Euler form.

$W = \langle s_i \rangle, i = 1, \ldots, r$ — the Weyl group and simple reflections.

$R = \{\alpha \in L \mid (\alpha, \alpha) = 2\} \subset L$ — the root system.

$R = R_+ \cup R_-$ — positive and negative roots.

Note that in the Dynkin case, both W and R are finite (see Theorem A.11).

3.3. Dynkin quivers

We will identify $L \simeq \mathbb{Z}^I$, $\sum n_i \alpha_i \mapsto (n_1, \ldots, n_r)$. By Theorem 1.15, this gives the identification

(3.4)
$$K(\vec{Q}) \simeq L$$
$$[V] \mapsto \mathbf{dim}(V) = \sum (\dim V_i) \alpha_i.$$

Under this identification, classes of simple representations $[S(i)]$ are identified with simple roots α_i.

The main result of this section is the following theorem.

Theorem 3.18. *Let $\vec{Q}$ be a Dynkin quiver. Then the map $[V] \mapsto \mathbf{dim}\, V$ gives a bijection between the set $\mathrm{Ind}(\vec{Q})$ of isomorphism classes of nonzero indecomposable representations of $\vec{Q}$ and the set R_+ of positive roots.*

We will denote by I_α the indecomposable representation of graded dimension $\alpha \in R_+$.

Remark 3.19. This theorem has a generalization to non-Dynkin quivers, due to Kac; it will be discussed in Section 7.11.

Before proving this theorem, note that it immediately implies the second part of Gabriel's theorem.

Corollary 3.20. *Any Dynkin quiver is of finite type; moreover, there are only finitely many isomorphism classes of indecomposable representations. This number and their dimensions do not depend on the choice of ground field* **k**.

Example 3.21. Let $\vec{Q}$ be the quiver below, with $n-1$ vertices:

$$\bullet \longrightarrow \bullet \longrightarrow \bullet \longrightarrow \cdots \longrightarrow \bullet \longrightarrow \bullet$$

The corresponding graph Q is the Dynkin diagram A_{n-1}. As is well known, the corresponding root lattice can be described as the subgroup in $\mathbb{Z}^n$ generated by $\alpha_i = e_i - e_{i+1}$ (where e_i is the standard basis in $\mathbb{Z}^n$). Positive roots are $\alpha_{ij} = e_i - e_j = \alpha_i + \alpha_{i+1} + \cdots + \alpha_{j-1}$, $1 \leq i < j \leq n$. For each such root α_{ij}, it is easy to construct an indecomposable representation I^{ij} with graded dimension α_{ij}:

$$\cdots \to \underset{i-1}{0} \to \underset{i}{\mathbf{k}} \to \underset{i+1}{\mathbf{k}} \to \cdots \to \underset{j-2}{\mathbf{k}} \to \underset{j-1}{\mathbf{k}} \to \underset{j}{0} \to \cdots$$

(all spaces not shown are zero; all maps $\mathbf{k} \to \mathbf{k}$ are identity maps).

Thus, Theorem 3.18 implies that representations $\{I^{ij}, 1 \leq i < j \leq n\}$ are the full list of nonzero indecomposable representations of $\vec{Q}$.

Example 3.22. Let

(3.5)

The corresponding graph is the Dynkin diagram of type D_4. It is known that the corresponding root system has 12 positive roots. On the other hand, we had given 12 nonisomorphic indecomposable representations of this quiver in Example 1.13. Thus, Theorem 3.18 implies that the list given in Example 1.13 is the full list of possible nonzero indecomposable representations. In particular, this gives a complete answer to the problem of classifying triples of subspaces in a vector space.

The proof of Theorem 3.18 will occupy the rest of this section. Our proof closely follows the arguments of [**BGP1973**].

Proof of Theorem 3.18. The idea of the proof is to construct indecomposable representations by applying the reflection functors to simple representations, much like one constructs all roots by applying simple reflections to simple roots. However, constructing representations is trickier than constructing roots: we can only apply the reflection functor $\Phi_i^\pm$ if i is a sink (respectively, source) for $\vec{Q}$. This justifies the following definition.

Definition 3.23. A sequence of indices $i_1, \ldots, i_k \in I$ (respectively, a reduced expression $w = s_{i_k} \ldots s_{i_1}$ for an element $w \in W$) is called *adapted* to an orientation Ω of Q if

i_1 is a sink for $\vec{Q} = (Q, \Omega)$,

i_2 is a sink for $s_{i_1}^+(\vec{Q})$,

$\vdots$

i_k is a sink for $s_{i_{k-1}}^+ \ldots s_{i_1}^+(\vec{Q})$.

Lemma 3.24. *Let* $\mathbf{v} \in \mathbb{Z}_+^I$, $\mathbf{v} \neq 0$. *Then there exists a sequence* $i_1, \ldots, i_{k+1} \in I$ *adapted to* $\vec{Q}$ *such that* $\mathbf{v} \geq 0$, $s_{i_1}(\mathbf{v}) \geq 0$, $\ldots$, $s_{i_k} \ldots s_{i_1}(\mathbf{v}) \geq 0$, *but* $s_{i_{k+1}} \ldots s_{i_1}(\mathbf{v}) \not\geq 0$.

The proof of this lemma will be given in Section 3.4.

Now, let V be an indecomposable representation of $\vec{Q}$. Denote $\mathbf{v} = \mathbf{dim}\, V$ and let $i_1, \ldots, i_{k+1}$ be a sequence of indices as in Lemma 3.24. By

Corollary 3.15, we see that

$\Phi^+_{i_1}(V)$ is a nonzero indecomposable representation of $s^+_{i_1}(\vec{Q})$,

$\Phi^+_{i_2}\Phi^+_{i_1}(V)$ is a nonzero indecomposable representation of $s^+_{i_2}s^+_{i_1}(\vec{Q})$,

$\vdots$

$V' = \Phi^+_{i_k} \ldots \Phi^+_{i_1}(V)$ is a nonzero indecomposable representation of $\vec{Q}' = s^+_{i_k} \ldots s^+_{i_1}(\vec{Q})$.

However, since $s_{i_{k+1}}(\mathbf{dim}\, V') \not\geq 0$, we must have $V' \simeq S(i_{k+1})$ — an irreducible representation of $\vec{Q}'$ (see Corollary 3.15). Thus, by Theorem 3.11, we see that we must have

(3.6) $$V \simeq \Phi^-_{i_1} \ldots \Phi^-_{i_k}(S(i_{k+1})).$$

In particular, this implies that $\mathbf{dim}\, V = s_{i_1} \ldots s_{i_k}(\alpha_{i_{k+1}})$ is a root and thus a positive root, and V is uniquely determined by $\mathbf{dim}\, V$.

Conversely, if $\mathbf{v}$ is a positive root, then choose a sequence $i_1, \ldots, i_{k+1}$ as in Lemma 3.24. Then, $\alpha' = s_{i_k} \ldots s_{i_1}(\mathbf{v})$ is a positive root, but $s_{i_{k+1}}(\alpha') \not\geq 0$. This implies that $\alpha' = \alpha_{i_{k+1}}$, so $\mathbf{v} = s_{i_1} \ldots s_{i_k}(\alpha_{i_{k+1}})$. Thus, if we define V by (3.6), then V will be an indecomposable representation with $\mathbf{dim}\, V = \mathbf{v}$. $\square$

This proof, together with Theorem 3.11, also gives the following result.

Theorem 3.25. *Let $\vec{Q}$ be a Dynkin quiver. Then for any nonzero indecomposable representation I, we have $\mathrm{End}_{\vec{Q}}(I) = \mathbf{k}$, $\mathrm{Ext}^1_{\vec{Q}}(I,I) = 0$.*

Indeed, for a simple representation $I = S(i)$ this was shown in Example 1.21; since any nonzero indecomposable representation is obtained from $S(i)$ by a sequence of reflection functors, Theorem 3.17 implies that the same holds for any I.

Corollary 3.26. *If $\vec{Q}$ is a Dynkin quiver, then for any nonzero indecomposable representation I_α, the corresponding orbit $\mathbb{O}_I \subset R(\vec{Q}, \alpha)$ is open.*

Indeed, this immediately follows from the previous theorem and the fact that the orbit is open iff $\mathrm{Ext}^1(V,V) = 0$ (Corollary 2.6).

3.4. Coxeter element

In this section, we give a proof of Lemma 3.24. Our proof is based on the use of a special element in the Weyl group, called the Coxeter element. We list the basic definitions and results here; details can be found in [**Hum1990**].

For future use, we consider not only Dynkin quivers but any quivers without edge loops. As before, we denote by $L = \mathbb{Z}^I$ the root lattice and by W the group generated by simple reflections (which in general is not necessarily finite).

Definition 3.27. An element $C \in W$ is called a *Coxeter element* if for some ordering of simple real roots $\Pi = \{\alpha_1, \ldots, \alpha_r\}$, we have

$$C = s_r \ldots s_1.$$

Example 3.28. For the root system of type A_{n-1} and the standard choice of simple roots, $\alpha_i = e_i - e_{i+1}$, we get $C = (1\ 2)(2\ 3)\ldots(n-1\ n) = (1\ 2\ \ldots\ n)$ — the cycle of length n.

Of course, there are many different Coxeter elements; however, it can be shown that in the Dynkin case, they are all conjugate in W (see [**Hum1990**, **Ste1959**]). In this case, the order of a Coxeter element is called the Coxeter number of the root system; we will denote it by h.

As in (1.23), for an element $\alpha = \sum n_i \alpha_i \in L$ we will write

$$\alpha \geqslant 0 \iff n_i \geq 0 \text{ for all } i.$$

Theorem 3.29. *Let $C \in W$ be a Coxeter element, $x \in L \otimes_{\mathbb{Z}} \mathbb{R}$. Then $Cx = x$ iff x is in the radical of the form $(\ ,\ )$.*

Proof. Assume $C = s_r \ldots s_1$ and $Cx = x$. Then $s_{r-1} \ldots s_1(x) = s_r(x)$, so $s_{r-1} \ldots s_1(x) - x = s_r(x) - x$. But by the definition of simple reflections, the right-hand side is a multiple of α_r, whereas the left-hand side is a linear combination of $\alpha_1, \ldots, \alpha_{r-1}$. Since simple roots are linearly independent, this implies $s_{r-1} \ldots s_1(x) = s_r(x) = x$, so $(x, \alpha_r) = 0$. Repeating the argument, we get $(x, \alpha_i) = 0$ for all i. □

Corollary 3.30. *Let Q be Dynkin. Then:*

(1) *The operator $C - 1$ is invertible in $L \otimes_{\mathbb{Z}} \mathbb{R}$.*

(2) *For any $\alpha \in L$, $\alpha \neq 0$, there exists $k \geq 0$ such that $C^k \alpha \not\geqslant 0$.*

Proof. The first part is obvious, since in the Dynkin case the form $(\ ,\ )$ is positive definite.

To prove the second statement, let h be the order of C; then $\frac{C^h - 1}{C - 1} = C^{h-1} + \cdots + 1 = 0$, so $C^{h-1}\alpha + \cdots + \alpha = 0$. Thus, it is impossible that all of $C^k \alpha \geqslant 0$. □

Now, given a quiver $\vec{Q}$ without oriented cycles we can order the vertices, writing $I = \{i_1, \ldots, i_r\}$ so that the sequence $i_1, \ldots, i_r$ is adapted to $\vec{Q}$, i.e.

(3.7)
$$i_1 \text{ is a sink for } \vec{Q},$$
$$i_2 \text{ is a sink for } s^+_{i_1}\vec{Q},$$
$$\vdots$$
$$i_r \text{ is a sink for } s^+_{i_{r-1}} \ldots s^+_{i_1}\vec{Q}$$

(for example, we can order vertices so that for any edge $h\colon i \to j$, we have $i > j$; since $\vec{Q}$ has no oriented cycles, this is always possible).

Consider the corresponding Coxeter element
$$C = s_{i_r} \ldots s_{i_1}.$$

Such a Coxeter element will be called *adapted* to $\vec{Q}$. In this case, we can define a sequence of quivers
$$\vec{Q}, \quad s^+_{i_1}\vec{Q}, \quad s^+_{i_2}s^+_{i_1}\vec{Q}, \quad \ldots, \quad C^+\vec{Q} = s^+_{i_r} \ldots s^+_{i_1}\vec{Q}$$

and the last quiver in this sequence is the same as the first one: $C^+\vec{Q} = \vec{Q}$. Indeed, for every edge h its orientation was reversed twice.

Now we can give a proof of Lemma 3.24.

Proof of Lemma 3.24. Let $\vec{Q}$ be a Dynkin quiver and let $\mathbf{v} \in \mathbb{Z}^I_+ \simeq L_+$, $\mathbf{v} \neq 0$. Choose an ordering of vertices $I = \{i_1, \ldots, i_r\}$ as in (3.7). Consider the sequence
$$\mathbf{v}, \; s_{i_1}(\mathbf{v}), \; \ldots, \; s_{i_r} \ldots s_{i_1}(\mathbf{v}) = C(\mathbf{v}),$$
$$s_{i_1} C(\mathbf{v}), \ldots, C^2(\mathbf{v}),$$
$$\vdots$$

Since this sequence contains $C^k(\mathbf{v})$ for all k, by Corollary 3.30 this sequence must contain an element $\mathbf{v}' \not\geq 0$. This completes the proof of the lemma. □

3.5. Longest element and ordering of positive roots

In this section $\vec{Q}$ is a Dynkin quiver; we keep the notation of Section 3.3.

We showed in Section 3.3 that for any positive root $\alpha \in R_+$, there is a unique up to an isomorphism indecomposable representation I_α of $\vec{Q}$, and this representation can be constructed by applying a sequence of reflection functors Φ^-_i to a simple representation. In this section, we refine this result, showing how one can get all indecomposable representations from a single sequence of reflection functors; in particular, it will give us a natural order on the set of positive roots.

Recall that the Weyl group W contains a unique element w_0 of length $l(w_0) = |R_+|$ (the longest element). This element satisfies $w_0(R_+) = R_-$. Moreover, if we choose a reduced expression

(3.8) $$w_0 = s_{i_l} \ldots s_{i_1}, \qquad l = |R_+|,$$

then one can write $R_+ = \{\gamma_1, \ldots, \gamma_l\}$, where

(3.9)
$$\gamma_1 = \alpha_{i_1},$$
$$\gamma_2 = s_{i_1}(\alpha_{i_2}),$$
$$\vdots$$
$$\gamma_l = s_{i_1} s_{i_2} \ldots s_{i_{l-1}}(\alpha_{i_l}).$$

So the defined γ_k satisfy $\gamma_k \in R_+$, $s_{i_1}\gamma_k \in R_+$, ..., $s_{i_{k-1}} \ldots s_{i_1}(\gamma_k) \in R_+$, but $s_{i_k} \ldots s_{i_1}(\gamma_k) \in R_-$ (see Appendix A and the references there).

The following theorem provides a relation between w_0 and the Coxeter element. Let us choose a partition $I = I_0 \sqcup I_1$ of vertices of Q so that for every edge, one of the endpoints will be in I_0 and the other in I_1 (since Q is a tree, it is easy to see that such a partition is possible). Define $c_0, c_1 \in W$ by $c_0 = \prod_{i \in I_0} s_i$, $c_1 = \prod_{i \in I_1} s_i$; note that the order of factors in c_0, c_1 is irrelevant, as all s_i, $i \in I_0$, commute, and similarly for $s_i, i \in I_1$. By definition, the product $C = c_1 c_0$ is a Coxeter element. The following result was proved in [**Ste1959**, Theorem 6.3].

Theorem 3.31. *If c_0, c_1 are as defined above, then*
$$w_0 = \ldots c_0 c_1 c_0 \quad (h \text{ factors})$$
$$= \ldots c_1 c_0 c_1 \quad (h \text{ factors})$$
are reduced expressions for the longest element $w_0 \in W$ (here h is the Coxeter number).

The proof is based on constructing a special two-dimensional real plane $P \subset L \otimes_{\mathbb{Z}} \mathbb{R}$ which is stable under c_0, c_1 and such that c_0, c_1 act on it by reflections around lines forming an angle of π/h. Details can be found in [**Hum1990**, Section 3.17] or in the original paper [**Ste1959**]. Note that in particular, if h is even, it gives $w_0 = C^{h/2}$ (for $C = c_1 c_0$); if h is odd, then it implies that $l(c_0) = l(c_1) = l(w_0)/h$.

Corollary 3.32. *One has the identity*
$$nh = |R|,$$
where $n = |I|$ is the rank of the root system and h is the Coxeter number.

Using this result, we can prove the following theorem.

3.5. Longest element and ordering of positive roots

Theorem 3.33 (Lusztig). *Given an orientation Ω of a Dynkin graph Q, there exists a reduced expression $w_0 = s_{i_l} \ldots s_{i_1}$ which is adapted to Ω (see Definition 3.23).*

Later we will show that such an adapted expression is unique up to exchanging $s_i s_j \leftrightarrow s_j s_i$ if i, j are not connected in the Dynkin diagram.

Proof. Let c_0, c_1 be as in Theorem 3.31. By Theorem 3.31,
$$w_0 = \ldots c_0 c_1 c_0 \quad (h \text{ factors})$$
is a reduced expression for w_0. This gives a reduced expression for w_0 adapted to the special orientation Ω_0, in which every $i \in I_0$ is a sink and every $i \in I_1$ is a source.

On the other hand, given a reduced expression $w_0 = s_{i_l} \ldots s_{i_1}$ which is adapted to orientation Ω, it is easy to see that
$$w_0 = s_j s_{i_l} \ldots s_{i_2},$$
where j is defined by $\alpha_j = -w_0(\alpha_{i_1})$, is a reduced expression adapted to orientation $s_{i_1}^+ \Omega$. Since any orientation can be obtained from Ω_0 by a sequence of sink$\leftrightarrow$source transformations (Lemma 3.6), this proves existence of an adapted reduced expression for any Ω. $\square$

Let us now choose a reduced expression (3.8) for w_0 adapted to $\vec{Q}$ and let $\gamma_1, \ldots, \gamma_l$ be defined by (3.9). By results of Section 3.3, this allows us to construct all indecomposable representations of $\vec{Q}$ by

(3.10)
$$\begin{aligned}
I_1 &= S(i_1), & \dim I_1 &= \gamma_1 = \alpha_{i_1}, \\
I_2 &= \Phi_{i_1}^-(S(i_2)), & \dim I_2 &= \gamma_2 = s_{i_1}(\alpha_{i_2}), \\
&\vdots & & \\
I_l &= \Phi_{i_1}^- \ldots \Phi_{i_{l-1}}^-(S(i_l)), & \dim I_l &= \gamma_l = s_{i_1} \ldots s_{i_{l-1}}(\alpha_{i_l}).
\end{aligned}$$

Example 3.34. Let $\vec{Q} = \underset{1}{\bullet} \longrightarrow \underset{2}{\bullet}$.

Then the adapted reduced expression for w_0 is $w_0 = s_2 s_1 s_2$, and the corresponding roots are
$$\begin{aligned}
\gamma_1 &= \alpha_2, \\
\gamma_2 &= s_2(\alpha_1) = \alpha_1 + \alpha_2, \\
\gamma_3 &= s_2 s_1(\alpha_2) = \alpha_1.
\end{aligned}$$

This illustrates the general fact: if α, β are positive roots such that $\alpha + \beta$ is also a positive root, then in the ordering of positive roots defined by a reduced expression for w_0 via (3.9), we will always have $\alpha + \beta$ between α and β (we will not need this fact so we skip the proof).

The corresponding indecomposables are

$$I_1 = S(2) = \underset{1}{\overset{0}{\bullet}} \longrightarrow \underset{2}{\overset{\mathbf{k}}{\bullet}},$$

$$I_2 = \Phi_2^-\left(\underset{1}{\overset{\mathbf{k}}{\bullet}} \longleftarrow \underset{2}{\overset{0}{\bullet}}\right) = \underset{1}{\overset{\mathbf{k}}{\bullet}} \overset{1}{\longrightarrow} \underset{2}{\overset{\mathbf{k}}{\bullet}} = P(1),$$

$$I_3 = \Phi_2^- \Phi_1^- \left(\underset{1}{\overset{0}{\bullet}} \longrightarrow \underset{2}{\overset{\mathbf{k}}{\bullet}}\right) = \Phi_2^-\left(\underset{1}{\overset{\mathbf{k}}{\bullet}} \overset{1}{\longleftarrow} \underset{2}{\overset{\mathbf{k}}{\bullet}}\right) = \underset{1}{\overset{\mathbf{k}}{\bullet}} \longrightarrow \underset{2}{\overset{0}{\bullet}} = S(1).$$

For future use, we state the following result.

Theorem 3.35. *Let $I_1, \ldots, I_l$ be indecomposable representations defined by (3.10). Then*

$$\operatorname{Hom}_{\vec{Q}}(I_a, I_b) = 0 \qquad \text{for } a > b,$$
$$\operatorname{Ext}^1_{\vec{Q}}(I_a, I_b) = 0 \qquad \text{for } a \leq b.$$

Proof. By Theorem 3.11 and Example 3.9, if i is a sink for $\vec{Q}$, then

$$\operatorname{Hom}(\Phi_i^-(V), S(i)) = \operatorname{Hom}(V, \Phi_i^+ S(i)) = 0$$

for any $V \in \operatorname{Rep}(s_i^+ \vec{Q})$. Using Theorem 3.17, this gives for $a > b$

$$\operatorname{Hom}(I_a, I_b) = \operatorname{Hom}(\Phi_{i_1}^- \ldots \Phi_{i_b}^- \ldots \Phi_{i_{a-1}}^-(S(i_a)), \Phi_{i_1}^- \ldots \Phi_{i_{b-1}}^-(S(i_b)))$$
$$= \operatorname{Hom}(\Phi_{i_b}^- \ldots \Phi_{i_{a-1}}^-(S(i_a)), S(i_b)) = 0.$$

The second part is proved similarly, using as the starting point the equality $\operatorname{Ext}^1(S(i), V) = 0$ for all V (if i is a sink), which is immediate since in this case $S(i) = P(i)$ is projective. □

In Chapter 6, we will show that the nonzero indecomposable representations $I_1, \ldots, I_l$ constructed above are in bijection with the vertices of a certain quiver, called the Auslander–Reiten quiver of $\vec{Q}$.

Chapter 4

Hall Algebras

In the previous chapters, we showed that in the case of Dynkin quivers one can recover the corresponding root system (or, rather, the positive part of the root system) using the representation theory of the quiver. It is natural to ask whether one can also recover the corresponding Lie algebra (or at least its positive part) from quiver representations. In this section we give a positive answer to this question in the case of Dynkin quivers, following works of Ringel [**Rin1990a, Rin1990b**] and Lusztig [**Lus1998a**].

4.1. Definition of Hall algebra

Let $\vec{Q}$ be a quiver. Consider the category of representations of this quiver over a finite field $\mathbb{F}_q$ with q elements. For any triple of representations L, M_1, M_2 of $\vec{Q}$ over $\mathbb{F}_q$, define the set

(4.1) $\quad \mathcal{F}^L_{M_1 M_2} = \{X \subset L \text{ — a subrepresentation} \mid X \simeq M_2,\ L/X \simeq M_1\}.$

Since L is a finite set, $\mathcal{F}$ is also finite, so we can define numbers

(4.2) $$F^L_{M_1 M_2} = |\mathcal{F}^L_{M_1 M_2}|.$$

This definition can be reformulated in the language of short exact sequences. Namely, let $X^L_{M_1 M_2} \subset \operatorname{Hom}_{\vec{Q}}(M_2, L) \times \operatorname{Hom}_{\vec{Q}}(L, M_1)$ be the set of pairs (f, g) such that

$$0 \to M_2 \xrightarrow{f} L \xrightarrow{g} M_1 \to 0$$

is a short exact sequence. Then one easily sees that the group $\operatorname{Aut}_{\vec{Q}}(M_1) \times \operatorname{Aut}_{\vec{Q}}(M_2)$ acts freely on $X^L_{M_1 M_2}$, and we have

$$\mathcal{F}^L_{M_1 M_2} = X^L_{M_1 M_2} / \operatorname{Aut}_{\vec{Q}}(M_1) \times \operatorname{Aut}_{\vec{Q}}(M_2),$$

$$F^L_{M_1 M_2} = \frac{|X^L_{M_1 M_2}|}{|\operatorname{Aut}_{\vec{Q}}(M_1) \times \operatorname{Aut}_{\vec{Q}}(M_2)|}.$$

More generally, we can define, for the collection $L, M_1, \ldots, M_k$ of representations of $\vec{Q}$ over $\mathbb{F}_q$,

(4.3)
$$\mathcal{F}^L_{M_1,\ldots,M_k} = \{L = X_0 \supset X_1 \supset \cdots \supset X_{k-1} \supset X_k = 0 \mid X_{i-1}/X_i \simeq M_i\},$$
$$F^L_{M_1,\ldots,M_k} = |\mathcal{F}^L_{M_1,\ldots,M_k}|.$$

It is obvious that $F^L_{M_1,\ldots,M_k}$ only depends on isomorphism classes of L, M_i and that $\mathcal{F}^L_{M_1,\ldots,M_n} = \varnothing$ (and thus $F^L_{M_1,\ldots,M_k} = 0$) unless

(4.4) $$\dim M_1 + \cdots + \dim M_k = \dim L.$$

Definition 4.1. Let $\vec{Q}$ be a quiver. For any natural number $q = p^r$, p prime, define the Hall algebra $H(\vec{Q}, \mathbb{F}_q)$ as the algebra over $\mathbb{C}$ with the basis given by isomorphism classes of representations of $\vec{Q}$ over $\mathbb{F}_q$ and multiplication defined by

(4.5) $$[M_1] * [M_2] = \sum_{[L]} F^L_{M_1 M_2}[L],$$

where the sum is over all isomorphism classes $[L]$ of representations of $\vec{Q}$ (condition (4.4) easily implies that the sum is actually finite).

Note that we define H as an algebra over $\mathbb{C}$, not over $\mathbb{F}_q$. One can also consider Hall algebras over $\mathbb{Q}$ or in fact over any commutative ring R.

The following theorem is due to Ringel [**Rin1990a**].

Theorem 4.2. *For any quiver $\vec{Q}$, $H(\vec{Q}, \mathbb{F}_q)$ is an associative algebra over $\mathbb{C}$ with unit $1 = [0]$.*

Proof. We will prove that

$$([M_1] \cdot [M_2]) \cdot [M_3] = [M_1] \cdot ([M_2] \cdot [M_3])$$

by proving that each side is in fact equal to

$$\sum_{[L]} F^L_{M_1 M_2 M_3}[L].$$

Indeed,

$$\mathcal{F}^L_{M_1 M_2 M_3} = \{L \supset X_1 \supset X_2 \supset 0 \mid L/X_1 \simeq M_1, X_1/X_2 \simeq M_2, X_2 \simeq M_3\}.$$

4.1. Definition of Hall algebra

This set has an obvious projection to $\bigcup_{[X_1]} \mathcal{F}^L_{M_1 X_1}$ given by
$$(L \supset X_1 \supset X_2 \supset 0) \mapsto (L \supset X_1 \supset 0).$$
The fiber of this projection is the set
$$\mathcal{F}^{X_1}_{M_2 M_3} = \{X_2 \mid X_1 \supset X_2 \supset 0, X_1/X_2 \simeq M_2, X_2 \simeq M_3\}.$$
Thus,
$$F^L_{M_1 M_2 M_3} = \sum_{[X_1]} F^L_{M_1 X_1} F^{X_1}_{M_2 M_3}$$
which shows that $[M_1]([M_2][M_3]) = \sum F^L_{M_1 M_2 M_3}[L]$.

Identity $([M_1][M_2])[M_3] = \sum F^L_{M_1 M_2 M_3}[L]$ is proved in a similar way, using projection $(L \supset X_1 \supset X_2 \supset 0) \mapsto (L \supset X_2 \supset 0)$. $\square$

It is obvious from (4.4) that the Hall algebra is naturally graded by $\mathbb{Z}^I_+$:
$$(4.6) \qquad H(\vec{Q}, \mathbb{F}_q) = \bigoplus_{\mathbf{v} \in \mathbb{Z}^I_+} H_{\mathbf{v}}(\vec{Q}, \mathbb{F}_q),$$
where $H_{\mathbf{v}}(\vec{Q}, \mathbb{F}_q)$ is spanned by isomorphism classes $[M]$ with $\dim M = \mathbf{v}$.

Finally, we note that many of the formulas will look more symmetric if we slightly twist the definition of the product in the Hall algebra, defining
$$(4.7) \qquad v \circ w = q^{\langle \mathbf{v}, \mathbf{w} \rangle / 2} v \cdot w, \quad v \in H_{\mathbf{v}}(\vec{Q}), w \in H_{\mathbf{w}}(\vec{Q}),$$
where $\langle \mathbf{v}, \mathbf{w} \rangle$ is the Euler form (1.15). Of course, this only makes sense if we have fixed a choice of $q^{1/2} \in \mathbb{C}$. It is easy to see that the twisted multiplication is still associative.

Example 4.3. Let $\vec{Q} = \bullet$. Then the only representations of $\vec{Q}$ are nS, where S is the one-dimensional vector space, and from the definition we get
$$[nS] \cdot [S] = [(n+1)S] |\mathbb{P}^n(\mathbb{F}_q)| = [(n+1)S] \frac{q^{n+1} - 1}{q - 1} = [(n+1)S][n+1]_q,$$
where we denoted $[m]_q = \frac{q^m - 1}{q - 1} = q^{m-1} + q^{m-2} + \cdots + q + 1$ (these are sometimes called q-numbers and are extensively used in representation theory of quantum groups).

Using associativity, it is easy to prove by induction that
$$[nS] \cdot [mS] = \binom{n+m}{m}_q [(n+m)S],$$
where $\binom{n+m}{m}_q$ is the q-binomial coefficient:
$$(4.8) \qquad \binom{n+m}{m}_q = \frac{[n+m]_q!}{[n]_q! [m]_q!},$$
$$[n]_q! = [1]_q [2]_q \ldots [n]_q.$$

Equivalently, this result can also be obtained by an explicit computation which requires computing the number of points in a Grassman variety over a finite field.

This shows that in this case the Hall algebra is isomorphic to the algebra $\mathbb{C}[x]$ of polynomials in one variable, with the isomorphism given by

$$[nS] \mapsto \frac{x^n}{[n]_q!}.$$

This result can be generalized: if S is any representation of a quiver $\vec{Q}$ satisfying $\operatorname{Ext}^1_{\vec{Q}}(S,S) = 0$ and $\operatorname{Hom}_{\vec{Q}}(S,S) = \mathbb{F}_q$, then the same argument as above shows that

(4.9) $$[nS] = \frac{[S]^n}{[n]_q!}.$$

It should be noted that the direct sum of representations is not related at all to addition in the Hall algebra: there is no relation between $[M \oplus N]$ and $[M] + [N]$. Instead, $[M \oplus N]$ is related to multiplication in the Hall algebra: the product $[M] \cdot [N]$ contains $[M \oplus N]$ with nonzero coefficient. The following proposition shows that under suitable conditions, one has $[M] \cdot [N] = [M \oplus N]$.

Proposition 4.4. *Let M_1, M_2 be representations of $\vec{Q}$ such that*

$$\operatorname{Hom}_{\vec{Q}}(M_2, M_1) = 0 = \operatorname{Ext}^1_{\vec{Q}}(M_1, M_2).$$

Then in the Hall algebra $H(\vec{Q}, \mathbb{F}_q)$ one has $[M_1] \cdot [M_2] = [M_1 \oplus M_2]$.

Proof. Indeed, $\operatorname{Ext}^1(M_1, M_2) = 0$ shows that if we have a short exact sequence $0 \to M_2 \to L \to M_1 \to 0$, then $L \simeq M_1 \oplus M_2$; thus, $[M_1][M_2]$ is a multiple of $M_1 \oplus M_2$. On the other hand, if $L = M_1 \oplus M_2$, then $\operatorname{Hom}(M_2, M_1) = 0$ implies that L contains a unique submodule isomorphic to M_2, so $F^L_{M_1 M_2} = 1$. □

Finally, note that in the simplest example of the trivial quiver (Example 4.3), the structure coefficients of the Hall algebra are polynomials in q. It turns out that the same holds for any Dynkin quiver.

Recall that if $\vec{Q}$ is a Dynkin quiver, then the set $\operatorname{Ind}(\vec{Q})$ of isomorphism classes of nonzero indecomposable representations over any field is identified with the set R_+ of positive roots of the corresponding root system. Thus, isomorphism classes of all representations are in bijection with the set

$$\mathcal{B} = \mathbb{Z}_+^{R_+} = \{n \colon R_+ \to \mathbb{Z}_+\}$$

4.1. Definition of Hall algebra

with the bijection defined by

$$\mathbf{n} \mapsto M_\mathbf{n} = \bigoplus_{\alpha \in R_+} n(\alpha) I_\alpha,$$

where I_α is the indecomposable representation with $\dim I_\alpha = \alpha$.

The following result is again due to Ringel.

Theorem 4.5. *Let $\vec{Q}$ be a Dynkin quiver. Then for any $\mathbf{n}, \mathbf{m}, \mathbf{k} \in \mathbb{Z}_+^{R_+}$, there is a polynomial $\varphi_{\mathbf{nk}}^{\mathbf{m}} \in \mathbb{Z}[t]$ such that the structure coefficients of the Hall algebra $H(\vec{Q}, \mathbb{F}_q)$ are given by*

$$F_{M_\mathbf{n} M_\mathbf{k}}^{M_\mathbf{m}} = \varphi_{\mathbf{nk}}^{\mathbf{m}}(q).$$

In other words, the structure constants of the Hall algebra are polynomials with integer coefficients in q.

The proof of this theorem can be found in [**Rin1990a**]; we will not repeat it here, describing instead just the main idea of the proof. The proof is based on defining the polynomials φ recursively using a certain partial order $\preceq$ on the set $\mathrm{Ind}(\vec{Q})$; we will describe this partial order, using the Auslander–Reiten quiver, in Section 6.4. In this recursive construction, the only dependence on q comes from counting the number of points in the vector space $\mathbb{F}_q^n$ and in the general linear group $\mathrm{GL}(n, \mathbb{F}_q)$, both of which depend polynomially on q. From this one gets that the structure coefficients are rational functions of the form $\varphi(q)/\psi(q)$, $\varphi, \psi \in \mathbb{Z}[t]$, and the leading coefficient of ψ is equal to 1. Since these rational functions take integer values at all $q = p^r$, it is easy to prove that in fact they are polynomials.

This theorem shows that in the case of Dynkin quivers, we can define the "universal" Hall algebra $H(\vec{Q})_\mathcal{A}$ over the ring $\mathcal{A} = \mathbb{Z}[t]$ as the algebra with basis $M_\mathbf{n}$, $\mathbf{n} \in \mathbb{Z}_+^{R_+}$, and multiplication defined by

$$(4.10) \qquad [M_\mathbf{n}] \cdot [M_\mathbf{k}] = \sum_\mathbf{m} \varphi_{\mathbf{nk}}^{\mathbf{m}}(t) [M_\mathbf{m}].$$

This algebra also has a natural grading: $H(\vec{Q})_\mathcal{A} = \bigoplus_{\mathbf{v} \in \mathbb{Z}_+^I} H_\mathbf{v}(\vec{Q})_\mathcal{A}$.

Theorem 4.5 shows that for any finite field $\mathbb{F}$, the algebra $H(\vec{Q}, \mathbb{F})$ is obtained from this universal Hall algebra by specialization:

$$H(\vec{Q}, \mathbb{F}) = H(\vec{Q})_\mathcal{A} \otimes_\mathcal{A} \mathbb{C},$$

where $\mathbb{C}$ is considered as an $\mathcal{A}$-module via $t \mapsto |\mathbb{F}|$. This, in particular, implies that $H(\vec{Q})_\mathcal{A}$ is asssociative: if some relation between polynomials holds for all values $t = p^r$, then it holds in the ring $\mathbb{Z}[t]$.

Moreover, existence of the universal algebra also shows that we can define specializations for any numeric value of t, not necessarily for $t = p^r$.

In particular, we will be especially interested in specialization at $t = 1$ which we will denote just by $H(\vec{Q})$:

(4.11) $$H(\vec{Q}) = H(\vec{Q})_{\mathcal{A}} \otimes_{\mathcal{A}} \mathbb{C}, \qquad t \mapsto 1.$$

4.2. Serre relations and Ringel's theorem

In this section, we will describe the Hall algebra of a Dynkin quiver by generators and relations, establishing the celebrated result of Ringel: the Hall algebra at $t = 1$ is isomorphic to the universal enveloping algebra of the positive part of the Lie algebra corresponding to Dynkin diagram Q.

Throughout this section, $\vec{Q}$ is a Dynkin quiver.

Define, for any $i \in I$, element $\theta_i \in H(\vec{Q})_{\mathcal{A}}$ (and thus in any $H(\vec{Q}, \mathbb{F}_q)$) by

(4.12) $$\theta_i = [S(i)],$$

where $S(i)$ is the simple representation supported at vertex i (see (1.5)).

Theorem 4.6. *Let $\tilde{\mathcal{A}}$ be the algebra obtained by joining to $\mathbb{Z}[t]$ all inverses of elements $[n]_t = 1 + t + \cdots + t^{n-1}$, $n \geq 1$. Then elements θ_i, $i \in I$, generate $H(\vec{Q})_{\tilde{\mathcal{A}}} = H(\vec{Q})_{\mathcal{A}} \otimes_{\mathcal{A}} \tilde{\mathcal{A}}$.*

Proof. Denote temporarily by $H' \subset H(\vec{Q})_{\tilde{\mathcal{A}}}$ the subalgebra in $H(\vec{Q})_{\tilde{\mathcal{A}}}$ generated by θ_i. We need to show that for any representation V of $\vec{Q}$, we have $[V] \in H'$.

As in Section 3.5, choose a reduced expression $w_0 = s_{i_l} \ldots s_{i_1}$ compatible with $\vec{Q}$. By results of Section 3.5, this gives a complete order on the set of positive roots: $R_+ = \{\gamma_1, \ldots, \gamma_l\}$; it also gives a construction of all indecomposable representations I_k of $\vec{Q}$ (see (3.10)).

Lemma 4.7. *If $V \simeq \bigoplus_{k=1}^{l} n_k I_k$, where I_k are indecomposable representations defined by (3.10), then in the Hall algebra $H(\vec{Q}, \mathbb{F}_q)$ we have*

(4.13) $$[V] = [I_1]^{(n_1)} \ldots [I_l]^{(n_l)},$$

where we denote $x^{(n)} = \frac{x^n}{[n]_q!}$ (compare with (4.9)).

Proof. Indeed, by Theorem 3.35, we have $\mathrm{Hom}_{\vec{Q}}(I_a, I_b) = 0$ for $a > b$ and $\mathrm{Ext}^1_{\vec{Q}}(I_a, I_b) = 0$ for $a \leq b$. By Proposition 4.4, this implies

$$[V] = [n_1 I_1] \cdot [n_2 I_2] \cdot \cdots \cdot [n_l I_l].$$

On the other hand, since $\mathrm{End}_{\vec{Q}}(I_k) = \mathbf{k}$, $\mathrm{Ext}^1_{\vec{Q}}(I_k, I_k) = 0$ (see Theorem 3.25), computation in Example 4.3 shows that $[n_k I_k] = [I_k]^{(n_k)}$. □

4.2. Serre relations and Ringel's theorem

Since relations (4.13) hold for any $q = p^r$, they must also hold in the universal algebra $H(\vec{Q})_{\tilde{\mathcal{A}}}$.

We can now give a proof of the theorem using induction. Define partial order on $\mathbb{Z}_+^I$ by writing $\mathbf{v} \leqslant \mathbf{w}$ if $\mathbf{v}_i \leq \mathbf{w}_i$ for all $i \in I$ (cf. (1.23)). Let us write $\mathbf{v} < \mathbf{w}$ if $\mathbf{v} \leq \mathbf{w}$, $\mathbf{v} \neq \mathbf{w}$.

Define the partial order on the set of isomorphism classes of representations by

$$[V] \leq [W] \text{ if } (\dim V < \dim W) \text{ or } (\dim V = \dim W \text{ and } \mathbb{O}_V \subset \overline{\mathbb{O}}_W),$$

where $\mathbb{O}_V$, $\mathbb{O}_W$ are the orbits in the representation space $R(\vec{Q})$ (over the algebraic closure $\overline{\mathbb{F}_q}$) corresponding to representations V, W respectively, and $\overline{\mathbb{O}}$ is the Zariski closure.

We will now use induction in the partial order on $[V]$ to prove that for any representation V of $\vec{Q}$, we have $[V] \in H'$. Indeed, for $V = 0$ the statement is obvious. Assume now that the statement is true for all $V' < V$. If V is decomposable, then Lemma 4.7 implies that $[V] = c[V'][V'']$ for some $c \in \tilde{\mathcal{A}}$ and representations V', V'' with $\dim V' < \dim V$, $\dim V'' < \dim V$; thus, by the induction assumption, $[V'], [V''] \in H'$, so $[V] \in H'$.

If V is indecomposable, then choose a vertex i such that $V_i \neq 0$, but for any edge $h \colon i \to j$, the corresponding operator $x_h \colon V_i \to V_j$ is zero. Since $\vec{Q}$ has no oriented loops, it is possible. Then we have a short exact sequence

$$0 \to V' \to V \to V/V' \to 0,$$

where $V'_i = V_i$ and $V'_j = 0$ for $j \neq i$, so $V' \simeq (\dim V_i)S(i)$. Therefore, in the Hall algebra we have $[V/V'] \cdot [V'] = [V] + \sum c_a[V''_a]$, where V''_a are other representations of dimension $\dim V''_a = \dim V$. But since V is indecomposable, the corresponding orbit is open (see Corollary 3.26), so $\overline{\mathbb{O}}_V = R(\dim V)$, and for any V''_a, we have $[V''_a] < [V]$. Therefore, by the induction assumption, $[V''_a] \in H'$ and $[V] = [V/V'] \cdot [V'] - \sum c_a[V''_a] \in H'$. □

Remark 4.8. This theorem is not true for non-Dynkin quivers. Even for Euclidean quivers, elements θ_i do not generate the Hall algebra.

Corollary 4.9. *Elements θ_i, $i \in I$, generate the Hall algebra $H(\vec{Q})$.*

Now let us study relations between the generators θ_i. Let us begin by considering the quiver $\vec{Q} = \underset{1}{\bullet} \longrightarrow \underset{2}{\bullet}$. As before, denote $\theta_1 = [S(1)]$, $\theta_2 = [S(2)] \in H(\vec{Q}, \mathbb{F}_q)$. Let us compute $\theta_1 \cdot \theta_2$, $\theta_2 \cdot \theta_1$.

Since the only representations of $\vec{Q}$ of dimension $\mathbf{d} = (1,1)$ are

$$S_{12} = \underset{1}{\overset{\mathbf{k}}{\bullet}} \xrightarrow{1} \underset{2}{\overset{\mathbf{k}}{\bullet}}, \qquad \tilde{S}_{12} = \underset{1}{\overset{\mathbf{k}}{\bullet}} \xrightarrow{0} \underset{2}{\overset{\mathbf{k}}{\bullet}},$$

an easy computation shows that

(4.14)
$$\theta_1 \cdot \theta_2 = [S_{12}] + [\tilde{S}_{12}],$$
$$\theta_2 \cdot \theta_1 = [\tilde{S}_{12}].$$

Next, let us compute products $\theta_1^2 \theta_2$, $\theta_1 \theta_2 \theta_1$, $\theta_2 \theta_1^2$. Each of them must be a linear combination of isomorphism classes of

$$N^1 = \underset{1}{\bullet} \xrightarrow{\mathbf{k}^2 \; x \; \mathbf{k}} \underset{2}{\bullet}, \qquad N^0 = \underset{1}{\bullet} \xrightarrow{\mathbf{k}^2 \; 0 \; \mathbf{k}} \underset{2}{\bullet},$$

where $x \colon \mathbf{k}^2 \to \mathbf{k}$ is surjective.

Using results of Example 4.3, we get

(4.15)
$$\theta_1^2 \cdot \theta_2 = (q+1)[2S(1)][S(2)] = (q+1)([N^0] + [N^1]),$$
$$\theta_2 \cdot \theta_1^2 = (q+1)[S(2)][2S(1)] = (q+1)[N^0],$$
$$\theta_1 \cdot \theta_2 \cdot \theta_1 = \theta_1 \cdot [\tilde{S}_{12}] = |\mathbb{P}^1(\mathbb{F}_q)|[N^0] + [N^1] = (q+1)[N^0] + [N^1],$$

since the only submodule in N^1 isomorphic to $\tilde{S}_{12}$ is $\underset{1}{\bullet} \xrightarrow{\mathrm{Ker}\, x \; 0 \; \mathbf{k}} \underset{2}{\bullet}$.

Thus, we get the following result.

Theorem 4.10. *For* $\vec{Q} = \underset{1}{\bullet} \longrightarrow \underset{2}{\bullet}$, *elements* $\theta_1 = [S(1)]$, $\theta_2 = [S(2)]$ *in the Hall algebra* $H(\vec{Q})_\mathcal{A}$ *satisfy the following relations:*

(4.16)
$$\theta_1^2 \cdot \theta_2 - (t+1)\theta_1 \cdot \theta_2 \cdot \theta_1 + t\theta_2 \cdot \theta_1^2 = 0,$$
$$\theta_2^2 \cdot \theta_1 - (t+1)\theta_2 \cdot \theta_1 \cdot \theta_2 + t\theta_1 \cdot \theta_2^2 = 0.$$

Proof. Indeed, for any $t = q = |\mathbb{F}|$, the first relation immediately follows from (4.15); thus, it must also hold in the universal algebra $H(\vec{Q})_\mathcal{A}$. The second relation is proved similarly. □

Now let $\vec{Q}$ be a Dynkin quiver; let R be the corresponding root system, and let $\mathfrak{g}(\vec{Q}) = \mathfrak{n}_- \oplus \mathfrak{h} \oplus \mathfrak{n}_+$ be the corresponding finite-dimensional semisimple complex Lie algebra (see Appendix A). Recall that the positive part $\mathfrak{n}_+$ can be described as the Lie algebra with generators e_i, $i \in I$, and Serre relations

$$(\mathrm{ad}\, e_i)^{1-c_{ij}} e_j = 0, \quad i \neq j,$$

where c_{ij} is the Cartan matrix. For a quiver without multiple edges (in particular, for all Dynkin quivers), c_{ij} is either 0 (if i, j are not connected in Q) or -1 (if i, j are connected in Q). Thus, in this case Serre relations take the form

(4.17)
$$[e_i, e_j] = 0, \quad \text{if } i, j \text{ are not connected in } Q,$$
$$[e_i, [e_i, e_j]] = 0, \quad \text{if } i, j \text{ are connected in } Q.$$

4.2. Serre relations and Ringel's theorem

Note that the last equation is equivalent to the relation
$$e_i^2 e_j - 2e_i e_j e_i + e_j e_i^2 = 0$$
in the universal enveloping algebra of $\mathfrak{g}$.

We are now ready to formulate Ringel's theorem.

Theorem 4.11 (Ringel). *Let $\vec{Q}$ be a Dynkin quiver and let $\mathfrak{n}_+$ be the positive part of the corresponding semisimple Lie algebra. Then there exists a unique homomorphism $\Psi\colon U\mathfrak{n}_+ \to H(\vec{Q})$ of associative algebras over $\mathbb{C}$ such that $\Psi(e_i) = [S(i)]$; moreover, Ψ is an isomorphism.*

Proof. First, let us check that Ψ does define a homomorphism of algebras. Since $U\mathfrak{n}_+$ is generated by elements e_i, $i \in I$, with generating relations (4.17), it suffices to check that elements $\theta_i = \Psi(e_i)$ satisfy relations (4.17). But these relations immediately follow from Theorem 4.10 by letting $t = 1$.

Since elements $\theta_i = [S(i)]$ generate $H(\vec{Q})$ (Theorem 4.6), Ψ is surjective.

Finally, to prove the injectivity of Ψ, denote by $(U\mathfrak{n}_+)_\mathbf{d}$ the component of the universal enveloping algebra with weight $\mathbf{d} \in \mathbb{Z}_+^I$. Then
$$\dim(U\mathfrak{n}_+)_\mathbf{d} = \left| \{(n_\alpha)_{\alpha \in R_+} \mid \sum n_\alpha \alpha = \mathbf{d}\} \right| = \dim H_\mathbf{d}(\vec{Q}).$$
The first identity follows from the Poincaré–Birkhoff–Witt theorem, the second is immediate from the definitions and the fact that any representation of $\vec{Q}$ can be uniquely written in the form $V \simeq \sum n_\alpha I_\alpha$. Therefore, since the dimensions are equal and the map is surjective, it is also injective. $\square$

This theorem generalizes to the $q \neq 1$ case. Namely, let $U_v\mathfrak{n}_+$ be the associative algebra over the field $\mathbb{C}(v)$ of rational functions in variable v with generators e_i, $i \in I$, and defining relations

(4.18)
$$[e_i, e_j] = 0, \quad \text{if } i, j \text{ are not connected in } Q,$$
$$e_i^2 e_j - (v + v^{-1}) e_i e_j e_i + e_j e_i^2 = 0, \quad \text{if } i, j \text{ are connected in } Q.$$

This algebra is a subalgebra in the "quantum group", or more formally the quantized universal enveloping algebra $U_v(\mathfrak{g})$; we refer the reader to books of Lusztig [**Lus1993**] and Jantzen [**Jan1996**] for the theory of quantum groups. (Note that normally the deformation parameter is denoted by q, not v; however, this would conflict with previously introduced notation $q = p^r$ for the cardinality of a finite field.)

Theorem 4.12. *Let $\vec{Q}$ be a Dynkin quiver, and let $\tilde{H}(\vec{Q}, v^2) = H(\vec{Q})_\mathcal{A} \otimes_\mathcal{A} \mathbb{C}(v)$ be the Hall algebra defined over the field $\mathbb{C}(v)$ by letting $t = v^2$ with the twisted multiplication (see (4.7)). Then there exists a unique homomorphism $\Psi\colon U_v\mathfrak{n}_+ \to H(\vec{Q}, v^2)$ of associative algebras over $\mathbb{C}(v)$ such that $\Psi(e_i) = [S(i)]$. Moreover, Ψ is an isomorphism.*

The proof is quite similar to the proof of Theorem 4.11 and will not be repeated here. We only note that in fact, it is also possible to construct the comultiplication in $U_v(\mathfrak{n}_+)$ (or rather in $U_v(\mathfrak{b}_+)$, where $\mathfrak{b}_+ = \mathfrak{n}_+ \oplus \mathfrak{h}$ is the Borel subalgebra) in terms of quiver representations (see [**Gre1995**]).

Remark 4.13. There is another method of proving the injectivity of the map Ψ which is not based on a computation of dimensions. Instead, it uses the fact that the map Ψ preserves a certain bilinear form on $U_v(\mathfrak{n}_+)$ which is known to be nondegenerate; thus, Ψ must be injective. Note, however, that this form as a function of v has poles at $v = 1$ and thus this method cannot be directly applied in the $q = 1$ case. This proof can be found in [**Lus1998a**, Theorem 1.20] or in [**Sch2006**].

Recently Ringel's construction was extended by Bridgeland [**Bri2013**] to give a construction of the whole $U\mathfrak{g}$, not just the positive part. Not surprisingly, this requires the use of derived category $D^b(\operatorname{Rep} \vec{Q})$. However, this construction presents additional challenges (naive extension of the notion of Hall algebras to derived categories fails) and will not be discussed here.

4.3. PBW basis

In this section, we let $\vec{Q}$ be a Dynkin quiver and study in detail the homomorphism $\Psi\colon U\mathfrak{n}_+ \to H(\vec{Q})$ defined in Theorem 4.11. In particular, we will show that it identifies the basis of isomorphism classes of objects in the Hall algebra with the so-called Poincaré–Birkhoff–Witt (PBW) basis in $U\mathfrak{n}_+$.

Our first goal is defining the counterpart of the reflection functors $\Phi^\pm$ for the algebra $U\mathfrak{g}$. This can be done as follows.

First, for any $i \in I$ we have an embedding of Lie algebras

$$\rho_i \colon \mathfrak{sl}(2, \mathbb{C}) \to \mathfrak{g},$$

$$e = \begin{pmatrix} 0 & 1 \\ 0 & 0 \end{pmatrix} \mapsto e_i,$$

$$f = \begin{pmatrix} 0 & 0 \\ 1 & 0 \end{pmatrix} \mapsto f_i,$$

$$h = \begin{pmatrix} 1 & 0 \\ 0 & -1 \end{pmatrix} \mapsto h_i.$$

By general theory of Lie groups and algebras, this lifts to a morphism of Lie groups $\rho_i \colon \operatorname{SL}(2, \mathbb{C}) \to G$, where G is the connected, simply-connected Lie

4.3. PBW basis

group with Lie algebra $\mathfrak{g}$. In particular, it gives elements

$$S_i^+ = \rho_i \begin{pmatrix} 0 & 1 \\ -1 & 0 \end{pmatrix} \in G,$$

$$S_i^- = \rho_i \begin{pmatrix} 0 & -1 \\ 1 & 0 \end{pmatrix} \in G.$$

Define operators $T_i^\pm \colon U\mathfrak{g} \to U\mathfrak{g}$ by

(4.19) $$T_i^\pm = \operatorname{Ad} S_i^\pm,$$

where Ad stands for the adjoint action of the group G on its Lie algebra (and thus on $U\mathfrak{g}$). It follows from the identity

$$\begin{pmatrix} 0 & 1 \\ -1 & 0 \end{pmatrix} = \begin{pmatrix} 1 & 1 \\ 0 & 1 \end{pmatrix} \begin{pmatrix} 1 & 0 \\ -1 & 1 \end{pmatrix} \begin{pmatrix} 1 & 1 \\ 0 & 1 \end{pmatrix}$$

that

(4.20)
$$\begin{aligned}
S_i^+ &= \exp(e_i)\exp(-f_i)\exp(e_i), \\
T_i^+ &= \exp(\operatorname{ad} e_i)\exp(-\operatorname{ad} f_i)\exp(\operatorname{ad} e_i), \\
S_i^- &= \exp(-e_i)\exp(f_i)\exp(-e_i), \\
T_i^- &= \exp(-\operatorname{ad} e_i)\exp(\operatorname{ad} f_i)\exp(-\operatorname{ad} e_i).
\end{aligned}$$

The following theorem lists some properties of these operators.

Theorem 4.14.

(1) $T_i^\pm$ are automorphisms of $U\mathfrak{g}$.
(2) $T_i^+ T_i^- = T_i^- T_i^+ = 1$.
(3) If $x \in \mathfrak{g}_\alpha$, then $T_i^\pm(x) \in \mathfrak{g}_{s_i(\alpha)}$, where $s_i \in W$ is the simple reflection corresponding to i.
(4) If $(\alpha_i, \alpha_j) = 0$, then $T_i^\pm, T_j^\pm$ commute; if $(\alpha_i, \alpha_j) = -1$, so that in the Weyl group we have $s_i s_j s_i = s_j s_i s_j$, then we also have $T_i^+ T_j^+ T_i^+ = T_j^+ T_i^+ T_j^+$, $T_i^- T_j^- T_i^- = T_j^- T_i^- T_j^-$.

The proof of all of these relations is easily obtained by explicit computation which can be found in [**Lus1993**].

Explicit computation shows that

(4.21)
$$\begin{aligned}
T_i^\pm(e_i) &= -f_i, \quad T_i^\pm(e_j) = e_j \text{ if } (\alpha_i, \alpha_j) = 0, \\
T_i^\pm(e_j) &= \pm[e_i, e_j] \text{ if } i \neq j,\ (\alpha_i, \alpha_j) = -1.
\end{aligned}$$

It turns out that the isomorphism $\Psi \colon U\mathfrak{n}_+ \to H(\vec{Q})$ identifies these automorphisms with the action of the reflection functors on the Hall algebra. Literally, this statement does not make sense, as T_i does not preserve the subalgebra $U\mathfrak{n}_+$. However, it can be corrected as follows.

Define, for any $i \in I$, subalgebras $U^{i,\pm} \subset U\mathfrak{n}_+$ by

(4.22) $\quad U^{i,+} = U\mathfrak{n}_+ \cap T_i^+(U\mathfrak{n}_+), \qquad U^{i,-} = U\mathfrak{n}_+ \cap T_i^-(U\mathfrak{n}_+).$

It is immediate from Theorem 4.14 that $U^{i,\pm}$ are subalgebras in $U\mathfrak{n}_+$ and that $T_i^+ : U^{i,-} \to U^{i,+}$, $T_i^- : U^{i,+} \to U^{i,-}$ are mutually inverse algebra isomorphisms.

Lemma 4.15. *Subalgebra* $U^{i,+} = U^{i,-}$ *is generated by elements* e_j, $j \neq i$, *and* $[e_j, e_i]$, *for* $j \neq i$, j *connected with* i *in* Q.

Proof. Let $\mathfrak{n}' \subset \mathfrak{n}_+$ be defined by $\mathfrak{n}' = \mathfrak{n}_+ \cap T_i^+ \mathfrak{n}_+ = \mathfrak{n}_+ \cap T_i^- \mathfrak{n}_+$. It is easy to see from (4.21) that $\mathfrak{n}' = \bigoplus_{\alpha \in R_+ \setminus \{\alpha_i\}} \mathfrak{g}_\alpha$.

First prove that $U^{i,+} = U\mathfrak{n}'$. Indeed, it follows from the Poincaré–Birkhoff–Witt theorem that any $x \in U\mathfrak{n}_+$ can be uniquely written in the form $x = \sum_{k \geq 0} x_k e_i^k$, where $x_k \in U\mathfrak{n}'$. By (4.21), $T_i^-(x) = \sum T_i^-(x_k)(-f_i)^k$. Since $T_i^-(x_k) \in U\mathfrak{n}_+$, using the PBW theorem once more we see that $T_i^- x \in U\mathfrak{n}_+$ iff $x_k = 0$ for all $k > 0$, i.e. if $x \in U\mathfrak{n}'$. In a similar way one proves that $U^{i,-} = U\mathfrak{n}'$.

Next, by Serre's theorem, $U\mathfrak{n}_+$ is spanned as a vector space by elements e_j and commutators of the form $[e_{j_1}, [\ldots, [e_{j_k}, e_{j_{k+1}}] \ldots]]$. It easily follows from this that $\mathfrak{n}'$ is generated as a Lie algebra by e_j, $j \neq i$, and $[e_i, e_j]$; thus, the same elements also generate $U\mathfrak{n}' = U^{i,+}$ as an associative algebra. $\square$

Remark 4.16. The operators $T_i^\pm$ can also be defined for the quantum group $U_v\mathfrak{g}$ (see [**Lus1993**]). Most of the results above generalize to the quantum group case with appropriate changes. The most important change is that in the quantum group case, $U^{i,+} \neq U^{i,-}$; however, we never use this identity in what follows.

Subalgebras $U^{i,\pm}$ have their counterpart on the Hall algebra side. Namely, assume that $i \in I$ is a sink for $\vec{Q}$; then we have the category $\text{Rep}^{i,-}(\vec{Q})$ (see Theorem 3.11). It was shown in the proof of Corollary 3.16 that $\text{Rep}^{i,-}(\vec{Q})$ is closed under extensions: if we have a short exact sequence $0 \to V \to A \to W \to 0$ of representations of $\vec{Q}$, and $V, W \in \text{Rep}^{i,-}(\vec{Q})$, then $A \in \text{Rep}^{i,-}(\vec{Q})$. Therefore, the subspace in the Hall algebra spanned by the isomorphism classes of objects $M \in \text{Rep}^{i,-}(\vec{Q})$ is a subalgebra; we will denote this subalgebra by $H^{i,-}(\vec{Q})_\mathcal{A}$:

(4.23) $\quad H^{i,-}(\vec{Q})_\mathcal{A} = (\text{span of } [M], M \in \text{Rep}^{i,-}(\vec{Q})) \subset H(\vec{Q})_\mathcal{A}.$

In a similar way, if i is a source for $\vec{Q}$, we define the subalgebra

(4.24) $\quad H^{i,+}(\vec{Q})_\mathcal{A} = (\text{span of } [M], M \in \text{Rep}^{i,+}(\vec{Q})) \subset H(\vec{Q})_\mathcal{A}.$

4.3. PBW basis

As in (4.11), we can specialize $t = 1$ to get the subalgebras $H^{i,\pm}(\vec{Q}) \subset H(\vec{Q})$.

We can now formulate the first theorem of this section.

Theorem 4.17. *Let $\vec{Q}$ be a Dynkin quiver and let $i \in I$ be a sink for $\vec{Q}$. Define $\vec{Q}' = s_i^+(\vec{Q})$. Then:*

(1) *For any $x \in U^{i,-}$, $\Psi_{\vec{Q}}(x) \in H^{i,-}(\vec{Q})$. Similarly, for any $x \in U^{i,+}$, $\Psi_{\vec{Q}'}(x) \in H^{i,+}(\vec{Q}')$.*

(2) *The following diagram is commutative:*

$$\begin{array}{ccc} U^{i,-} & \xrightarrow{\Psi_{\vec{Q}}} & H^{i,-}(\vec{Q}) \\ T_i^+ \Big\Updownarrow T_i^- & & \Phi_i^+ \Big\Updownarrow \Phi_i^- \\ U^{i,+} & \xrightarrow{\Psi_{\vec{Q}'}} & H^{i,+}(\vec{Q}) \end{array}$$

Proof. To prove the first part, it suffices to prove it in the case when x is a generator of $U^{i,\pm}$. By Lemma 4.15, the generators are $x = e_j$, $x = [e_i, e_j]$. For $x = e_j$, $j \neq i$, we have $\Psi_{\vec{Q}}(e_j) = \theta_j \in H^{i,-}(\vec{Q})$ and similarly for $\Psi_{\vec{Q}'}$. For $x = [e_j, e_i]$, it follows from calculations in the A_2 case (see (4.14)) that $\Psi([e_j, e_i]) = \theta_j \theta_i - \theta_i \theta_j = [S_{ji}]$, where

$$S_{ji} = \underset{j}{\overset{\mathbf{k}}{\bullet}} \longrightarrow \underset{i}{\overset{\mathbf{k}}{\bullet}}$$

and thus $\Psi([e_j, e_i]) \in H^{i,-}(\vec{Q})$. The proof for $\vec{Q}'$ is similar.

To prove the second part, it suffices to check that $\Psi(T_i^+(x)) = \Phi_i^+ \Psi(x)$ for each generator x of $U^{i,-}$. We check it for $x = [e_j, e_i]$; the check for $x = e_j$ is similar.

For $x = [e_j, e_i]$, we get $\Phi_i^+ \Psi([e_j, e_i]) = [\Phi_i^+(S_{ji})] = [S(j)] = \theta_j$.

On the other hand, since $[e_j, e_i] = T_i^-(e_j)$, we get $T_i^+[e_j, e_i] = T_i^+ T_i^-(e_j) = e_j$, so $\Psi T_i^+[e_j, e_i] = \Psi(e_j) = \theta_j$. Thus,

$$\Psi T_i^+[e_j, e_i] = \Phi_i^+ \Psi[e_j, e_i].$$

The proof for T_i^- is similar. $\square$

We are now ready to describe the basis in $U\mathfrak{n}_+$ corresponding to the basis of isomorphism classes of representations of $\vec{Q}$. As before, let $w_0 \in W$ be the longest element in the Weyl group and let $w_0 = s_{i_l} \ldots s_{i_1}$ be a reduced expression adapted to $\vec{Q}$ (see Section 3.5). Recall that this gives a complete order on the set of positive roots: $R_+ = \{\gamma_1, \ldots, \gamma_l\}$; it also gives a construction of all indecomposable representations I_k of $\vec{Q}$ (see (3.10)).

Define the corresponding "root basis" $E_{\gamma_k} \in \mathfrak{g}_{\gamma_k}$ by

(4.25)
$$\begin{aligned} E_{\gamma_1} &= e_{i_1}, \\ E_{\gamma_2} &= T_{i_1}^- e_{i_2}, \\ &\vdots \\ E_{\gamma_k} &= T_{i_1}^- T_{i_2}^- \ldots T_{i_{k-1}}^- (e_{i_k}). \end{aligned}$$

It is immediate that so-defined elements form a basis of $\mathfrak{n}_+$ (indeed, each $\mathfrak{g}_\gamma$ is one-dimensional, so E_γ is a basis in $\mathfrak{g}_\gamma$).

Now, define the Poincaré–Birkhoff–Witt (PBW) basis in $U\mathfrak{n}_+$ as follows. Recall notation $\mathcal{B} = (\mathbb{Z}_+)^{R_+}$. For any $\mathbf{n} = (n_\alpha)_{\alpha \in R_+} \in \mathcal{B}$, define

(4.26)
$$X_\mathbf{n} = E_{\gamma_1}^{(n_{\gamma_1})} \ldots E_{\gamma_l}^{(n_{\gamma_l})} \in U\mathfrak{n}_+,$$

where we used notation $E^{(n)} = \frac{E^n}{n!}$.

It follows immediately from the Poincaré–Birkhoff–Witt theorem that elements $X_\mathbf{n}$, $\mathbf{n} \in \mathcal{B}$, form a basis of $U\mathfrak{n}_+$.

Example 4.18. Let $\vec{Q} = \underset{1}{\bullet} \longrightarrow \underset{2}{\bullet}$. Then we have $w_0 = s_2 s_1 s_2$ and $E_1 = e_2$, $E_2 = T_2^-(e_1) = [e_2, e_1]$, and $E_3 = T_2^- T_1^-(e_2) = e_1$ (cf. Example 3.34). Thus, in this case the PBW basis has the form

$$e_2^{(m)} [e_2, e_1]^{(n)} e_1^{(k)}.$$

Theorem 4.19. *Let $\vec{Q}$ be a Dynkin quiver, and let $\Psi\colon U\mathfrak{n}_+ \to H(\vec{Q})$ be the homomorphism defined in Theorem 4.11. Then for any $\mathbf{n} \in \mathcal{B}$,*

$$\Psi(X_\mathbf{n}) = [\bigoplus_{\alpha \in R_+} n_\alpha I_\alpha].$$

Proof. First, we prove that for any positive root we have $\Psi(E_{\gamma_k}) = I_k$. Indeed, by construction $I_k = \Phi_{i_1}^- \ldots \Phi_{i_{k-1}}^-(S(i_k))$; thus, by Theorem 4.17, we have

$$\Psi(E_{\gamma_k}) = \Psi\left(T_{i_1}^- T_{i_2}^- \ldots T_{i_{k-1}}^- (e_{i_k})\right) = \Phi_{i_1}^- \ldots \Phi_{i_{k-1}}^- (\Psi(e_{i_k})) = I_k.$$

The general result immediately follows from this and Lemma 4.7. $\square$

4.4. Hall algebra of constructible functions

So far we have defined the Hall algebra by counting the number of points of certain varieties $\mathcal{F}_{M_1 M_2}^L$ over the finite field $\mathbb{F}_q$; to get the universal enveloping algebra $U\mathfrak{n}_+$ over $\mathbb{C}$, we have to let $q=1$. It would seem more natural if we had a construction of the Hall algebra which would only use complex geometry and would not require algebraic varieties over $\mathbb{F}_q$. Of course then we can no longer count the number of points. A natural suggestion would be to replace the number of points by a suitably defined volume of a manifold; however, the varieties that appear in the construction of the Hall algebra do not have a distinguished volume form.

Instead, following ideas of Lusztig and Riedtmann, we use the Euler characteristic for complex algebraic varieties and their subsets.

Definition 4.20. Let X be an algebraic variety over $\mathbb{C}$. A subset $Y \subset X$ is called *constructible* if it is obtained from algebraic subvarieties in X by a finite sequence of set-theoretic operations (union, intersection, complement in X).

It is easy to see that any constructible set is automatically locally closed in Zariski topology in X and thus also locally closed in the analytic topology.

For any constructible $Y \subset X$, denote by $\chi(Y)$ the Euler characteristic of Y, considering Y as a topological space with analytic (not Zariski!) topology. We also denote by $\chi_c(Y)$ the Euler characteristic with compact support.

Theorem 4.21. *Let X be an algebraic variety over $\mathbb{C}$.*

(1) *For any constructible $Y \subset X$, we have $\chi(Y) = \chi_c(Y)$.*
(2) *If $Y', Y'' \subset X$ are constructible and $Y' \cap Y'' = \varnothing$, then $\chi(Y' \cup Y'') = \chi(Y') + \chi(Y'')$.*
(3) *For any algebraic fiber bundle $M \to B$ with fiber F, we have $\chi(M) = \chi(B)\chi(F)$.*

The statement of the first result for algebraic varieties over $\mathbb{C}$ can be found in [**Lau1981**], [**Ful1993**, p. 141]; it should be noted that it is false for varieties over $\mathbb{R}$, as is clear from the example $X = \mathbb{R}$. The rest follows easily from this; details can be found, for example, in [**KP1985**].

This theorem shows that χ enjoys properties similar to the properties of volume or to the number of points in an algebraic variety over $\mathbb{F}_q$. In Section 4.5, we will give some arguments why Euler characteristic should be considered a proper complex analog of the number of points over $\mathbb{F}_q$. These arguments are related to some very deep results in arithmetic algebraic geometry.

Definition 4.22. For a Dynkin quiver $\vec{Q}$, its Hall algebra of constructible functions $H^{\mathrm{con}}(\vec{Q})$ is the algebra over $\mathbb{C}$ with the basis given by isomorphism classes of representations of $\vec{Q}$ over $\mathbb{C}$ and multiplication defined by

$$[M_1][M_2] = \sum_{[L]} \chi(\mathcal{F}^L_{M_1 M_2})[L],$$

where $\mathcal{F}^L_{M_1 M_2}$ is defined by (4.1) (it is easy to see that $\mathcal{F}$ is an algebraic variety).

Theorem 4.23 (Riedtmann). *For any Dynkin quiver $\vec{Q}$, the Hall algebra $H^{\mathrm{con}}(\vec{Q})$ is a unital associative algebra.*

A proof of this theorem can be obtained by suitably modifying the proof of Theorem 4.2; this requires the properties of Euler characteristic listed in Theorem 4.21. Details of this proof can be found in [**Rie1994**]. We will not give the proof here; instead, we will give an alternative definition of the Hall algebra using the notion of constructible function (thus explaining the name) and will use this alternative definition to prove associativity of H (see (4.35) below).

Definition 4.24. Let X be an algebraic variety over $\mathbb{C}$. A function $\varphi \colon X \to \mathbb{C}$ is called *constructible* if it takes only finitely many values, and for each $a \in \mathbb{C}$, the level set $\varphi^{-1}(a)$ is constructible. We denote the space of constructible functions on X by $\mathrm{Con}(X)$.

It is easy to see that the set of constructible functions is indeed a vector space (and, in fact, an algebra).

For any constructible function $\varphi \in \mathrm{Con}(X)$, define its integral over X by

$$(4.27) \qquad \int_X \varphi = \sum_{a \in \mathbb{C}} a\, \chi(\varphi^{-1}(a)).$$

It follows from Theorem 4.21 that $\int_X \colon \mathrm{Con}(X) \to \mathbb{C}$ is a linear map.

If $f \colon X \to Y$ is a morphism of algebraic varieties, then it is easy to see that for any constructible $Z \subset Y$, $f^{-1}(Z)$ is constructible. Thus, we can define maps

$$(4.28) \qquad \begin{aligned} f^* &\colon \mathrm{Con}(Y) \to \mathrm{Con}(X), \\ f_! &\colon \mathrm{Con}(X) \to \mathrm{Con}(Y) \end{aligned}$$

by

$$(4.29) \qquad \begin{aligned} (f^*\varphi)(x) &= \varphi(f(x)), \\ (f_!\varphi)(y) &= \int_{f^{-1}(y)} \varphi. \end{aligned}$$

4.4. Hall algebra of constructible functions

Using Theorem 4.21, one can check that this is functorial: for $g\colon X \to Y$, $f\colon Y \to Z$, we have
$$(4.30) \qquad \begin{aligned} (fg)^* &= g^*f^*, \\ (fg)_! &= f_!g_!. \end{aligned}$$

In particular, for any morphism of algebraic varieties $\pi\colon M \to B$ and a constructible function φ on M, we have
$$\int_M \varphi = \int_B \pi_!(\varphi) = \int_B \left(\int_{\pi^{-1}(b)} \varphi \right).$$

Remark 4.25. These operations on constructible functions are nothing but shadows of corresponding pullback and pushforward functors on the category of constructible sheaves. We will return to this later in Section 4.5.

We can now give a definition of the Hall algebra of a Dynkin quiver in terms of constructible functions. Consider the category $\mathrm{Rep}(\vec{Q})$ of complex representations of a quiver $\vec{Q}$. Recall that the isomorphism classes of representations of dimension $\mathbf{v} \in \mathbb{Z}_+^I$ are in bijection with $\mathrm{GL}(\mathbf{v})$-orbits in the representation space $R(\vec{Q}, \mathbf{v})$ (see Section 2.1). Define, for a Dynkin quiver $\vec{Q}$,
$$(4.31) \qquad H_{\mathbf{v}}^{\mathrm{con}}(\vec{Q}) = \mathrm{Con}(R(\vec{Q}, \mathbf{v}))^{\mathrm{GL}(\mathbf{v})}.$$

It is clear that each $\mathrm{GL}(\mathbf{v})$-orbit is a constructible subset in $R(\vec{Q}, \mathbf{v})$ and that characteristic functions of orbits form a basis of the space $\mathrm{Con}(R(\mathbf{v}, \vec{Q}))^{\mathrm{GL}(\mathbf{v})}$ of invariant constructible functions (this is based on the fact that there are only finitely many orbits and thus fails for non-Dynkin quivers).

To define the multiplication $H_{\mathbf{v}}^{\mathrm{con}} \otimes H_{\mathbf{w}}^{\mathrm{con}} \to H_{\mathbf{v}+\mathbf{w}}^{\mathrm{con}}$, consider the following diagram:

$$(4.32) \qquad \begin{array}{c} R' \\ {}^{\beta}\swarrow \quad \searrow^{\beta'} \\ R(\mathbf{v}) \times R(\mathbf{w}) \qquad R'' \\ \qquad \qquad \searrow^{\beta''} \\ \qquad \qquad \quad R(\mathbf{v}+\mathbf{w}), \end{array}$$

where
$$R'' = \{(x, W) \mid x \in R(\mathbf{v}+\mathbf{w}), W \subset \mathbb{C}^{\mathbf{v}+\mathbf{w}}, \dim W = \mathbf{w}, xW \subset W\},$$
$$R' = \{(x, W) \in R'', \varphi\colon W \to \mathbb{C}^{\mathbf{w}}, \psi\colon \mathbb{C}^{\mathbf{v}+\mathbf{w}}/W \to \mathbb{C}^{\mathbf{v}}\}.$$

Here φ, ψ are isomorphisms of graded vector spaces; obviously, fixing φ, ψ is equivalent to fixing a choice of basis in each W_i, $\mathbb{C}^{\mathbf{v}_i+\mathbf{w}_i}/W_i$.

The maps β', β'' are the obvious forgetting maps; the map $\beta\colon R' \to R(\mathbf{v}) \times R(\mathbf{w})$ is given by restriction of x to $\mathbb{C}^{\mathbf{v}+\mathbf{w}}/W \simeq \mathbb{C}^{\mathbf{v}}$ and $W \simeq \mathbb{C}^{\mathbf{w}}$.

Lemma 4.26.

(1) R', R'' are algebraic varieties, and β, β', β'' are morphisms of algebraic varieties.

(2) β commutes with the action of $\mathrm{GL}(\mathbf{v}) \times \mathrm{GL}(\mathbf{w})$ on $R(\mathbf{v}) \times R(\mathbf{w})$, R'.

(3) β' is a principal $\mathrm{GL}(\mathbf{v}) \times \mathrm{GL}(\mathbf{w})$-bundle over R''.

(4) β'' is proper.

The proof of this lemma is left as an exercise to the reader.

Now define multiplication in the Hall algebra of constructible functions by

(4.33) $$f_1 \cdot f_2 = \beta''_!(\beta'_\flat \beta^*(f_1 f_2)),$$

where $\beta'_\flat : \mathrm{Con}(R')^{\mathrm{GL}(\mathbf{v}) \times \mathrm{GL}(\mathbf{w})} \to \mathrm{Con}(R'')$ is defined as follows: since β' is a principal $\mathrm{GL}(\mathbf{v}) \times \mathrm{GL}(\mathbf{w})$-bundle, every $\mathrm{GL}(\mathbf{v}) \times \mathrm{GL}(\mathbf{w})$-invariant function f on R' is obtained by a pull-back of a unique function $\beta'_\flat(f)$ on R'':

$$f = \beta'^*(\beta'_\flat(f)).$$

This definition can also be rewritten as follows:

(4.34) $$f_1 \cdot f_2(x) = \int_W f_1(x|_{\mathbb{C}^\mathbf{d}/W}) f_2(x|_W), \qquad \mathbf{d} = \mathbf{v} + \mathbf{w},$$

where the integral is over the set $\beta''^{-1}(x)$ of all subrepresentations $W \subset \mathbb{C}^{\mathbf{v}+\mathbf{w}}$, $xW \subset W$ with $\dim W = \mathbf{w}$.

In particular, in the case when f_1, f_2 are characteristic functions of orbits, we get

$$\mathbf{1}_{\mathbb{O}_1} \cdot \mathbf{1}_{\mathbb{O}_2}(x) = \chi(\{W \subset \mathbb{C}^\mathbf{d} \mid xW \subset W, x|_{\mathbb{C}^\mathbf{d}/W} \in \mathbb{O}_1, x|_W \in \mathbb{O}_2\})$$

so

$$\mathbf{1}_{\mathbb{O}_1} \cdot \mathbf{1}_{\mathbb{O}_2} = \sum_\mathbb{O} \chi(\mathcal{F}^\mathbb{O}_{\mathbb{O}_1 \mathbb{O}_2}) \mathbf{1}_\mathbb{O},$$

where the sum is over all orbits $\mathbb{O} \subset R(\mathbf{d})$, and $\mathcal{F}^\mathbb{O}_{\mathbb{O}_1 \mathbb{O}_2} = \{W \subset \mathbb{C}^\mathbf{d} \mid xW \subset W, x|_{\mathbb{C}^\mathbf{d}/W} \in \mathbb{O}_1, x|_W \in \mathbb{O}_2\}$, which coincides with Definition 4.22.

Now in this new language one can easily prove associativity. Indeed, if $f_i \in H_{\mathbf{v}_i}(\vec{Q})$ and $x \in R(\mathbf{d})$, $\mathbf{d} = \mathbf{v}_1 + \mathbf{v}_2 + \mathbf{v}_3$, then

(4.35) $$(f_1 f_2) f_3(x) = \int f_1(x|_{\mathbb{C}^\mathbf{d}/V_2}) f_2(x|_{V_2/V_1}) f_3(x|_{V_1}) = f_1(f_2 f_3)(x),$$

where the integral is taken over the set of all flags $0 \subset V_1 \subset V_2 \subset \mathbb{C}^\mathbf{d}$ satisfying $\dim V_1 = \mathbf{v}_3$, $\dim V_2 = \mathbf{v}_3 + \mathbf{v}_2$, $xV_i \subset V_i$.

4.4. Hall algebra of constructible functions

Example 4.27. Let $\vec{Q} = \bullet$. Then, similar to computations in Example 4.3, we get the following relations in $H^{\mathrm{con}}(\vec{Q})$:

$$[nS] \cdot [S] = [(n+1)S]\chi(\mathbb{P}^n(\mathbb{C})) = (n+1)[(n+1)S]$$

which implies

$$[nS] \cdot [mS] = \binom{n+m}{m}[(n+m)S].$$

In particular,

$$[nS] = \frac{[S]^n}{n!}.$$

Example 4.28. Let $\vec{Q} = \bullet_1 \longrightarrow \bullet_2$. Then the same computations as in Section 4.2 show that in $H^{\mathrm{con}}(\vec{Q})$ we have:

(4.36)
$$\theta_1 \cdot \theta_2 = [S_{12}] + [\tilde{S}_{12}],$$
$$\theta_2 \cdot \theta_1 = [\tilde{S}_{12}]$$

and

(4.37)
$$\theta_1^2 \cdot \theta_2 = 2[2S(1)][S(2)] = 2([N^0] + [N^1]),$$
$$\theta_2 \cdot \theta_1^2 = 2[S(2)][2S(1)] = 2[N^0],$$
$$\theta_1 \cdot \theta_2 \cdot \theta_1 = \theta_1 \cdot [\tilde{S}_{12}] = \chi(\mathbb{P}^1(\mathbb{C}))[N^0] + [N^1] = 2[N^0] + [N^1]$$

(same notation as in Section 4.2). This implies the Serre relations:

$$\theta_1^2 \theta_2 - 2\theta_1\theta_2\theta_1 + \theta_2\theta_1^2 = 0,$$
$$\theta_2^2 \theta_1 - 2\theta_2\theta_1\theta_2 + \theta_1\theta_2^2 = 0.$$

Theorem 4.29 (Riedtmann). *If $\vec{Q}$ is a Dynkin quiver, then the Hall algebra $H^{\mathrm{con}}(\vec{Q})$ is isomorphic to $U\mathfrak{n}_+$, with the isomorphism given by $\Psi(e_i) = \theta_i = $ characteristic function of orbit $\mathbb{O}_{S(i)}$.*

The proof is similar to the proof of Theorem 4.10. The only difference is in the proof of Serre relations, which is given in Example 4.28 above.

Corollary 4.30. *For a Dynkin quiver $\vec{Q}$, the Hall algebra of constructible functions is isomorphic to the Hall algebra $H(\vec{Q})$ obtained by specialization at $t = 1$ of the universal Hall algebra $H(\vec{Q})_{\mathcal{A}}$.*

Note that the theorem makes no sense for non-Dynkin quivers, as in the general case it is impossible to define specialization $q = 1$ of the Hall algebra $H(\vec{Q}, \mathbb{F}_q)$.

4.5. Finite fields vs. complex numbers

In this section, we will give some explanation of why the definition of Hall algebra obtained by counting points over finite field $\mathbb{F}_q$ and then letting $q = 1$ gives the same answer as computing the Euler characteristic over $\mathbb{C}$. This is based on the following result, which we give here in the form due to N. Katz.

Theorem 4.31. *Let X be an algebraic variety defined over $\mathbb{Z}$. Assume that there exists a polynomial $P_X(t) \in \mathbb{C}[t]$ such that for any finite field $\mathbb{F}_q$, the number of points of X over $\mathbb{F}_q$ is given by*

$$|X(\mathbb{F}_q)| = P_X(q).$$

Let $X(\mathbb{C})$ be the corresponding algebraic variety over $\mathbb{C}$, considered with the analytic topology. Then its Euler characteristic is given by

$$\chi(X(\mathbb{C})) = P_X(1).$$

The proof of this theorem can be found in the Appendix of [**HRV2008**] (it is a special case of Theorem 6.1.2(3), with $x = y = 1$). It is based on a number of very deep results of Grothendieck and Deligne, used to prove the Weil conjectures in arithmetic algebraic geometry. Detailed discussion of them goes far beyond the limits of this book; however, it is one of the most remarkable results in all of modern mathematics, so we do give a brief overview. We refer interested readers to the book [**FK1988**] for details.

The starting point of Grothendieck's work are two observations:

- Let X be an algebraic variety defined over a finite field $\mathbb{F}_q$; we will denote by $X(\overline{\mathbb{F}_q})$ the set of points of X over the algebraic closure $\overline{\mathbb{F}_q}$. Then the Frobenius automorphism $\mathrm{Fr}\colon \overline{\mathbb{F}_q} \to \overline{\mathbb{F}_q}\colon x \mapsto x^q$ defines an automorphism of $X(\overline{\mathbb{F}_q})$, and the points of X over $\mathbb{F}_q$ are exactly the fixed points of Frobenius automorphism:
$$X(\mathbb{F}_q) = (X(\overline{\mathbb{F}_q}))^{\mathrm{Fr}}.$$

- For a compact complex manifold M, one can compute the number of fixed points of a morphism $f\colon M \to M$ by the Lefschetz fixed point formula:
$$|M^f| = \sum_i (-1)^i \operatorname{tr}_{H^i(M)}(f^*)$$

(assuming that the graph of f intersects the diagonal transversally).

Developing these ideas, Grothendieck proved an analog of the Lefshetz fixed point formula for the Frobenius automorphism. This requires introducing a new cohomology theory for algebraic varieties over $\overline{\mathbb{F}_q}$, namely *étale* cohomology with l-adic coefficients.

4.5. Finite fields vs. complex numbers

To relate this with the Euler characteristic over $\mathbb{C}$, we need to relate the Euler characteristic over $\mathbb{C}$ with the étale Euler characteristic over $\overline{\mathbb{F}_p}$. It turns out that for a large class of algebraic varieties defined over $\mathbb{Z}$ (but not for all of them), these two coincide: $\chi_c(X(\mathbb{C})) = \chi_c^{et}(X(\overline{\mathbb{F}_q}))$; precise conditions can be found in [**FK1988**]. Thus, for such varieties we see that counting the number of points over a finite field and then letting $q = 1$ gives exactly the Euler characteristic over $\mathbb{C}$. This is the primary motivation for the use of the Euler characteristic in Definition 4.22.

Finally it should be noted that constructible functions (both over the finite field and over $\mathbb{C}$) are but a shadow of a much more interesting theory, namely the derived category of constructible sheaves. A description of these categories can be found, for example, in Lusztig's book [**Lus1993**] or in [**KS1990**]; here we just mention that each constructible sheaf $\mathcal{F}$ defines a constructible function whose value at point x is equal to $\mathrm{tr}_{\mathcal{F}_x}(\mathrm{Fr})$, where $\mathcal{F}_x$ is the stalk at point x and Fr is the Frobenius automorphism; for $\mathbb{C}$, we just take the dimensions of stalks. In the Dynkin case, the Hall algebra can be described as the Grothendieck group of the category of $GL(\mathbf{v})$-invariant constructible sheaves. Moreover, all operations we defined for constructible functions can actually be extended to suitable functors on the derived category of constructible sheaves. In particular, using this approach Lusztig was able to define a remarkable basis in the Hall algebra, and thus in the positive part of the (quantum) universal enveloping algebra, called the *canonical basis*. We refer the reader to the original papers [**Lus1990a**, **Lus1990b**, **Lus1991**] or to the book [**Lus1993**] for the description of this theory.

Chapter 5

Double Quivers

In this chapter, we provide another approach to the geometric construction of the positive part of the universal enveloping algebra corresponding to a Dynkin quiver. This is based on the notions of a double quiver and of a preprojective algebra, both of which will also be extensively used later.

5.1. The double quiver

Definition 5.1. Let Q be a graph, with a set of vertices I and a set of edges E. We define the corresponding *double quiver* $Q^\sharp$ as the quiver with the same vertices and with the set of oriented edges

$$H = \{(e, o(e))\}, \qquad e \text{ — an edge of } Q, o(e) \text{ — an orientation of } e.$$

Thus, each edge e connecting vertices i and j in Q gives rise to two oriented edges $h\colon i \to j$, $\bar h\colon j \to i$ in $Q^\sharp$. An example of a double quiver is shown in Figure 5.1.

The set H of edges of $Q^\sharp$ has a natural involution $h \mapsto \bar h$ which reverses edge orientation: for an edge $h\colon i \to j$, $\bar h$ is the corresponding edge $j \to i$.

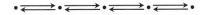

Figure 5.1. A double quiver.

Note that in this language, any quiver $\vec Q$ with the underlying graph Q is naturally a subquiver in $Q^\sharp$: namely, the set Ω of edges of $\vec Q$ is a subset in the set H of edges of $Q^\sharp$ such that $H = \Omega \cup \bar\Omega$.

The category of representations of a double quiver is defined in the same way as for any quiver. By results of Section 2.1, isomorphism classes of representations of graded dimension $\mathbf{v}$ are in bijection with $\mathrm{GL}(\mathbf{v})$-orbits in the representation space

(5.1) $$R(Q^\sharp, \mathbf{v}) = \bigoplus_{h \in H} \mathrm{Hom}_{\mathbf{k}}(\mathbf{k}^{\mathbf{v}_{s(h)}}, \mathbf{k}^{\mathbf{v}_{t(h)}}).$$

Note that for any pair of finite-dimensional vector spaces V, W we have a canonical pairing $\mathrm{Hom}(V, W) \otimes \mathrm{Hom}(W, V) \to \mathbf{k}$, given by $(f, g) = \mathrm{tr}(fg)$. This pairing is nondegenerate and therefore defines an isomorphism $\mathrm{Hom}(W, V) \simeq \mathrm{Hom}(V, W)^*$. This gives us the relation between the representation spaces $R(\vec{Q}, \mathbf{v})$ and $R(Q^\sharp, \mathbf{v})$:

(5.2) $$R(Q^\sharp, \mathbf{v}) = R(\vec{Q}, \mathbf{v}) \oplus R(\vec{Q}, \mathbf{v})^* = T^*(R(\vec{Q}, \mathbf{v})).$$

The last identity follows from the observation that for any vector space E, the cotangent bundle is $T^*E = E \oplus E^*$.

5.2. Preprojective algebra

In general, classifying representations of double quivers is a problem which does not have a satisfactory answer (as we will discuss in Chapter 7, such classification problem are called *wild*). Instead, we will consider representations satisfying some additional restrictions.

In what follows, we will need to choose a function $\varepsilon \colon H \to \mathbf{k}^\times$ such that

(5.3) $$\varepsilon(h) + \varepsilon(\bar{h}) = 0 \quad \text{for any edge } h \in H.$$

Of course, there are many such functions. For example, one can choose an orientation Ω of Q and define ε by

(5.4) $$\varepsilon_\Omega(h) = \begin{cases} 1, & h \in \Omega, \\ -1, & h \in \bar{\Omega}. \end{cases}$$

However, up to suitable isomorphisms, all our subsequent constructions will not depend on ε.

Definition 5.2. Let $Q^\sharp$ be the double quiver of a graph Q, and let $\varepsilon \colon H \to \mathbf{k}^\times$ be a function satisfying (5.3). Define the *preprojective algebra*

(5.5) $$\Pi = \mathbf{k}Q^\sharp/J,$$

where J is the two-sided ideal generated by elements θ_i, $i \in I$, defined by

$$\theta_i = \sum_{t(h)=i} \varepsilon(h) h \bar{h}.$$

The name "preprojective algebra" will be explained in Chapter 6, where we introduce the notion of a preprojective representation of $\vec{Q}$ and show that the preprojective algebra is exactly the direct sum of all indecomposable preprojective representations (see Corollary 6.32).

5.2. Preprojective algebra

Preprojective algebra also naturally appears in the study of McKay correspondence, which will be discussed in Section 8.3.

The definition of preprojective algebra is motivated by symplectic geometry. Namely, define the map

(5.6)
$$\mu_{\mathbf{v}} \colon R(Q^\sharp, \mathbf{v}) \to \bigoplus_i \mathfrak{gl}(\mathbf{v}_i, \mathbf{k})$$
$$z \mapsto \bigoplus_i \sum_{t(h)=i} \varepsilon(h) z_h z_{\bar{h}}.$$

This map naturally appears as the moment map for the action of $\mathrm{GL}(\mathbf{v})$ on $R(Q^\sharp, \mathbf{v})$. This will be discussed in detail in Section 10.2. In particular, if ε is determined by a choice of orientation Ω of Q (see (5.4)), then

$$\mu(z) = \sum_{h \in \Omega} [z_h, z_{\bar{h}}].$$

Consider now the set

(5.7) $\quad \mu_{\mathbf{v}}^{-1}(0) = \{ z \in R(Q^\sharp, \mathbf{v}) \mid \sum_{t(h)=i} \varepsilon(h) z_h z_{\bar{h}} = 0 \text{ for any } i \in I \}.$

The following lemma is immediate from the definitions.

Lemma 5.3. *$\mu_{\mathbf{v}}^{-1}(0)$ is a closed algebraic subvariety in $R(Q^\sharp, \mathbf{v})$, which is invariant under the action of $\mathrm{GL}(\mathbf{v})$. The $\mathrm{GL}(\mathbf{v})$-orbits in $\mu_{\mathbf{v}}^{-1}(0)$ are in bijection with the isomorphism classes of representations of the preprojective algebra Π.*

We now study some basic properties of the preprojective algebra.

Lemma 5.4.

(1) *Up to an isomorphism, the preprojective algebra Π and the variety $\mu_{\mathbf{v}}^{-1}(0)$ do not depend on the choice of function ε.*

(2) *For any choice of orientation Ω of Q, the path algebra $\mathbf{k}\vec{Q}$, where $\vec{Q} = (Q, \Omega)$, is a subalgebra in Π.*

Proof. Let $\varepsilon', \varepsilon''$ be two functions satisfying (5.3). Let Π', Π'' be the corresponding preprojective algebras. Obviously, in this case we have $\varepsilon''(h) = \lambda(h)\varepsilon'(h)$ for some function $\lambda \colon H \to \mathbf{k}^\times$ such that $\lambda(\bar{h}) = \lambda(h)$. Choose an orientation Ω of Q. Then the map

$$z_h \mapsto \begin{cases} \lambda(h) z_h, & h \in \Omega, \\ z_h, & h \notin \Omega, \end{cases}$$

is an isomorphism $\Pi'' \to \Pi'$. The statement about $\mu_{\mathbf{v}}^{-1}(0)$ is proved similarly.

To prove the second part, consider the path algebra $A = \mathbf{k}Q^\sharp$. Obviously, the algebra $\mathbf{k}\vec{Q}$ is a subalgebra in A and $J \cap \mathbf{k}\vec{Q} = \{0\}$ (every element of J is a linear combination of paths which contain the product $h\bar{h}$, and $\mathbf{k}\vec{Q}$ does not contain such paths). Thus, the composition $\mathbf{k}\vec{Q} \to \mathbf{k}Q^\sharp \to \Pi$ is injective. $\square$

It is easy to see from the definition that the preprojective algebra is naturally graded:

$$(5.8) \qquad \Pi = \bigoplus_{l \geq 0} \Pi^l = \bigoplus_{l \geq 0, i, j \in I} {}_j\Pi^l_i,$$

where Π^l is spanned by paths of length l, and ${}_j\Pi^l_i = e_j \Pi^l e_i$ is spanned by paths of length l from i to j.

5.3. Varieties $\Lambda(\mathbf{v})$

Recall that the isomorphism classes of representations of the preprojective algebra are in bijection with $GL(\mathbf{v})$-orbits in $\mu_\mathbf{v}^{-1}(0)$, where $\mu_\mathbf{v} \colon R(Q^\sharp, \mathbf{v}) \to \mathfrak{gl}(\mathbf{v})$ is defined by (5.6). In this section we study the geometry of these spaces.

Example 5.5. Let Q be the quiver of type A_2 and $\mathbf{v} = (m, n)$. Then $Q^\sharp$ is shown in Figure 5.2, and the representation space $R(Q^\sharp, \mathbf{v}) = \{(x, y) \mid x \colon \mathbf{k}^m \to \mathbf{k}^n, y \colon \mathbf{k}^n \to \mathbf{k}^m\}$. In this case the variety $\mu_\mathbf{v}^{-1}(0) \subset R(Q^\sharp, \mathbf{v})$ is given by equations $xy = 0$, $yx = 0$. In particular, if $m = n = 1$, then $R(Q^\sharp, \mathbf{v}) \simeq \mathbf{k}^2$, and $\mu_\mathbf{v}^{-1}(0) = \{(x, y) \in \mathbf{k}^2 \mid xy = 0\}$ is the coordinate cross. Thus, in this case it is reducible and has two irreducible components, both of which are nonsingular one-dimensional varieties.

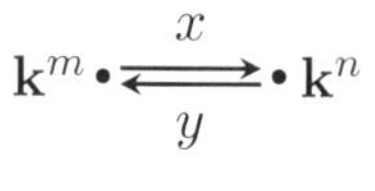

$$\mathbf{k}^m \bullet \underset{y}{\overset{x}{\rightleftarrows}} \bullet \mathbf{k}^n$$

Figure 5.2. Double of quiver A_2.

The following lemma gives an explicit description of $\mu_\mathbf{v}^{-1}(0)$.

Lemma 5.6. *Let Ω be an orientation of Q and let $\vec{Q} = (Q, \Omega)$ be the corresponding quiver so that*

$$R(Q^\sharp, \mathbf{v}) = R(\vec{Q}, \mathbf{v}) \oplus R(\vec{Q}^{opp}, \mathbf{v})$$

(compare with (5.2)). Then for $x \in R(\vec{Q}, \mathbf{v})$, $y \in R(\vec{Q}^{opp}, \mathbf{v}) = R(\vec{Q}, \mathbf{v})^*$, we have

$$x + y \in \mu_\mathbf{v}^{-1}(0) \text{ iff } \langle T_x \mathbb{O}_x, y \rangle = 0,$$

where $\mathbb{O}_x$ is the orbit of x in $R(\vec{Q}, \mathbf{v})$ and $\langle \, , \, \rangle$ stands for the pairing $R(\vec{Q}, \mathbf{v}) \otimes R(\vec{Q}^{opp}, \mathbf{v}) \to \mathbf{k}$ given by $\langle a, b \rangle = \mathrm{tr}_{\mathbf{k}^\mathbf{v}}(ab)$.

5.3. Varieties $\Lambda(\mathbf{v})$

Proof. Indeed,

$$\langle \mu(x+y), a \rangle = \sum_{h \in \Omega} \operatorname{tr}([x_h, y_{\bar{h}}]a) = \sum_{h \in \Omega} \operatorname{tr}([a, x_h]y_{\bar{h}}) = \langle [a, x], y \rangle.$$

Thus, $\mu(x+y) = 0$ iff y is orthogonal to every element of the form $[a, x]$, $a \in \mathfrak{gl}(\mathbf{v})$. But $T_x \mathbb{O}_x = \{[a, x], a \in \mathfrak{gl}(\mathbf{v})\}$. $\square$

In other words, we see that $\mu_{\mathbf{v}}^{-1}(0)$ is the union of conormal bundles to $\operatorname{GL}(\mathbf{v})$-orbits in $R(\vec{Q}, \mathbf{v})$.

Corollary 5.7. *Let Q be Dynkin. Then:*

(1) *Irreducible components of $\mu_{\mathbf{v}}^{-1}(0)$ are exactly the closures of conormal bundles to $\operatorname{GL}(\mathbf{v})$-orbits in $R(\vec{Q}, \mathbf{v})$:*

$$Z = \overline{N^*\mathbb{O}},$$

where $N_x^* = (T_x R/T_x \mathbb{O}_x)^*$.

(2) *Each irreducible component is a Lagrangian subvariety in $R(Q^\sharp, \mathbf{v})$; in particular, it has dimension equal to $\dim R(Q^\sharp, \mathbf{v})/2$.*

Proof. Since $\operatorname{GL}(\mathbf{v})$ is irreducible, we see that $\overline{N^*\mathbb{O}}$ is an irreducible subvariety in $\mu_{\mathbf{v}}^{-1}(0)$. By Lemma 5.6, $\mu_{\mathbf{v}}^{-1}(0)$ is the union of these conormal bundles. Since there are only finitely many of them (Gabriel's theorem), this implies that each Z is an irreducible component.

The second part of the theorem is immediate from the standard arguments of symplectic geometry: if $Y \subset X$ is a submanifold, then $N^*Y \subset T^*X$ is Lagrangian (see Example 9.42 in Chapter 9). $\square$

This result fails if Q is not Dynkin. In this case, it is useful to modify the definition and consider the subvariety of nilpotent elements in $\mu_{\mathbf{v}}^{-1}(0)$.

Definition 5.8. An element $z \in R(Q^\sharp, \mathbf{v})$ is called nilpotent if there exists N such that $z_{h_l} \ldots z_{h_1} = 0$ for any path $h_l \ldots h_1$ in $Q^\sharp$ with $l \geq N$.

Now define

(5.9) $\quad \Lambda(\mathbf{v}) = \{z \in R(Q^\sharp, \mathbf{v}) \mid \mu_{\mathbf{v}}(z) = 0, \; z \text{ is nilpotent } \} \subset \mu_{\mathbf{v}}^{-1}(0).$

It is easy to prove that Λ is a closed subvariety of $R(Q^\sharp, \mathbf{v})$; it is also obvious that Λ is stable under the action of the group $\operatorname{GL}(\mathbf{v})$.

Theorem 5.9 (Lusztig)**.** *If Q is Dynkin, then any $z \in \mu_{\mathbf{v}}^{-1}(0)$ is nilpotent: $\Lambda(\mathbf{v}) = \mu_{\mathbf{v}}^{-1}(0)$.*

Proof. Let us choose an orientation Ω and let $\vec{Q} = (Q, \Omega)$ be the corresponding quiver. Then we can write $z = x + y$, $x \in R(\vec{Q}, \mathbf{v})$, $y \in R(\vec{Q}^{opp}, \mathbf{v})$.

Recall that by Theorem 3.18, indecomposable irreducible representations of $\vec{Q}$ are classified by positive roots of the corresponding root system. Moreover, it is possible to order the positive roots in such a way that

$$\operatorname{Hom}_{\vec{Q}}(I_a, I_b) = 0 \quad \text{for } a > b,$$
$$\operatorname{Ext}^1_{\vec{Q}}(I_a, I_b) = 0 \quad \text{for } a \leq b$$

(see Theorem 3.35).

Let $V = (\mathbb{C}^{\mathbf{v}}, x)$ be the representation of $\vec{Q}$ corresponding to $x \in R(\vec{Q}, \mathbf{v})$. We can write $V = \bigoplus_k n_k I_k$, where I_k are indecomposable representations chosen as above. Let $V^k = n_k I_k$. Then

$$R(\vec{Q}, \mathbf{v}) = \bigoplus_{k,l} H_\Omega^{kl}, \qquad H_\Omega^{kl} = \bigoplus_{h \in \Omega} \operatorname{Hom}_{\mathbf{k}}(V^k_{s(h)}, V^l_{t(h)}).$$

Since each of V^k is stable under x, we see that $x \in \bigoplus_k H_\Omega^{kk}$.

From the description of a tangent space to the orbit given in Theorem 2.4, we easily see that in this case the space $T_x \mathbb{O}_x \subset R(\vec{Q}, \mathbf{v})$ is graded:

$$T_x \mathbb{O}_x = \bigoplus (T_x \mathbb{O}_x \cap H_\Omega^{kl})$$

and we have a short exact sequence

$$0 \to T_x \mathbb{O}_x \cap H_\Omega^{kl} \to H_\Omega^{kl} \to \operatorname{Ext}^1_{\vec{Q}}(V^k, V^l) \to 0.$$

By Theorem 3.35, $\operatorname{Ext}^1_{\vec{Q}}(V^k, V^l) = 0$ for $k \leq l$; thus,

$$T_x \supset \bigoplus_{k \leq l} H_\Omega^{kl}.$$

Since we have $\langle y, T_x \rangle = 0$ (Lemma 5.6), we see that $y \in \bigoplus_{k < l} H_{\bar{\Omega}}^{kl}$, where

$$H_{\bar{\Omega}}^{kl} = \bigoplus_{h \in \bar{\Omega}} \operatorname{Hom}_{\mathbf{k}}(V^k_{s(h)}, V^l_{t(h)}).$$

Therefore, we see that

$$z = x + y \in \bigoplus_{k \leq l} \operatorname{Hom}_{\mathbf{k}}(V^k, V^l)$$

and the block-diagonal part of z coincides with x. Since $\vec{Q}$ has no oriented cycles, x is nilpotent. This easily implies that z is nilpotent. $\square$

Corollary 5.10. *If Q is Dynkin, then the only irreducible representations of the preprojective algebra Π are the representations $S(i)$ defined in (1.5).*

Proof. Let V be an irreducible representation of Π. Consider the spaces $\Pi^l V$, where Π^l is the subspace in the preprojective algebra spanned by paths of length l. By Theorem 5.9, $\Pi^l V = 0$ for l large enough. Therefore, $\Pi^1 V$ is a proper subspace in V. Since it is also a subrepresentation, we see that we must have $\Pi^1 V = 0$, which means that all operators $z_h, h \in H$, act by zero in V. Therefore, $V \simeq \bigoplus \mathbf{v}_i S(i)$. □

In general, however, $\Lambda(\mathbf{v})$ is a proper subset in $\mu_{\mathbf{v}}^{-1}(0)$ (we will give an example in the next section).

Theorem 5.11. *Assume that Q has no edge loops. Then*:
  (1) $\Lambda(\mathbf{v})$ *is a variety of pure dimension* $\dim R(Q^\sharp, \mathbf{v})/2$ (*i.e. each irreducible component of $\Lambda(\mathbf{v})$ has this dimension*).
  (2) $\Lambda(\mathbf{v})$ *is a Lagrangian subvariety in* $R(Q^\sharp, \mathbf{v})$.

A proof of this theorem can be found in [**Lus1991**, Theorem 12.3].

Note that this fails if Q has edge loops, as is easy to see in the example of the Jordan quiver. A possible replacement for the varieties $\Lambda(\mathbf{v})$ for graphs with edge loops was suggested in [**Boz**].

5.4. Composition algebra of the double quiver

In this section we will give another realization of the positive part of the universal enveloping algebra $U\mathfrak{g}$ in terms of quivers. Throughout this section, we assume that Q is a connected graph without edge loops. We also assume that the ground field is the field of complex numbers: $\mathbf{k} = \mathbb{C}$.

Consider the variety $\Lambda(\mathbf{v})$ (note that in the Dynkin case, by Theorem 5.9, it coincides with $\mu_{\mathbf{v}}^{-1}(0)$). Consider the space of complex-valued $\mathrm{GL}(\mathbf{v})$-invariant constructible functions on $\Lambda(\mathbf{v})$ (see Definition 4.24 for the definition of constructible function):
$$C(\mathbf{v}) = \mathrm{Con}(\Lambda(\mathbf{v}))^{\mathrm{GL}(\mathbf{v})}.$$
We will define a structure of a graded associative algebra on $C = \bigoplus_{\mathbf{v}} C(\mathbf{v})$. To do this, consider the following diagram (compare with (4.32)):

(5.10)
$$\begin{array}{ccc}
 & \Lambda' & \\
{}^{\beta}\swarrow & & \searrow^{\beta'} \\
\Lambda(\mathbf{v}) \times \Lambda(\mathbf{w}) & & \Lambda'' \\
 & & \downarrow^{\beta''} \\
 & & \Lambda(\mathbf{v} + \mathbf{w}),
\end{array}$$

where
$$\Lambda'' = \{(z, W) \mid z \in \Lambda(\mathbf{v} + \mathbf{w}), W \subset \mathbb{C}^{\mathbf{v}+\mathbf{w}}, \dim W = \mathbf{w}, zW \subset W\},$$
$$\Lambda' = \{(z, W) \in \Lambda'', \varphi \colon W \to \mathbb{C}^{\mathbf{w}}, \psi \colon \mathbb{C}^{\mathbf{v}+\mathbf{w}}/W \to \mathbb{C}^{\mathbf{v}}\}.$$

Here φ, ψ are isomorphisms of graded vector spaces; obviously, fixing φ, ψ is equivalent to fixing a choice of basis in each W_i, $\mathbb{C}^{\mathbf{v}_i + \mathbf{w}_i}/W_i$.

The maps β', β'' are the obvious forgetting maps; the map $\beta \colon \Lambda' \to \Lambda(\mathbf{v}) \times \Lambda(\mathbf{w})$ is given by restriction of z to $\mathbb{C}^{\mathbf{v}+\mathbf{w}}/W \simeq \mathbb{C}^{\mathbf{v}}$ and $W \simeq \mathbb{C}^{\mathbf{w}}$. Note that since z is nilpotent, its restrictions to $\mathbb{C}^{\mathbf{v}+\mathbf{w}}/W \simeq \mathbb{C}^{\mathbf{v}}$ and $W \simeq \mathbb{C}^{\mathbf{w}}$ are also nilpotent; similarly, condition $\sum_h \varepsilon(h) z_h z_{\overline{h}} = 0$ in $\mathfrak{gl}(\mathbf{v}+\mathbf{w}, \mathbb{C})$ implies that we also have the same condition for the restrictions. Thus, map β is well defined.

As in Section 4.4, we see that Λ', Λ'' are algebraic varieties, β, β', β'' are morphisms of algebraic varieties such that β commutes with the action of $\mathrm{GL}(\mathbf{v}) \times \mathrm{GL}(\mathbf{w})$ on $\Lambda(\mathbf{v}) \times \Lambda(\mathbf{w})$, Λ'; moreover, β' is a principal $\mathrm{GL}(\mathbf{v}) \times \mathrm{GL}(\mathbf{w})$-bundle over Λ'' and β'' is proper. Thus, we can define the multiplication map

$$(5.11) \quad \begin{aligned} C(\mathbf{v}) \otimes C(\mathbf{w}) &\to C(\mathbf{v}+\mathbf{w}) \\ f_1 \cdot f_2 &= \beta''_!(\beta'_\flat \beta^*(f_1 f_2)), \end{aligned}$$

where $\beta'_\flat$ is defined in the same way as in (4.33).

As in Section 4.4, this definition can also be rewritten as follows:

$$(5.12) \qquad f_1 \cdot f_2(z) = \int_W f_1(z|_{\mathbb{C}^\mathbf{d}/W}) f_2(z|_W), \qquad \mathbf{d} = \mathbf{v}+\mathbf{w},$$

where the integral is over the set $\beta''^{-1}(z)$ of all subrepresentations $W \subset \mathbb{C}^{\mathbf{v}+\mathbf{w}}$, $zW \subset W$ with $\dim W = \mathbf{w}$.

Theorem 5.12. *Multiplication (5.11) defines on $C = \bigoplus_\mathbf{v} C(\mathbf{v})$ a structure of an associative algebra with unit.*

The proof of this theorem is identical to the proof of Theorem 4.23 (see (4.35)) and will not be repeated here.

Example 5.13. Let $\vec{Q} = \bullet$. Then $\Lambda(\mathbf{v}) = pt$ for any $\mathbf{v}$. If we denote by f_n the characteristic function of $\Lambda(n)$, then it follows from the definition that

$$f_n f_m = \chi(G(n, n+m)) f_{n+m} = \binom{n+m}{n} f_{n+m},$$

where $G(n, d)$ is the Grassman variety of n-planes in $\mathbb{C}^d$. Thus, the algebra C is isomorphic to the algebra $\mathbb{C}[x]$ of polynomials in one variable, with the isomorphism given by

$$f_n \mapsto \frac{x^n}{n!}$$

(compare with Example 4.3, where the same result was obtained for the Hall algebra of Q).

5.4. Composition algebra of the double quiver

Example 5.14. Let Q be the quiver of type A_2 so that $Q^\sharp$ is shown below:

$$\bullet \rightleftarrows \bullet$$

If we denote

$$f_1 = \text{characteristic function of } \Lambda(1,0),$$
$$f_2 = \text{characteristic function of } \Lambda(0,1),$$

then

$$f_1 f_2(x,y) = \begin{cases} 1, & y = 0, \\ 0 & \text{otherwise.} \end{cases}$$

Indeed, a representation with graded dimension $(1,1)$ has a subrepresentation of dimension $(0,1)$ iff $y = 0$, in which case this subrepresentation is unique. Thus, $f_1 f_2 = f_{Z_1}$ is the characteristic function of $Z_1 = \{(x,y) \mid y = 0\} \subset \Lambda(1,1) = \{(x,y) \mid xy = 0\}$. Note that Z_1 is one of the irreducible components of $\Lambda(1,1)$ in this case.

In a similar way one can show that $f_2 f_1 = f_{Z_2}$ is the characteristic function of $Z_2 = \{(x,y) \mid x = 0\} \subset \Lambda(1,1)$.

An explicit computation similar to the one done in Example 4.28 (and also in the proof of Theorem 4.10) shows that we have the following relations in the algebra C:

$$f_1^2 f_2 - 2 f_1 f_2 f_1 + f_2 f_1^2 = 0,$$
$$f_2^2 f_1 - 2 f_2 f_1 f_2 + f_1 f_2^2 = 0.$$

The following lemma provides some link between the algebra C and the Hall algebra of constructible functions defined in Section 4.4. Let Ω be an orientation of a Dynkin graph Q and let $\vec{Q} = (Q, \Omega)$ be the corresponding quiver. Recall that then $R(\vec{Q}, \mathbf{v})$ is a Lagrangian subspace in $R(Q^\sharp, \mathbf{v})$; moreover, we have $R(\vec{Q}, \mathbf{v}) \subset \mu_\mathbf{v}^{-1}(0)$. In addition, since Ω has no oriented cycles (every Dynkin graph is a tree), $R(\vec{Q}, \mathbf{v}) \subset \Lambda(\mathbf{v})$. Thus, we have the restriction map

(5.13) $$C(\mathbf{v}) \to H^{con}(\vec{Q}, \mathbf{v}),$$

where H^{con} is the Hall algebra of constructible functions: $H^{con}(\vec{Q}, \mathbf{v}) = \text{Con}(R(\vec{Q}, \mathbf{v}))^{\text{GL}(\mathbf{v})}$.

Lemma 5.15. *Let $\vec{Q} = (Q, \Omega)$ be a Dynkin quiver. Then the restriction map (5.13) induces an algebra homomorphism $C \to H^{con}(\vec{Q})$.*

This immediately follows from the definition of multiplication in C and H^{con} respectively.

Definition 5.16. For any quiver Q, the composition algebra $\mathcal{F}$ is the subalgebra in the algebra C generated by elements

$$f_i = \text{the characteristic function of } R(Q^\sharp, \alpha_i),$$

where $\alpha_i = (0,0,\ldots,1,0,\ldots,0) \in \mathbb{Z}_+^I$ is the usual basis in $\mathbb{Z}^I$ so that $R(Q^\sharp, \alpha_i)$ is a point.

Theorem 5.17. *Let Q be a graph without edge loops. Let $\mathfrak{n}_+$ be the positive part of the corresponding Kac–Moody Lie algebra $\mathfrak{g}(Q)$ (see Appendix A), and let $\mathcal{F}$ be the composition algebra defined in Definition 5.16. Then there is a unique algebra homomorphism $\Psi \colon U\mathfrak{n}_+ \to \mathcal{F}$ defined by*

$$\Psi(e_i) = f_i.$$

Moreover, Ψ is an isomorphism.

Proof. We give here a proof in the case when Q is Dynkin. A general proof can be found in [**Lus2000**, Theorem 2.7].

To show that Ψ lifts to an algebra homomorphism we need to check that $\Psi(e_i)$ satisfy the defining relations of $U\mathfrak{n}_+$, i.e. the Serre relations (see Appendix A). This follows from computations in Example 5.14.

Surjectivity of Ψ is exactly the definition of the composition algebra.

To prove that Ψ is injective, choose an orientation Ω of Q. Then we can define the chain of algebra morphisms

$$U\mathfrak{n}_+ \xrightarrow{\Psi} \mathcal{F} \xrightarrow{res} H(\vec{Q}),$$

where the second map is the restriction morphism defined in Lemma 5.15.

By Ringel's theorem (Theorem 4.11), the composition $res \circ \Psi \colon U\mathfrak{n}_+ \to H(\vec{Q})$ is an isomorphism. Thus, Ψ is injective. $\square$

It turns out that this construction also gives a basis in the composition algebra $\mathcal{F}$.

Theorem 5.18 (Lusztig)**.** *Denote by $\mathrm{Irr}(\Lambda)$ the set of irreducible components of an algebraic variety Λ. Then in the assumptions of Theorem 5.17, there exists a unique basis $f_Z \in \mathcal{F}(\mathbf{v})$, $Z \in \mathrm{Irr}(\Lambda(\mathbf{v}))$, such that:*

- *$f_Z = 1$ on an open dense subset of Z.*
- *For every $Z' \in \mathrm{Irr}(\Lambda(\mathbf{v}))$, $Z' \neq Z$, $f_Z = 0$ on an open dense subset of Z'.*

The proof of this theorem can be found in [**Lus2000**, Theorem 2.7]. This basis is called the *semicanonical* basis; we note for experts that it does not coincide with Lusztig's canonical basis (see [**KS1997**]).

5.4. Composition algebra of the double quiver

Example 5.19. Let Q be of type A_2, $\mathbf{v} = (1,1)$ (see Example 5.14). Then $\Lambda(\mathbf{v})$ has two irreducible components, $Z_1 = \{(x,y) \mid x = 0\}$ and $Z_2 = \{(x,y) \mid y = 0\}$; it follows from the computations in Example 5.14 that in this case the basis decribed in Theorem 5.18 is f_{Z_1} = characteristic function of Z_1, f_{Z_2} = characteristic function of Z_2. Note that in this case the space of invariant constructible functions $C(\mathbf{v})$ is 3-dimensional, with the basis given by characteristic functions of $Z_1 \setminus (0,0)$, $Z_2 \setminus (0,0)$, and $\{(0,0)\}$, so the composition algebra $\mathcal{F}$ is a proper subalgebra in C.

Part 2

Quivers of Infinite Type

Chapter 6

Coxeter Functor and Preprojective Representations

In this chapter, we continue the study of reflection functors introduced in Section 3.3 and their use for the construction of indecomposable representations. We consider a more general situation, allowing Q to be non-Dynkin; we show that in this case, one can use reflection functors to construct a large class of indecomposable representations of $\vec{Q}$, called preprojective representations (in the Dynkin case, every representation is preprojective; for general quivers, it is not so). We also give an elementary discussion of the so-called Auslander–Reiten quiver, which provides a new perspective on the indecomposables constructed using reflection functors, even in the Dynkin case. Our exposition is heavily influenced by [**CB1992**].

The results of this chapter will be used later, when we classify representations of Euclidean quivers.

Throughout this chapter, Q is a graph with no edge loops (in some sections, we impose more conditions). All orientations of Q we consider will be such that there are no oriented cycles. We keep the notation of the previous chapters; in particular, we denote by L the root lattice of Q and by W the Weyl group (see Section 1.7). We also denote by R (respectively, R^{re}) the root system (respectively, set of real roots); see Appendix A for details. As before, all representations are considered over the ground field $\mathbf{k}$, which is assumed to be algebraically closed and of characteristic 0.

6.1. Coxeter functor

In this section, we study some further properties of the Coxeter element and the corresponding reflection functor. As in Section 3.4, we let $\vec{Q}$ be a quiver without oriented cycles and denote by $C = s_{i_r} \ldots s_{i_1}$ a Coxeter element adapted to $\vec{Q}$, $r = |I|$.

First, note that if we consider the corresponding sequence of reflections $C^+ = s_{i_r}^+ \ldots s_{i_1}^+$, then $C^+ \vec{Q} = \vec{Q}$ (orientation of every edge is reversed twice).

For any $k = 1, \ldots, r$ define roots

(6.1)
$$p_k = s_{i_1} \ldots s_{i_{k-1}}(\alpha_{i_k}),$$
$$q_k = s_{i_r} \ldots s_{i_{k+1}}(\alpha_{i_k}).$$

Lemma 6.1. *The elements p_k are distinct positive real roots, and*
$$\{p_1, \ldots, p_r\} = \{\alpha \in R_+^{re} \mid C\alpha \in R_-^{re}\}.$$

Similarly, the elements q_k are distinct positive real roots, and
$$\{q_1, \ldots, q_r\} = \{\alpha \in R_+^{re} \mid C^{-1}\alpha \in R_-^{re}\}.$$

This immediately follows from the geometric interpretation of the length function on the Weyl group (see Theorem A.23 in Appendix A) and the fact that $C = s_{i_r} \ldots s_{i_1}$ is a reduced expression for C (similarly, $C^{-1} = s_{i_1} \ldots s_{i_r}$ is a reduced expression for C^{-1}).

Note also that it is immediate from the definitions that

(6.2)
$$C(p_k) = -q_k.$$

We can define the corresponding indecomposable representations, as in (3.6):

(6.3)
$$P_k = \Phi_{i_1}^- \ldots \Phi_{i_{k-1}}^-(S(i_k))$$

(note that here $S(i_k)$ is the simple representation of the quiver $s_{i_{k-1}}^+ \ldots s_{i_1}^+ \vec{Q}$), and similarly for q's:

(6.4)
$$Q_k = \Phi_{i_r}^+ \ldots \Phi_{i_{k+1}}^+(S(i_k)).$$

Obviously, $\dim P_k = p_k$, $\dim Q_k = q_k$.

Lemma 6.2. *Representations P_k defined by (6.3) are exactly the indecomposable projective representations:*
$$P_k = P(i_k).$$

Similarly,
$$Q_k = Q(i_k)$$
are exactly the indecomposable injective representations.

6.1. Coxeter functor

Proof. It is immediate from the definition that if $\vec{Q}' = s_k^+(\vec{Q})$ and $P'(i)$ is an indecomposable projective representation of $\vec{Q}'$, $i \ne k$, then $\Phi_k^-(P'(i)) = P(i)$ is the indecomposable projective representation of $\vec{Q}$. Since $S(i_k) = P(i_k)$ is the projective representation of the quiver $s_{i_{k-1}}^+ \ldots s_{i_1}^+ \vec{Q}$ (since i_k is the sink for this quiver), this implies that $P_k = P(i_k)$ is a projective representation of $\vec{Q}$. A similar argument applies to injective modules. □

Corollary 6.3. *For any $\alpha, \beta \in L$, we have*
$$\langle \alpha, \beta \rangle = -\langle \beta, C\alpha \rangle = \langle C\alpha, C\beta \rangle,$$
where $\langle \alpha, \beta \rangle$ is the Euler form (1.15).

Proof. It suffices to prove the corollary in the case $\alpha = \dim P(i) = p(i)$. In this case $\langle p(i), \beta \rangle = \beta_i$ (see Example 1.24) and $C(p(i)) = -q(i)$, so $\langle \beta, C(p(i)) \rangle = -\langle \beta, q(i) \rangle = -\beta_i$ (see Example 1.24). □

Consider now the composition of the reflection functors
$$(6.5) \qquad \mathbf{C}^+ = \Phi_{i_r}^+ \ldots \Phi_{i_1}^+ \colon \operatorname{Rep}(\vec{Q}) \to \operatorname{Rep}(\vec{Q}).$$

Similarly, we can also define
$$(6.6) \qquad \mathbf{C}^- = \Phi_{i_1}^- \ldots \Phi_{i_r}^- \colon \operatorname{Rep}(\vec{Q}) \to \operatorname{Rep}(\vec{Q}).$$

Note that the corresponding elements of W are C and C^{-1} respectively.

Functors $\mathbf{C}^\pm$ will be called the *Coxeter functors*. They play an important role in representation theory of quivers. In particular, they are crucial ingredients in the theory of Auslander–Reiten sequences: $\mathbf{C}^+$ coincides with the translation functor $\tau = DTr$. We will not discuss this theory here, referring the reader to [**ARS1997**] or the review [**Gab1980**]. We just list some of the more important properties of these functors.

Theorem 6.4. *Let I be an indecomposable representation of $\vec{Q}$. Then $\mathbf{C}^+(I)$, $\mathbf{C}^-(I)$ are also indecomposable (possibly zero). If $I, \mathbf{C}^+(I)$ are nonzero, then $\mathbf{C}^-\mathbf{C}^+(I) \simeq I$; similarly, if $I, \mathbf{C}^-(I)$ are nonzero, then $\mathbf{C}^+\mathbf{C}^-(I) \simeq I$.*

Proof. This immediately follows from Theorem 3.14. □

We can now look at indecomposables which satisfy $\mathbf{C}^\pm(I) = 0$.

Theorem 6.5. *Let V be a nonzero indecomposable representation of $\vec{Q}$. Then the following conditions are equivalent:*

(1) $\mathbf{C}^+(V) = 0$.
(2) $V \simeq P_k$ for some $k = 1, \ldots, r$.
(3) V is projective.

Similarly, the following are also equivalent:

(1′) $\mathbf{C}^-(V) = 0$.

(2′) $V \simeq Q_k$ for some $k = 1, \ldots, r$.

(3′) V is injective.

Proof. Equivalence of (1) and (2) follows from Corollary 3.15 and Lemma 6.1. Projectivity of P_k was proved in Lemma 6.2; since $\{i_1, \ldots, i_r\} = I$ and $P(i), i \in I$, is the full set of indecomposable projectives (see Theorem 1.18), we see that any indecomposable projective is of the form P_k.

The second part is proved similarly, using Theorem 1.23. □

Corollary 6.6. *For a representation V of $\vec{Q}$:*

(1) $\mathbf{C}^+(V) = 0$ iff V is projective.

(2) $\mathbf{C}^-(V) = 0$ iff V is injective.

Finally, we have the following lemma, which will be useful later.

Lemma 6.7. *For any representations X, Y of $\vec{Q}$,*
$$\mathrm{Ext}^1(X, Y) \simeq \mathrm{Hom}(Y, \mathbf{C}^+(X))^*.$$

This lemma can be thought of as an analog of Serre duality in algebraic geometry. We will not prove this here; interested readers can find the proof in [**ARS1997**, Corollary 4.7].

6.2. Preprojective and preinjective representations

As before, we choose an orientation of Q without oriented cycles; we choose a Coxeter element C adapted to this orientation, and we let $\mathbf{C}^\pm$ be the corresponding Coxeter functors (6.5).

Our goal in this section is to prove an analog of Theorem 3.18, establishing a bijection between positive roots and indecomposable representations. It is clear that we cannot expect this to hold in general (for example, the Kronecker quiver has infinitely many representations of given dimension). However, it turns out that this theorem still holds for special classes of representations.

Definition 6.8. An indecomposable representation V of $\vec{Q}$ is called *preprojective* if $(\mathbf{C}^+)^n V = 0$ for $n \gg 0$, and *preinjective* if $(\mathbf{C}^-)^n V = 0$ for $n \gg 0$.

An indecomposable representation V is called *regular* if for any $n \in \mathbb{Z}_+$, we have $(\mathbf{C}^+)^n V \neq 0$, $(\mathbf{C}^-)^n V \neq 0$.

A representation V is called preprojective, preinjective, or regular if each of its indecomposable summands is of the corresponding type.

6.2. Preprojective and preinjective representations

We will call a positive real root $\alpha \in R_+^{re}$ *preprojective* if it satisfies the condition

(6.7) $$C^n \alpha \not> 0 \text{ for some } n > 0.$$

Similarly, a positive real root $\alpha \in R_+^{re}$ is called *preinjective* if it satisfies the condition

(6.8) $$C^n \alpha \not> 0 \text{ for some } n < 0.$$

For example, roots p_k defined by (6.1) are preprojective, and roots q_k are preinjective.

In the Dynkin case it follows from Corollary 3.30 that any positive root is both preprojective and preinjective.

Theorem 6.9. *Let $\vec{Q}$ be a quiver without oriented cycles, $C = s_{i_r} \ldots s_{i_1}$ a Coxeter element adapted to $\vec{Q}$, and $\mathbf{C}^\pm$ the corresponding Coxeter functors.*

(1) *If I is a nonzero preprojective indecomposable representation, then $I \simeq (\mathbf{C}^-)^n(P(i))$ for some $n \geq 0, i \in I$. In this case, $\dim I$ is a preprojective real positive root. Conversely, for every preprojective real positive root α, there is a unique indecomposable representation I_α of graded dimension α; this representation is preprojective.*

(2) *If I is a nonzero preinjective indecomposable representation, then $I \simeq (\mathbf{C}^+)^n(Q(i))$ for some $n \geq 0, i \in I$. In this case, $\dim I$ is a preinjective real positive root. Conversely, for every preinjective real positive root α, there is a unique indecomposable representation I_α of graded dimension α; this representation is preinjective.*

Proof. The proof is similar to the Dynkin case (Theorem 3.18).

If I is a nonzero indecomposable preprojective representation, then by Theorem 6.5, we see that $(\mathbf{C}^+)^n I \simeq P(i)$ for some $i \in I$, $n \geq 0$. Thus, $I \simeq (\mathbf{C}^-)^n(P(i))$. Since $i = i_k$ for some $k = 1, \ldots, r$, $\dim P(i) = p_k$ is a preprojective real positive root (see Lemma 6.1); thus, $\dim I = C^{-n} p_k$ is also preprojective. It also shows that I is the unique indecomposable of the given dimension.

Conversely, let $\alpha \in R_+^{re}$ be a positive preprojective real root. Take n to be maximal such that $C^n \alpha \in R_+^{re}$; then, by Lemma 6.1, $C^n(\alpha) = p_k$ for some k, so $\alpha = C^{-n} p_k$. Taking $I = (\mathbf{C}^-)^n(P(i_k))$, we get an indecomposable preprojective representation; since $p_k, C^{-1} p_k, \ldots, C^{-n}(p_k)$ are all positive, we see that all representations $P(i_k), \mathbf{C}^-(P(i_k)), \ldots, (\mathbf{C}^-)^n(P(i_k))$ are nonzero.

The second part is proved similarly. $\square$

Corollary 6.10. *If $\vec{Q}$ is Dynkin, then any indecomposable representation is both preprojective and preinjective, and there are no nonzero regular representations.*

6.3. Auslander–Reiten quiver: Combinatorics

In this section, we introduce a tool which makes the combinatorics of orientation reversal operations s_i: (sink $\leftrightarrow$ source) much more transparent. The key object of our discussion will be the so-called Auslander–Reiten (AR) quiver which plays an important role in representation theory of artin algebras and, in particular, in representation theory of quivers (see [**ARS1997**]). This quiver is usually defined in terms of almost split sequences which goes beyond the limits of this book. However, it is possible to give a purely elementary combinatorial construction of a part of the AR quiver, corresponding to preprojective representations; for Dynkin quivers, this actually gives the whole AR quiver, since in this case any representation is preprojective.

Our exposition follows paper [**KT**].

From now on, we assume that Q is a connected bipartite graph: $I = I_0 \sqcup I_1$, and for every edge of Q, one of the endpoints is in I_0 and the other is in I_1. Such a partition is possible if and only if Q has no cycles of odd length; in particular, any tree graph is bipartite. We will also assume that Q is not the single vertex graph $\bullet$.

For a vertex $i \in I$, we define its parity $p(i) \in \mathbb{Z}_2$ by

(6.9) $$p(i) = \begin{cases} 0, & i \in I_0, \\ 1, & i \in I_1. \end{cases}$$

Define the quiver

$$Q \times \mathbb{Z}$$

with the set of vertices $I \times \mathbb{Z}$ and (oriented) edges $(i, n) \to (j, n+1)$, $(j, n) \to (i, n+1)$ for every edge $i - j$ in Q and every $n \in \mathbb{Z}$. It is easy to check that $Q \times \mathbb{Z}$ is not connected: (i, n) is connected with (j, m) iff $m - n \equiv d(i, j)$ mod 2, where $d(i, j)$ is the length of a path connecting i and j in Q. Let

(6.10) $$\mathbb{Z}Q = \{(i, n) \mid p(i) + n \equiv 0 \mod 2\} \subset Q \times \mathbb{Z}.$$

The quiver $\mathbb{Z}Q$ is called the translation quiver; one easily sees that $\mathbb{Z}Q$ is connected unless Q is a single point. It has a canonical automorphism τ defined by

(6.11) $$\tau \colon (i, n) \mapsto (i, n - 2).$$

An example of the quiver $\mathbb{Z}Q$ for the graph of type A_5 is shown in Figure 6.1.

6.3. Auslander–Reiten quiver: Combinatorics

Remark 6.11. In [**KT**], $\mathbb{Z}Q$ was shown in the figures with the $\mathbb{Z}$-direction going vertically. Here we adopt the more traditional convention, with the $\mathbb{Z}$-direction being horizontal.

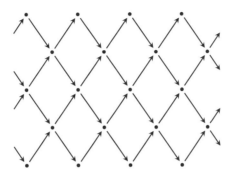

Figure 6.1. Quiver $\mathbb{Z}Q$ for the graph of type A_5.

The quiver $\mathbb{Z}Q$ has a natural projection to Q; we can consider sections of this projection, i.e. embeddings of Q into $\mathbb{Z}Q$. The following definition makes this precise.

Definition 6.12. A subset $T \subset \mathbb{Z}Q$ is called a *slice* if, for any $i \in I$, there is a unique vertex $q = (i, h_i) \in T$ and if, whenever vertices i, j are connected by an edge in Q, we have $h_i = h_j \pm 1$.

Note that every choice of slice T automatically gives a choice of orientation Ω_T on Q and thus a quiver $\vec{Q}_T$ with the underlying graph Q. Namely, for every edge e connecting vertices i, j in Q, we orient it by

(6.12)
$$e\colon i \to j \text{ if } h_i = h_j + 1,$$
$$e\colon j \to i \text{ if } h_i = h_j - 1.$$

Note that this orientation is exactly opposite to the orientation one would get considering T as a subquiver in $\mathbb{Z}Q$. The reason for this orientation reversal will be made clear later.

Figure 6.2 shows an example of a slice and the corresponding orientation of the graph Q.

Lemma 6.13. *If Q is a tree, any orientation of Q can be obtained from a slice in $\mathbb{Z}Q$. Two slices T, T' give the same orientation iff $T' = \tau^k(T)$ for some $k \in \mathbb{Z}$.*

The proof is obvious and left to the reader.

Thus, choosing a slice is closely related to choosing an orientation. It turns out that the operations $s_i^\pm$ which relate different orientations of the same graph have their counterpart for slices.

6. Coxeter Functor and Preprojective Representations

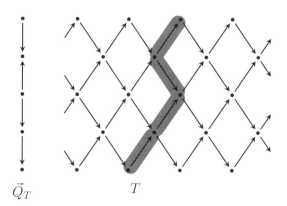

Figure 6.2. A slice and corresponding orientation. The slice is highlighted.

Definition 6.14. Let $T = \{(i, h_i)\}$ be a slice and let $k \in I$ be a sink for the corresponding orientation Ω_T (which means that the function $i \mapsto h_i$ has a local minimum at k). We define a new slice $T' = s_k^+ T$ by $T' = \{(i, h_i')\}$, where

$$h_i' = \begin{cases} h_i + 2 & \text{if } i = k, \\ h_i & \text{if } i \neq k. \end{cases}$$

Similarly, if $k \in I$ is a source for the orientation Ω_T (which means that the function $i \mapsto h_i$ has a local maximum at k), we define a new slice $T' = s_k^- T$ by $T' = \{(i, h_i')\}$, where

$$h_i' = \begin{cases} h_i - 2 & \text{if } i = k, \\ h_i & \text{if } i \neq k. \end{cases}$$

Figure 6.3 illustrates these operations.

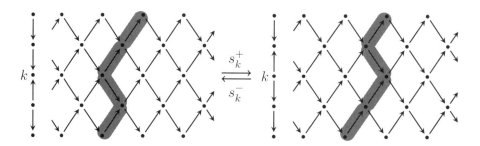

Figure 6.3. Operations $s_i^\pm$ for slices.

It is easy to see that these operations agree with the operations $s_i^\pm$ for orientations: $\Omega_{s_i^\pm T} = s_i^\pm \Omega_T$.

We can now give a proof of the fact that any two orientations can be obtained from each other by a sequence of operations $s_i^{\pm}$. In fact, we can make a more precise statement.

Definition 6.15. Let $T = \{(i, h_i)\} \subset \mathbb{Z}Q$ be a slice. We say that a vertex $q = (j, n) \in \mathbb{Z}Q$ is above (respectively, strictly above) T and write $q \succcurlyeq T$ (respectively, $q \succ T$) if we have $n \geq h_j$ (respectively, $n > h_j$). We say that a slice T' is above a slice T if every vertex $q \in T'$ is above T.

In a similar way, one defines what it means for a vertex q to be below or strictly below T.

Exercise 6.16.

(1) Show that $p \succcurlyeq T$ if and only if there exists a path in $\mathbb{Z}Q$ from some vertex $q \in T$ to p.

(2) Show that if $p \succ T$, $q \in T$, then there are no paths from p to q in $\mathbb{Z}Q$.

Theorem 6.17. *If T, T' are two slices such that $T' \succcurlyeq T$, then one can obtain T' by applying a sequence of operations s_i^+ to T. Moreover, this sequence is defined uniquely up to interchanging $s_i^+ s_j^+ \leftrightarrow s_j^+ s_i^+$ for i, j not connected in Q.*

Proof. Let us define the distance $d(T, T')$ between T and T' as the number of vertices q between T, T', i.e. vertices satisfying $T \preccurlyeq q \prec T'$. We will prove the theorem by induction in $d(T, T')$.

If $d(T, T') = 0$, then $T = T'$, so the theorem is proved. Otherwise, consider all vertices of T which are strictly below T'. It is easy to see that if (i, h_i) is such a vertex, and we have an edge $i \to j$ in Ω_T, then (j, h_j) is also strictly below T'. Thus, among all vertices of T which are strictly below T' there is a vertex k which is a sink for Ω_T. Let us consider $\tilde{T} = s_k^+ T$. It is immediate that it is a slice satisfying $\tilde{T} \preccurlyeq T'$, and $d(\tilde{T}, T') = d(T, T') - 1$. Thus, by the induction assumption, T' can be obtained from $\tilde{T}$ (and thus from T) by a sequence of operations s_i^+.

To prove the uniqueness of such a sequence of reflections, we will again argue by induction in $d(T, T')$. Indeed, assume that $T' = s_{i_l}^+ \ldots s_{i_1}^+(T) = s_{j_n}^+ \ldots s_{j_1}^+(T)$. Then i_1 is a sink for T; since the only reflection that can move vertex (i, h_i) up is s_i^+, we see that the second sequence of reflections must also contain s_{i_1}: we must have $j_a = i_1$ for some a. Let us take the smallest such value of a. Thus, the second sequence has the form $s_{j_n}^+ \ldots s_{i_1}^+ s_{j_{a-1}}^+ \ldots s_{j_1}^+$. But then i_1 is a sink for $s_{j_1}^+ T$, $s_{j_2}^+ s_{j_1}^+ T$, $\ldots$, $s_{j_{a-1}}^+ \ldots s_{j_1}^+ T$. Therefore, since a vertex connected to a sink cannot itself be a sink, we see that none of $j_1, \ldots, j_{a-1}$ are connected to i_1. Therefore, by permuting $s_i^+ s_j^+ \leftrightarrow s_j^+ s_i^+$ for

i, j not connected in Q, we can rewrite the second sequence of reflections as $s_{j_n}^+ \ldots s_{j_{a-1}}^+ \ldots s_{j_1}^+ s_{i_1}^+$. Now the statement easily follows by induction. □

Corollary 6.18. *For any two orientations Ω, Ω' of a tree Q, one can obtain Ω' by applying a sequence of operations s_i^+ to Ω. One can also get Ω' by applying a sequence of operations s_i^- to Ω.*

This refines the statement of Lemma 3.6.

6.4. Auslander–Reiten quiver: Representation theory

In the previous section, we introduced the quiver $\mathbb{Z}Q$ and introduced the notion of the slice as a convenient tool for visualizing sink-to-source transformations. In this section, we show that this quiver has a natural interpretation in terms of representation theory of quivers: there is a natural correspondence between (some) vertices of $\mathbb{Z}Q$ and indecomposable preprojective representations of $\vec{Q}$.

Theorem 6.19. *Let Q be a bipartite graph. Then one can assign to every pair (T, q), where T is a slice of $\mathbb{Z}Q$ and q is a vertex of $\mathbb{Z}Q$, a representation I_q^T of $\vec{Q}_T$ so that the following conditions hold:*

(1) *If q is strictly below T, then $I_q^T = 0$.*
(2) *If $q = (i, h_i) \in T$ is a sink for T, then $I_q^T = S(i)$.*
(3) *If $T' = s_i^+ T$ and $q \succcurlyeq T'$, then $I_q^T = \Phi_i^- I_q^{T'}$, $I_q^{T'} = \Phi_i^+ I_q^T$.*

Moreover, these conditions determine I_q^T uniquely up to isomorphism.

Condition (3) is illustrated in Figure 6.4.

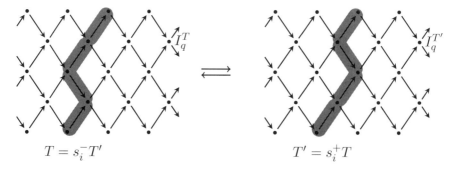

Figure 6.4. Defining relation for I_q^T: $I_q^T = \Phi_i^- I_q^{T'}$, $I_q^{T'} = \Phi_i^+ I_q^T$.

Proof. First, note that if the collection of representations I_q^T described in the theorem exists, it is unique. Indeed, given $q \succcurlyeq T$, choose a slice $T' \succcurlyeq T$

containing q such that q is a sink for T'. By Theorem 6.17, one can write $T' = s_{i_l}^+ \ldots s_{i_1}^+(T)$ for some sequence of reflections $s_{i_l}^+ \ldots s_{i_1}^+$. Then we must have

(6.13) $$I_q^T = \Phi_{i_1}^- \ldots \Phi_{i_l}^- S(i),$$

where $S(i)$ is the simple representation of $\vec{Q}_{T'}$. This proves uniqueness of I_q^T.

To prove existence, choose, for every pair (q, T), $q \succcurlyeq T$, a slice $T' \succcurlyeq T$ containing q such that q is a sink for T' and a sequence of reflections $s_{i_l}^+ \ldots s_{i_1}^+$ such that $T' = s_{i_l}^+ \ldots s_{i_1}^+(T)$. Define I_q^T by (6.13). We claim that the so-defined representation does not depend on the choice of the slice T' or the sequence of reflections $s_{i_l}^+ \ldots s_{i_1}^+$. Indeed, by Theorem 6.17, for a fixed choice of T', the sequence of reflections is defined uniquely up to permuting s_i^+ and s_j^+, where vertices i, j are not connected in Q. But, in this case, functors Φ_i^- and Φ_j^- commute: one has a functorial isomorphism $\Phi_i^- \Phi_j^- \simeq \Phi_j^- \Phi_i^-$. Thus, I_q^T does not depend on the choice of the sequence of reflections.

Now, let us show that nothing depends on the choice of T'. Namely, if $T' = \{(j, h_j')\}, T'' = \{(j, h_j'')\}$ are two slices both having $q = (i, n)$ as a sink, then these two slices coincide at i and at all neighbors of i: $h_j' = h_j''$ for any j connected by an edge to i in Q. In this case, an easy modification of arguments in the proof of Theorem 6.17 shows that one can get T'' from T' by a sequence of operations $T' \mapsto s_k^+ T'$, k not connected to i, and their inverses. But T' and $s_k^+ T'$ give canonically isomorphic I_q^T: if $T' = s_{i_l}^+ \ldots s_{i_1}^+(T)$, $T'' = s_k^+ s_{i_l}^+ \ldots s_{i_1}^+ T$ and we get

$$I_q^T = \Phi_{i_1}^- \ldots \Phi_{i_l}^- S(i) = \Phi_{i_1}^- \ldots \Phi_{i_l}^- \Phi_k^- S(i)$$

since $\Phi_k^-(S(i)) = S(i)$.

Thus, we see that representation I_q^T defined by (6.13) only depends on q, T and does not depend on the choice of T' or the sequence of reflections. The fact that so-defined representations satisfy conditions of the theorem is now obvious. $\square$

The following theorem lists some of the properties of the constructed representations.

Theorem 6.20.

(1) If $q = (i, n) \in T$, then $I_q^T = P(i)$ is the indecomposable projective representations.

(2) The construction is invariant under the translation τ defined by (6.11): $I_{\tau q}^{\tau T} = I_q^T$.

(3) If $q \succ T$, then $I^T_{\tau^{-1}q} = I^{\tau T}_q = \mathbf{C}^-(I^T_q)$, where $\mathbf{C}^-$ is the Coxeter functor (6.5).

(4) Each representation I^T_q is indecomposable and preprojective (possibly zero).

Proof. To prove part (1), note that it is immediate from the definition that if $\vec{Q}' = s^+_k(Q)$ and $P'(i)$ is an indecomposable projective representation of $\vec{Q}'$, $i \neq k$, then $\Phi^-_k(P'(i)) = P(i)$ is the indecomposable projective representation of $\vec{Q}$. Thus, if $q = (i,n) \in T$, then $I^T_q = \Phi^-_{i_1} \ldots \Phi^-_{i_l} S(i) = \Phi^-_{i_1} \ldots \Phi^-_{i_l} P(i) = P(i)$, since all indices $i_1, \ldots, i_l$ are distinct from i.

Part (2) follows from the uniqueness of I^T_q, and part (3) follows from (2) and property (3) in the definition of I^T_q.

Part (4) is immediate from (1), (3): every $q \succeq T$ can be written as $q = \tau^{-n}q'$, $q' \in T$, $n \geq 0$. Thus, $I^T_q = (\mathbf{C}^-)^n I^T_{q'} = (\mathbf{C}^-)^n P(i)$. □

Note that representations I_q constructed in Theorem 6.19 can be zero. Let us study those of them that are nonzero.

Definition 6.21. Let Q be a bipartite graph and let $T \subset \mathbb{Z}Q$ be a slice. We define the *preprojective Auslander–Reiten quiver* of $\vec{Q}_T$ as the subquiver of $\mathbb{Z}Q$ with the set of vertices

(6.14) $$\Delta_T = \{q \in \mathbb{Z}Q \mid I^T_q \neq 0\} \subset \mathbb{Z}Q.$$

If Q is Dynkin, we will drop the word "preprojective" and call Δ the Auslander–Reiten quiver of $\vec{Q}_T$.

Theorem 6.22. *In the assumptions of Definition 6.21, the map*

$$q \mapsto I^T_q$$

is a bijection between Δ_T and the set of isomorphism classes of nonzero preprojective indecomposable representations of $\vec{Q}_T$.

Proof. We have already proved that each I^T_q is an indecomposable preprojective representation.

Conversely, let I be an indecomposable preprojective. As was proved in Theorem 6.9, we then have $I = (\mathbf{C}^-)^n(P(i)) = (\mathbf{C}^-)^n(I_q) = I_{\tau^{-n}q}$, $q = (i, h_i) \in T$. Therefore, each indecomposable preprojective can be written as I^T_q.

It remains to show that different representations of the form $(\mathbf{C}^-)^n P(k)$ are nonisomorphic. Indeed, if $I = (\mathbf{C}^-)^{n_1} P(i_1) = (\mathbf{C}^-)^{n_2} P(i_2)$, then $n_1 = \max\{n \mid (\mathbf{C}^+)^n I \neq 0\} = n_2$, so $P(i_1) \simeq P(i_2)$, which is only possible if $i_1 = i_2$. □

6.4. Auslander–Reiten quiver: Representation theory

Remark 6.23. In the next section, we will show that edges of Δ_T can also be interpreted in terms of representation theory: each edge gives a morphism between the corresponding indecomposable representations. A precise statement is given in Corollary 6.34.

Let us now consider the special case of Dynkin quivers.

Theorem 6.24. *Let Q be a Dynkin graph, and let T, Δ_T be as in Definition 6.21. Let $w_0 = s_{i_l} \ldots s_{i_1}$ be a reduced expression for the longest element of the Weyl group adapted to orientation Ω_T, and let $T' = s_{i_l}^+ \ldots s_{i_1}^+ T$. Then*

$$\Delta_T = \{q \in \mathbb{Z}Q \mid T \preccurlyeq q \prec T'\}.$$

Proof. Denote temporarily $\Delta_T' = \{q \in \mathbb{Z}Q \mid T \preccurlyeq q \prec T'\}$. Then $\{I_q^T, q \in \Delta_T'\}$ are exactly the indecomposable representations $I_1, \ldots, I_l$ defined in (3.10); thus, $q \mapsto I_q$ gives a bijection $\Delta_T' \to \mathrm{Ind}(Q_T)$. Comparing it with the previous theorem, we see that $\Delta_T = \Delta_T'$. □

Corollary 6.25. *The slice $T' = s_{i_l}^+ \ldots s_{i_1}^+ T$ used in the proof does not depend on the choice of the reduced expression $w_0 = s_{i_l} \ldots s_{i_1}$.*

In fact, it can be shown (see [**KT**]) that if $T = \{(i, h_i)\}$, then $T' = \{(i, h_i')\}$, where $h_i' = h_{i^\vee} + h$, h is the Coxeter number, and $i^\vee$ is defined by $\alpha_{i^\vee} = -w_0(\alpha_i)$.

Since in the Dynkin case nonzero indecomposable representations are in bijection with positive roots, we see that in the Dynkin case, we have a bijection

$$\Delta_T \simeq R_+.$$

Figure 6.5 shows an example of the Auslander–Reiten quiver Δ_T for the Dynkin graph of type A_4.

Theorem 6.26. *Let Q be a Dynkin graph, $T \subset \mathbb{Z}Q$ a slice, and Δ_T the Auslander–Reiten quiver. Define a partial order $\preccurlyeq$ on Δ_T (and thus on $\mathrm{Ind}(\vec{Q}_T)$) by*

$$q \preccurlyeq q' \iff \text{there is a path in } \Delta_T \text{ from } q \text{ to } q'.$$

(1) *For any choice of reduced expression $w_0 = s_{i_l} \ldots s_{i_1}$ adapted to $\vec{Q}_T$, the order on R_+ defined by (3.9) is compatible with the partial order $\preccurlyeq$: if $I_k \preccurlyeq I_n$, then $k \leq n$.*

(2) *Any linear order on R_+ compatible with $\preccurlyeq$ can be obtained from some reduced expression for w_0 adapted to $\vec{Q}_T$.*

(3)
$$\mathrm{Hom}_{\vec{Q}_T}(I_q, I_{q'}) \neq 0 \implies q \preccurlyeq q',$$
$$\mathrm{Ext}^1_{\vec{Q}_T}(I_q, I_{q'}) \neq 0 \implies q \succ q'.$$

96   6. Coxeter Functor and Preprojective Representations

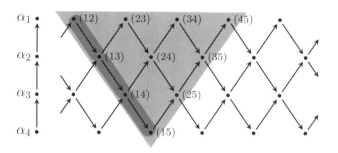

Figure 6.5. Auslander–Reiten quiver Δ_Q for quiver of type A_4. The slice T is highlighted in the darker shaded area; the lighter shaded area is the AR quiver $\Delta_T \subset \mathbb{Z}Q$. Next to each vertex $q \in \Delta_T$, we write the root $\alpha_q = \dim I_q$, using the notation (ij) for $e_i - e_j = \alpha_i + \cdots + \alpha_{j-1}$.

Proof. Consider the sequence of slice
$$T_0 = T,$$
$$T_1 = s_{i_1}^+ T,$$
$$\vdots$$
$$T_l = s_{i_l}^+ \ldots s_{i_1}^+ T.$$

Thus, each T_i differs from T_{i-1} in one place: there is a unique vertex $q_i \in \mathbb{Z}Q$ such that $q_i \succ T_{i-1}$, $q_i \in T_i$. It is immediate from definition that the set $\{q_1, \ldots, q_l\} = \Delta_T$ and that $I_{q_k}^T = I_k$ are exactly the indecomposable representations defined by (3.10). Thus, it suffices to prove that if $q_k \preccurlyeq q_n$, then $k \le n$, which follows from Exercise 6.16.

To prove part (2), note that having a linear order on Δ_T defines a sequence of slices $T_0 \preccurlyeq T_1 \preccurlyeq \cdots \preccurlyeq T_l = T'$, where each T_k is obtained from T_{k-1} by applying a reflection $s_{i_k}^+$; thus, it defines a reduced expression for w_0.

To prove the last part, note that $q \preccurlyeq q'$ iff for every linear order $\le$ compatible with $\preccurlyeq$, we have $q \le q'$. Now the result follows from part (2) and Theorem 3.35: if $\mathrm{Hom}(I_a, I_b) \ne 0$, then $a \le b$. □

6.5. Preprojective algebra and Auslander–Reiten quiver

Throughout this section, we keep the notation of the previous section. In particular, Q is a bipartite graph and $\mathbb{Z}Q$ is the corresponding translation quiver.

6.5. Preprojective algebra and Auslander–Reiten quiver

In this section, we will show that there is a close relation between the path algebra $\mathbf{k}(\mathbb{Z}Q)$ and the preprojective algebra Π defined in Section 5.2.

We begin by choosing a skew-symmetric function on edges of $\mathbb{Z}Q$. Namely, note that by construction of $\mathbb{Z}Q$, for every edge $h\colon (i,n) \to (j, n+1)$ in $\mathbb{Z}Q$, we have the corresponding edge $(j, n+1) \to (i, n+2)$. We will denote this edge by $h(1)$.

Let us now choose a function $\varepsilon\colon$ (Edges of $\mathbb{Z}Q) \to \mathbf{k}^\times$ so that for any edge h, we have
$$\varepsilon(h) + \varepsilon(h(1)) = 0.$$

It is easy to see that this is equivalent to choosing a skew-symmetric function ε on the set of edges of the double quiver $Q^\sharp$ as in (5.3).

Similar to the definition of the preprojective algebra, define, for every $q \in \mathbb{Z}Q$, an element θ_q in the path algebra $\mathbf{k}(\mathbb{Z}Q)$ by

(6.15) $$\theta_q = \sum_h \varepsilon(h) h(1) h \colon \tau q \to q,$$

where, for $q = (i, n)$, the sum is over all edges $h \colon \tau q = (i, n-2) \to (j, n-1)$:

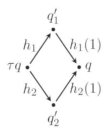

Figure 6.6. Definition of θ_q.

Definition 6.27. We define the algebra $\tilde{\Pi}$ as the quotient
$$\tilde{\Pi} = A/J, \qquad A = \mathbf{k}(\mathbb{Z}Q),$$
where J is the ideal generated by elements θ_q, $q \in \mathbb{Z}Q$.

It is obvious from the definition that $\tilde{\Pi}$ can be decomposed as

(6.16) $$\tilde{\Pi} = \bigoplus_{p,q \in \mathbb{Z}Q} {}_p\tilde{\Pi}_q,$$

where ${}_p\tilde{\Pi}_q$ is spanned by paths from q to p.

Theorem 6.28.

(1) For $q = (i, m)$, $p = (j, n)$, we have a canonical isomorphism
$$_p\tilde{\Pi}_q \simeq {_j\Pi_i^{n-m}},$$
where $_j\Pi_i^l$ is the component of the preprojective algebra Π of Q spanned by paths $i \to j$ of length l (see (5.8)).

(2) If $T \subset \mathbb{Z}Q$ is a slice, then for any $p, q \in T$, we have
$$_p\tilde{\Pi}_q = (\text{span of paths } p \to q \text{ in } T).$$

Proof. The first part immediately follows from the observation that we have a natural bijection

$$(\text{paths } q \to p \text{ in } \mathbb{Z}Q) \leftrightarrow (\text{paths } i \to j \text{ of length } l = n - m \text{ in } Q^\sharp).$$

The second part is proved in the same way as Lemma 5.4. $\square$

Thus, we see that $\tilde{\Pi}$ is almost the same as the preprojective algebra Π. The following result is the key to the study of the algebra $\tilde{\Pi}$.

Theorem 6.29. *Let $p, q \in \mathbb{Z}Q$. Consider the sequence*

(6.17) $$0 \to {_{\tau p}\tilde{\Pi}_q} \xrightarrow{d_1} \bigoplus_h {_{p'}\tilde{\Pi}_q} \xrightarrow{d_2} {_p\tilde{\Pi}_q} \to 0,$$

where the sum is over all edges $h\colon \tau(p) \to p'$ in $\mathbb{Z}Q$ with source $\tau(p)$, and the maps d_1, d_2 are given by

(6.18) $$d_1(x) = \sum_h \varepsilon(h) h x,$$

(6.19) $$d_2(x) = \sum_h h(1) x.$$

Then this sequence is a complex; if $p \neq q$, this complex is exact.

Similarly, consider the sequence

(6.20) $$0 \to {_p\tilde{\Pi}_q} \xrightarrow{d_1} \bigoplus_h {_p\tilde{\Pi}_{q'}} \xrightarrow{d_2} {_p\tilde{\Pi}_{\tau q}} \to 0,$$

where the sum is over all edges $\tau q \to q'$, and

(6.21) $$d_1(x) = \sum_h x h(1),$$

(6.22) $$d_2(x) = \sum_h \varepsilon(h) x h.$$

Then this sequence is a complex; if $p \neq \tau q$, this complex is exact.

6.5. Preprojective algebra and Auslander–Reiten quiver

A proof of this theorem will be given at the end of this section.

This theorem immediately gives a number of useful corollaries. First, let $T \subset \mathbb{Z}Q$ be a slice. Define

$$\tag{6.23} {}_p\tilde{\Pi}_T = \bigoplus_{q \in T} {}_p\tilde{\Pi}_q.$$

This space has a structure of a left $\vec{Q}_T$-module defined by $x_h \pi = \pi \circ \bar{h}$ (where, for an edge h of $\vec{Q}_T$, we denote by $\bar{h}$ the corresponding edge in T).

The following result is a reformulation of the theorem of Gelfand and Ponomarev [**GP1979**].

Theorem 6.30. *Let Q be a bipartite graph, and let $T \subset \mathbb{Z}Q$ be a slice. Then for every $p \in \mathbb{Z}Q$ we have an isomorphism of $\vec{Q}_T$-modules*

$$I_p^T \simeq {}_p\tilde{\Pi}_T,$$

where I_p^T is the (possibly zero) indecomposable module constructed in Theorem 6.19.

Proof. This immediately follows from the definition of I_p and the following statements:

(1) ${}_p\tilde{\Pi}_T = 0$ if p is strictly below T. This immediately follows from the definition: in this case, there are no paths from a vertex $q \in T$ to p.

(2) For $p = (i, h_i) \in T$, ${}_p\tilde{\Pi}_T = P(i)$. This follows from Theorem 6.28.

(3) If i is a sink for T and $T' = s_i^+(T)$, then for any $p \succcurlyeq T'$ we have

$${}_p\tilde{\Pi}_T = \Phi_i^-({}_p\tilde{\Pi}_{T'}).$$

Indeed, in this case vertices of T and T' coincide except for one: vertex (i, h_i) of T is replaced by vertex $(i, h_i + 2)$ of T'. Denoting $q = (i, h_i + 2)$ so that $(i, h_i) = \tau q$ and using (6.20) we see that

$${}_p\tilde{\Pi}_{\tau q} = \mathrm{Coker}\left({}_p\tilde{\Pi}_q \to \bigoplus {}_p\tilde{\Pi}_{q'}\right)$$

which is the definition of Φ^-. $\square$

Recall the Auslander–Reiten quiver $\Delta_T = \{q \in \mathbb{Z}Q \mid I_q \neq 0\}$ defined in Definition 6.21.

Corollary 6.31. *In the assumptions of Theorem 6.30, fix a slice T. Assume that $q = (i, n) \in \Delta_T$ is such that $\tau q = (i, n - 2) \in \Delta_T$. Then we have the following short exact sequence of $\vec{Q}_T$-modules:*

$$\tag{6.24} 0 \to I_{\tau q} \to \bigoplus_{q'} I_{q'} \to I_q \to 0,$$

where the sum is over all edges $h \colon \tau q \to q'$ in $\mathbb{Z}Q$.

This immediately follows from Theorem 6.30 and Theorem 6.29.

Short exact sequences of the form (6.24) are examples of *almost split sequences*; we refer the reader to [**ARS1997**] for definition and discussion of the theory of almost split sequences.

Corollary 6.32. *In the assumptions of Theorem 6.30, the preprojective algebra Π of Q, considered as a module over $\mathbf{k}\vec{Q}_T \subset \Pi$, is isomorphic to the direct sum of all preprojective indecomposable modules over $\vec{Q}_T$:*

$$\Pi \simeq \bigoplus_{q \in \Delta_T} I_q.$$

In particular, if Q is Dynkin, then Π is finite-dimensional.

Proof. The proof follows from Theorem 6.30 and the equality

$$\Pi = \bigoplus{}_i\Pi_j^l = \bigoplus_{q \in T, p \in \mathbb{Z}Q} {}_p\tilde{\Pi}_q = \bigoplus_{p \in \Delta_T} {}_p\tilde{\Pi}_T. \qquad \square$$

(In fact, the last statement of the theorem can be reversed: if Π is finite-dimensional, then Q is Dynkin.)

This corollary explains the name "preprojective algebra".

The results we have proved about the preprojective algebra can be used to get some information about the indecomposable modules I_q constructed in Section 6.3.

Remark 6.33. Replacing the abelian category $\operatorname{Rep}\vec{Q}$ by the corresponding derived category $D^b(\vec{Q})$, one can use (6.24) (or, rather, the corresponding exact triangle in $D^b(\vec{Q})$) to define an indecomposable object $I_q \in D^b(\vec{Q})$ for every vertex $q \in \mathbb{Z}Q$. It can be shown that in the Dynkin case, this gives a bijection between isomorphism classes of nonzero indecomposable objects in $D^b(\vec{Q})$ and vertices of $\mathbb{Z}Q$. A modification of this also allows one to construct a bijection

(indec. objects in $D^b(\vec{Q})/T^2$) $\leftrightarrow$ vertices of $\mathbb{Z}Q/\tau^h$ $\leftrightarrow$ roots of Q,

where T is the translation functor $X \mapsto X[1]$ and h is the Coxeter number.

Details can be found in [**KT**].

Corollary 6.34. *In the assumptions of Theorem 6.30, for any two vertices $q, p \in \Delta_T$, we have a natural isomorphism*

$$\operatorname{Hom}_{\vec{Q}_T}(I_q, I_p) = {}_p\tilde{\Pi}_q.$$

Proof. Without loss of generality we can assume that there are paths in $\mathbb{Z}Q$ from q to p; otherwise, the right-hand side is obviously zero, and the left-hand side is also zero by Theorem 6.26. So let us assume that $q \preccurlyeq p$ in the order defined by Theorem 6.26.

6.5. Preprojective algebra and Auslander–Reiten quiver

If $q = (i, n) \in T$, then by Theorem 6.20 $I_q = P(i)$ is the projective module, and by Theorem 1.16 $\operatorname{Hom}_{\vec{Q}_T}(I_q, V) = V_i$. Applying it to $V = I_p = {}_p\tilde{\Pi}_T$, we get the statement of the corollary in this case.

The proof in the general case is obtained by choosing a slice T' above T such that $q \in T'$. Since we assumed that $q \preccurlyeq p$, we will also have $T' \preccurlyeq p$, so $I_q^T = \Phi_{i_1}^- \ldots \Phi_{i_l}^-(I_q^{T'})$ and $I_p^T = \Phi_{i_1}^- \ldots \Phi_{i_l}^-(I_p^{T'})$. Using Theorem 3.17, we see that $\operatorname{Hom}_{\vec{Q}_T}(I_q^T, I_p^T) = \operatorname{Hom}_{\vec{Q}_{T'}}(I_q^{T'}, I_p^{T'}) = {}_p\Pi_q$. □

Chapter 7

Tame and Wild Quivers

In this section, we discuss representation theory of quivers which are not of finite type. This theory is significantly more complicated than the theory of finite type quivers, and full classification of representation is only possible in special cases (namely, for tame quivers as defined below).

Throughout this chapter, $\vec{Q}$ is a connected quiver and $\mathbf{k}$ is an algebraically closed field.

7.1. Tame-wild dichotomy

Definition 7.1. A quiver $\vec{Q}$ is called *tame* if for any $\mathbf{v} \in \mathbb{Z}_+^I$, the set of isomorphism classes of indecomposable representations of graded dimension $\mathbf{v}$ is a union of a finite number of one-parameter families and a finite number of isolated points (see [**CB1988**] for a more rigorous definition).

Note that any finite type quiver is automatically tame; however, it is usually convenient to consider separately finite type and tame but not finite type cases.

Example 7.2. The Jordan quiver ⟲ is tame but not of finite type (over an algebraically closed field, for every d there is exactly one one-parameter family of indecomposable representations, namely J_λ —the Jordan block of size d with eigenvalue $\lambda \in \mathbf{k}$).

For the Kronecker quiver , we have shown in Example 1.5 that for $\mathbf{v} = (1,1)$, the set of indecomposable representations of dimension $\mathbf{v}$ is parametrized by $\mathbb{P}^1$. It will be shown in Section 7.7 that this quiver is in fact also tame.

Example 7.3. Consider the double loop quiver shown below:

Its path algebra is the algebra of polynomials in two noncommuting variables: $A = \mathbf{k}\langle x, y\rangle$. Classifying its representations is equivalent to classifying pairs of square matrices up to conjugation. This problem is not tame: even for $d = 1$, isomorphism classes of representations are parametrized by pairs $(\lambda_1, \lambda_2) \in \mathbf{k}^2$.

The problem of classifying representations of the double loop quiver from the last example (or, equivalently, classifying pairs of matrices up to conjugation) is a classical problem of linear algebra which so far has no satisfactory answer: while there exist algorithms which, for any d, construct a set of invariants which allow one to distinguish any two nonisomorphic representations of dimension d (see [**Fri1983**], [**Bel2000**]), the number of steps in these algorithms and the number of invariants they produce grow with d, so there is no uniform description. Moreover, it can be shown that classifying representations of $\mathbf{k}\langle x, y\rangle$ is in a certain sense a universal classification problem: for any finite-dimensional algebra A, the category of representations of A is a subcategory in the category of representations of the double loop quiver (see [**GP1969**]).

We will call a classification problem *wild* if it contains as a subproblem the problem of classifying pairs of matrices up to conjugation. More formally, we give the following definition, borrowed from [**CB1988**].

Definition 7.4. Let $W = \text{Rep}(\mathbf{k}\langle x, y\rangle)$ be the category of finite-dimensional representations of the double loop quiver from Example 7.3. We say that an associative algebra A is *wild* if there is a functor $F \colon W \to \text{Rep}(A)$, where $\text{Rep}(A)$ is the category of finite-dimensional representations of A, such that

(1) F preserves indecomposability and isomorphism classes: $F(X)$ is indecomposable in $\text{Rep}(A)$ iff X is indecomposable in W, and $F(X) \simeq F(Y)$ iff $X \simeq Y$.

(2) F is given by $F(X) = M \otimes_{\mathbf{k}\langle x,y \rangle} X$ for some $A-\mathbf{k}\langle x,y\rangle$ bimodule M, which is free and finitely generated as a right module over $\mathbf{k}\langle x, y\rangle$.

A quiver $\vec{Q}$ is wild if its path algebra is wild.

Note that the first condition is automatically satisfied if F is fully faithful, i.e. if $\operatorname{Hom}_W(X,Y) = \operatorname{Hom}_A(F(X), F(Y))$ (indeed, X is decomposable if and only if there exist $e_1, e_2 \in \operatorname{End}(X)$ such that $e_i^2 = e_i$, $e_1 e_2 = e_2 e_1 = 0$, $e_i \neq 0$). The second condition is technical and will be automatically satisfied for all the functors we construct in this book.

It is clear from Example 7.3 that wild quivers cannot be tame. The following "Tame-Wild" theorem was proved by Drozd.

Theorem 7.5. *Any algebra which is not tame is wild.*

We will not give a proof of this theorem here. Interested readers can find it in [**Drozd1980, CB1988**].

The main result of this chapter is the following theorem, the proof of which occupies several sections. Recall that in Section 1.6 we defined a Dynkin (respectively, Euclidean) graph as a connected graph such that the corresponding Tits form is positive definite (respectively, positive semidefinite), and gave a full classification of such graphs.

Theorem. *A connected quiver $\vec{Q}$ is tame if and only if the underlying graph Q is Dynkin or Euclidean.*

This theorem was proved independently by Nazarova [**Naz1973**] and by Donovan and Freislich [**DF1973**] (see also [**DR1976**]).

The plan of the proof is as follows. We begin by considering a simple example of representations of the cyclic quiver. After this, we develop the general theory of representations of Euclidean quivers, using the tools developed in the previous sections (most importantly, the Coxeter functor). The final result is given in Theorem 7.41, which classifies all indecomposable representations of Euclidean quivers. In particular, it implies that every Euclidean quiver is tame.

In Section 7.10, we show that every non-Euclidean quiver is wild (Theorem 7.46). Taken together, Theorem 7.41 and Theorem 7.46 imply the theorem above, which is restated as Theorem 7.47.

7.2. Representations of the cyclic quiver

We begin with a simple example: representations of the cyclic quiver.

Consider the cyclic quiver C_n with n vertices and with clockwise orientation shown in Figure 7.1. The corresponding graph is the Euclidean graph $\widehat{A}_{n-1}$.

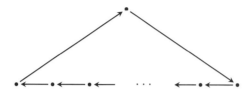

Figure 7.1. The cyclic quiver.

A representation of this quiver is the same as a $\mathbb{Z}_n$-graded vector space $V = \bigoplus_{k \in \mathbb{Z}_n} V_k$ together with a linear operator $A\colon V \to V$ such that $AV_k \subset V_{k+1}$. Note that, in particular, this shows that $A^n V_k \subset V_k$. We will call a representation *nilpotent* if the operator A is nilpotent.

Theorem 7.6.

(1) *For any $i \in \mathbb{Z}_n$, $k \geq 1$, define a representation $V = X_i^{(k)}$ as a vector space with the basis $v_0, v_1, \ldots, v_{k-1}$. The grading is given by $v_a \in V_{i+a}$, and the action of A is given by $Av_a = v_{a+1}$, $Av_{k-1} = 0$. Then V is an indecomposable representation of the cyclic quiver; it has dimension k, is nilpotent, and has the simple representation $S(i)$ as a quotient.*

(2) *For any $l \geq 1$, $\lambda \in \mathbf{k}^\times$, there exists a unique up to isomorphism representation V such that for every $a \in \mathbb{Z}_n$, we have $\dim V_a = l$, and the operator $A^n\colon V_a \to V_a$ is the Jordan block of size l with eigenvalue λ. This representation is indecomposable; it will be denoted $X_{\lambda,l}$.*

(3) *Representations $X_i^{(k)}$, $X_{\lambda,l}$ defined above give a full list of nonzero indecomposable representations of the cyclic quiver.*

Proof. The proof is left to the reader as an exercise. □

Corollary 7.7. *The cyclic quiver C_n is tame.*

7.3. Affine root systems

Our first goal is to show that any Dynkin or Euclidean quiver is tame. For Dynkin quivers, this follows from Gabriel's theorem (Theorem 3.3), since any finite type quiver is automatically tame. Thus, we will concentrate on the study of Euclidean quivers. Since the only Euclidean quiver which has oriented loops is the cyclic quiver, which we have already studied in Section 7.2, from now on we also assume that $\vec{Q}$ has no oriented cycles (which automatically excludes the Jordan quiver).

7.4. Affine Coxeter element

As in Section 1.7, let $L = \mathbb{Z}^I$ be the root lattice of Q and let $(\,,\,)$ be the symmetrized Euler form (1.17). We also denote by $\alpha_i, i \in I$, the standard basis in L and by W the Weyl group, generated by simple reflections s_i. As described in Appendix A, L, W can also be described as the root lattice and the Weyl group of the corresponding Kac–Moody Lie algebra $\mathfrak{g}(Q)$, which in this case is an (untwisted) affine Lie algebra. Thus, we will frequently refer to W as an affine Weyl group.

We denote by R the set of roots of $\mathfrak{g}(Q)$, and by R^{re}, R^{im}, the sets of real and imaginary roots. Recall that in the case of Euclidean quivers, the root system can be described as follows:

$$R^{re} = \{\alpha \in L - \{0\} \mid (\alpha, \alpha) = 2\},$$
$$R^{im} = \{\alpha \in L - \{0\} \mid (\alpha, \alpha) = 0\}.$$

In fact, one can give an even more explicit description. Let $\delta \in R_+$ be the generator of the radical of $(\,,\,)$ (see Theorem 1.28). Fix a choice of a vertex $i_0 \in I$ such that $Q_f = Q - \{i_0\}$ is a connected graph; such a vertex will be called an *extending vertex*. It follows from the proof of Theorem 1.28 that Q_f is Dynkin. We will denote by $L_f = \mathbb{Z}^{I-\{i_0\}}$ the corresponding root lattice and by R_f and W_f the corresponding root system and Weyl group respectively; note that in particular, both R_f and W_f are finite.

Theorem 7.8.

(1) $R^{re} = \{\alpha + n\delta, \alpha \in R_f, n \in \mathbb{Z}\}$ and $R^{im} = \{n\delta, n \in \mathbb{Z}, n \neq 0\}$.

(2) For an element $\alpha \in L_f$, define the automorphism $\tau_\alpha \colon L \to L$ by

$$\tau_\alpha(x) = x + (\alpha, x)\delta.$$

Then $\tau_{\alpha+\beta} = \tau_\alpha \tau_\beta$, so the group $T = \{\tau_\alpha, \alpha \in L_f\}$ is isomorphic to L_f. Moreover, we have an isomorphism

$$W = W_f \ltimes T.$$

(3) For any $w \in W$, $w(\delta) = \delta$.

A proof of this theorem can be found in [**Kac1990**].

An example of an affine root system and the corresponding Weyl group will be given in Example 7.15.

7.4. Affine Coxeter element

We will keep all the notation of the previous section. Namely, we assume that $\vec{Q}$ is a Euclidean quiver without oriented cycles, with extending vertex i_0. We denote by R the corresponding root system and by W the affine Weyl group. We also denote by R_f and W_f the root system and the Weyl group of the quiver $\vec{Q}_f = Q - \{i_0\}$ and denote by $r = |I| - 1$ the number of vertices of Q_f.

Recall that a Coxeter element in the affine Weyl group is an element

(7.1) $$C = s_{i_r} \ldots s_{i_1} s_{i_0},$$

where $\{i_0, i_1, \ldots, i_r\} = I$ is some ordering of the elements of I (compare with Definition 3.27).

It is easy to see that for a fixed choice of the set of positive roots $\alpha_i, i \in I$, all Coxeter elements obtained for different orderings of I are conjugate to each other. [Unlike the Dynkin case, if we just start with the root lattice and do not fix the positive roots, then not all Coxeter elements are conjugate to each other; this is related to the fact that for the graph $\widehat{A}_n$, not every two orientations can be obtained from each other by sink-to-source transformations.] It is also not true that C has finite order. Instead, we have the following results.

Theorem 7.9.

(1) *There exists a positive integer g such that $C^g \in T$, where $T = \tau(L_f)$ is defined in Theorem 7.8.*

(2) $C\alpha = \alpha$ *iff* $\alpha = n\delta$.

Proof. The first part is immediate from the fact that $W/T = W_f$ is finite.

To prove the second part, assume that $C(\alpha + n\delta) = \alpha + n\delta$ for some $\alpha \in L_f$; clearly this is equivalent to $C\alpha = \alpha$. Let us first consider the case when $C = C_f s_0$, where s_0 is the simple reflection corresponding to the extending vertex and $C_f = s_{i_1} \ldots s_{i_r}$ is a Coxeter element for W_f. Since C_f does not change the δ-component of a root, $C\alpha = \alpha$ is only possible if $(\alpha, \alpha_0) = 0$, so $s_0(\alpha) = \alpha$, $C_f(\alpha) = \alpha$. By Theorem 3.29, this implies $\alpha = 0$.

The general case (when s_0 appears somewhere in the middle of C rather than at the end) easily follows: any Coxeter element is conjugate to an element of the form $C_f s_0$, so

$$C\alpha = \alpha \implies (wC_f s_0 w^{-1}\alpha = \alpha) \implies (w^{-1}\alpha = n\delta) \implies \alpha = n\delta.$$

$\square$

Remark 7.10. It is possible to compute the number g explicitly. Namely, consider the Dynkin graph Q_f. Recall that by Theorem 1.30, Q_f must be a star diagram $\Gamma(p_1, p_2, p_3)$ consisting of three branches of lengths p_1, p_2, p_3 meeting at one branch point (lengths of branches include the branch point). Then it is shown in [**Ste1985**] that the image of C in $W_f = W/T$ is conjugate to a product of commuting elements C_i corresponding to branches of Q_f, and the order of C_i is p_i. (For type A, this is true if we choose

7.4. Affine Coxeter element

$p_1 = n, p_2 = p_3 = 1$). Thus, one can take $g = \gcd(p_i)$; for example, for $Q_f = D_n$ we have $p_1 = p_2 = 2$, $p_3 = n - 2$, so $g = \gcd(2, n-2)$.

From now on, we choose the Coxeter element C adapted to $\vec{Q}$ (see Definition 3.23); since we assumed that $\vec{Q}$ has no oriented cycles, this is possible. We will study the action of C on the set of roots. We begin by repeating the constructions of Section 6.1.

As before, we denote by $\langle\,,\rangle$ the Euler form of $\vec{Q}$. Recall that by Corollary 6.3, we have
$$\langle \alpha, \beta \rangle = -\langle \beta, C\alpha \rangle = \langle C\alpha, C\beta \rangle.$$
In particular, this implies
$$\langle \delta, x \rangle = \langle \delta, Cx \rangle = -\langle x, \delta \rangle.$$

For any $k = 0, \ldots, r$ define roots
$$
\begin{aligned}
p_k &= s_{i_0} \ldots s_{i_{k-1}}(\alpha_{i_k}), \\
q_k &= s_{i_r} \ldots s_{i_{k+1}}(\alpha_{i_k}).
\end{aligned}
\tag{7.2}
$$

Note that then $Cp_k = -q_k$. It follows from Lemma 6.1 that the elements p_k are distinct real positive roots, and

$$\{p_0, \ldots, p_r\} = \{\alpha \in R_+ \mid C\alpha \in R_-\}. \tag{7.3}$$

Similarly, the elements q_k are distinct real positive roots, and

$$\{q_0, \ldots, q_r\} = \{\alpha \in R_+ \mid C^{-1}\alpha \in R_-\}. \tag{7.4}$$

Lemma 7.11. *For any $k = 0, \ldots, r$, we have $C^n p_k \in R_-$ for $n > 0$ and $C^{-n} p_k \in R_+$ for $n \geq 0$.*

Similarly, for any $k = 0, \ldots, r$, we have $C^{-n} q_k \in R_-$ for $n > 0$ and $C^n q_k \in R_+$ for $n \geq 0$.

Proof. Assume that $C^{-n} p_k \in R_-$ for some $n > 0$. Choosing the smallest such n and using (7.4), we see that then $C^{-(n-1)} p_k = q_l$ for some l. Since $q_l = -Cp_l$ (see (6.2)), this gives $C^{-n} p_k = -p_l$. Taking the Euler pairing of both sides with δ gives $\langle C^{-n} p_k, \delta \rangle = -\langle p_l, \delta \rangle$. Using the C-invariance of $\langle\,,\rangle$, we see that the left-hand side is $\langle p_k, \delta \rangle = \delta_{i_k} > 0$ (by Example 1.24), whereas the right-hand side is $-\langle p_l, \delta \rangle = -\delta_{i_l} < 0$, which gives a contradiction.

The remaining statements are proved in a similar way. $\square$

We can now give a full classification of roots depending on the properties of their C-orbit.

Definition 7.12. Let g, λ be as in Theorem 7.9, so that $C^g = \tau_\lambda$. For any $\alpha \in L$, we define its *defect* by
$$\partial(\alpha) = (\alpha, \lambda) \in \mathbb{Z}$$

so that

(7.5) $$C^g(\alpha) = \alpha + \partial(\alpha)\delta.$$

For a representation V of the quiver $\vec{Q}$, we will write $\partial(V)$ for $\partial(\mathbf{dim}\, V)$.

Note that it is immediate from the definition that $\partial(C\alpha) = \partial(\alpha)$.

Theorem 7.13. *Let $\alpha \in R_+$. Then:*

(1)
$$\partial(\alpha) < 0 \iff C^N\alpha \in R_- \text{ for all } N \gg 0$$
$$\iff \alpha = C^{-n}p_k \text{ for some } n \geq 0, k = 0, \ldots, r.$$

(2)
$$\partial(\alpha) > 0 \iff C^{-N}\alpha \in R_- \text{ for all } N \gg 0$$
$$\iff \alpha = C^n q_k \text{ for some } n \geq 0, k = 0, \ldots, r.$$

(3)
$$\partial(\alpha) = 0 \iff C^g\alpha = \alpha \iff C^n\alpha \in R_+ \text{ for all } n \in \mathbb{Z}.$$

Proof. Equivalences
$$\partial(\alpha) = 0 \iff C^g\alpha = \alpha,$$
$$\partial(\alpha) < 0 \iff C^N\alpha \in R_- \text{ for all } N \gg 0,$$
$$\partial(\alpha) > 0 \iff C^{-N}\alpha \in R_- \text{ for all } N \gg 0$$

immediately follow from the equality $C^{ng}(\alpha) = \alpha + n\partial(\alpha)\delta$. Comparing it with Lemma 7.11, we get the remaining statements. □

We will also need one more lemma.

Lemma 7.14. *Let $\alpha \in L$ be an element with defect zero. Then $\langle \delta, \alpha \rangle = 0$.*

Proof. Since $\partial(\alpha) = 0$, we have $C^g(\alpha) = \alpha$. Thus, the element $\beta = \alpha + C\alpha + \cdots + C^{g-1}\alpha$ satisfies $C\beta = \beta$; by Theorem 7.9, we have $\beta = k\delta$. Therefore,
$$0 = \langle \delta, \beta \rangle = \sum_{i=0}^{g-1} \langle \delta, C^i\alpha \rangle = g\langle \delta, \alpha \rangle.$$
□

7.4. Affine Coxeter element

Example 7.15. Consider the Kronecker quiver:

$$\underset{1}{\bullet} \rightrightarrows \underset{0}{\bullet}$$

Then the corresponding root system is given by

$$R^{re} = \{\pm\alpha_1 + n\delta,\ n \in \mathbb{Z}\},$$
$$R^{im} = \{n\delta, n \neq 0\},$$

where $\delta = \alpha_0 + \alpha_1$ (see Figure 7.2). This is exactly the root system of the affine Lie algebra $\widehat{\mathfrak{sl}_2}$.

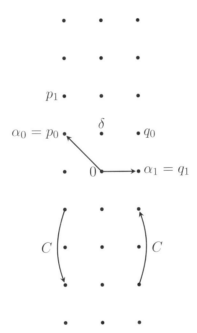

Figure 7.2. Root system $\widehat{A}_1$ and the action of the affine Coxeter element C.

The Coxeter element is $C = s_1 s_0$ (it is chosen to be adapted to the quiver $\vec{Q}$). It acts on the root lattice by

$$C(\alpha_1) = \alpha_1 + 2\delta, \qquad C(\alpha_0) = \alpha_0 - 2\delta, \qquad C(\delta) = \delta.$$

Comparing it with the definition of defect, we see that in this case $g = 1$, $\partial(\alpha_1) = 2$, $\partial(\alpha_0) = -2$. The roots p_i, q_i defined by (7.2) are

$$p_0 = \alpha_0, \qquad\qquad p_1 = s_0(\alpha_1) = \alpha_1 + 2\alpha_0,$$
$$q_0 = s_1(\alpha_0) = \alpha_0 + 2\alpha_1, \qquad q_1 = \alpha_1.$$

7.5. Preprojective, preinjective, and regular representations

In this section, we begin the study of indecomposable representations of Euclidean quivers. It will be based on the results of Chapter 6.

Throughout this section $\vec{Q}$ is a Euclidean quiver without oriented cycles (thus, it cannot be the Jordan quiver), and $C \in W$ is an adapted Coxeter element.

Recall that we have defined the Coxeter functors $\mathbf{C}^{\pm}$ on the category $\operatorname{Rep} \vec{Q}$ and introduced the notion of preprojective, preinjective, and regular indecomposable representations (see Definition 6.8): an indecomposable representation V of $\vec{Q}$ is called

- *preprojective* if $(\mathbf{C}^{+})^{n} V = 0$ for $n \gg 0$,
- *preinjective* if $(\mathbf{C}^{-})^{n} V = 0$ for $n \gg 0$,
- *regular* if for any $n \in \mathbb{Z}_{+}$, we have $(\mathbf{C}^{+})^{n} V \neq 0$, $(\mathbf{C}^{-})^{n} V \neq 0$.

We also have similar notions for positive roots: a positive root $\alpha \in R_{+}$ is

- *preprojective* if $C^{n} \alpha \not> 0$ for some $n > 0$,
- *preinjective* if $C^{n} \alpha \not> 0$ for some $n < 0$,
- *regular* if $C^{n} \alpha > 0$ for all n.

It is immediate from Theorem 7.13 that if α is a positive root, then it is preprojective (preinjective, regular) if and only if we have $\partial(\alpha) < 0$ (respectively, $\partial(\alpha) > 0$, $\partial(\alpha) = 0$). Combining it with Theorem 6.9, which gives a bijection between the set of indecomposable preprojective representations and preprojective positive roots, we get the following result.

Theorem 7.16. *Let I be an indecomposable preprojective representation of $\vec{Q}$. Then $\alpha = \dim I$ is a positive root satisfying $\partial(\alpha) < 0$. Conversely, for any positive root satisfying $\partial(\alpha) < 0$, there is a unique up to an isomorphism indecomposable representation I_{α} of dimension α. This representation is preprojective.*

A similar result holds for preinjective representations.

Theorem 7.17. *Let I be an indecomposable preinjective representation of $\vec{Q}$. Then $\alpha = \dim I$ is a positive root satisfying $\partial(\alpha) > 0$. Conversely, for any positive root α with $\partial(\alpha) > 0$, there is a unique up to an isomorphism indecomposable representation I_{α} of dimension α. This representation is preinjective.*

Corollary 7.18. *Let I be an indecomposable representation of $\vec{Q}$. Then I is*

- *preprojective iff $\partial(I) < 0$,*
- *preinjective iff $\partial(I) > 0$,*
- *regular iff $\partial(I) = 0$.*

Thus, the theory of preprojective and preinjective representations is quite similar to the Dynkin case. In particular, we have a bijection

(isomorphism classes of preprojective and preinjective indecomposables)
$$\leftrightarrow \text{(positive roots } \alpha \in R \text{ such that } \partial(\alpha) \neq 0).$$

Note that every such root must be real, as for imaginary roots, we have $\partial(\alpha) = 0$. However, not all real roots have nonzero defect: in general, there exist real roots with $\partial(\alpha) = 0$.

This gives a complete classification of preinjective and preprojective indecomposable representations. It remains to study the regular indecomposables.

7.6. Category of regular representations

In this section, we keep the assumptions and notation of Sections 7.4 and 7.5.

To complete the analysis of indecomposable representations of Euclidean quivers, we need to study regular indecomposable representations. Our exposition follows [**DR1976**], [**CB1992**].

Let $\mathcal{R}(\vec{Q})$ be the category of regular representations (see Definition 6.8). It follows from Theorem 6.4 that $\mathbf{C}^+, \mathbf{C}^-$ are mutually inverse autoequivalences of the category $\mathcal{R}$ and that restrictions of $\mathbf{C}^\pm$ to $\mathcal{R}$ are exact. In particular, this implies that
$$\mathbf{dim}(\mathbf{C}^\pm(V)) = C^{\pm 1} \mathbf{dim}(V)$$
for any $V \in \mathcal{R}(\vec{Q})$.

Lemma 7.19. *Let $V \in \mathrm{Rep}(\vec{Q})$ be such that $\partial(V) = 0$. Then the following are equivalent*:

(1) $V \in \mathcal{R}(\vec{Q})$.
(2) *For any subrepresentation $V' \subset V$, $\partial(V') \leq 0$.*
(3) *For any quotient $V'' = V/V'$, $\partial(V'') \geq 0$.*

Proof. We show that (1) implies (2). Indeed, assume that V is regular and $V' \subset V$, $\partial(V') > 0$. Without loss of generality, we can assume that V' is indecomposable; thus, by Corollary 7.18, V' is preinjective, so for any $n \geq 0$, we have $\mathbf{dim}(\mathbf{C}^+)^n V' = C^n \mathbf{v}'$, $\mathbf{v}' = \mathbf{dim}\, V'$.

Since the functor $\mathbf{C}^+$ is left exact, we have $(\mathbf{C}^+)^n V' \subset (\mathbf{C}^+)^n(V)$, so
$$C^n \mathbf{v}' \leq C^n \mathbf{v} \qquad \text{for any } n \geq 0.$$

But by definition of the defect, $C^{gm}\mathbf{v}' = \mathbf{v}' + m\partial(\mathbf{v}')\delta$, while $C^{gm}\mathbf{v} = \mathbf{v}$. Thus, using $\partial(\mathbf{v}') > 0$, for large m we get a contradiction.

Conversely, assume that V satisfies condition (2). Writing $V = \bigoplus V_i$, where V_i are indecomposable, we get that $\partial(V_i) \leq 0$. Since $\partial(V) = \sum \partial(V_i) = 0$, this implies $\partial(V_i) = 0$, so $V \in \mathcal{R}(\vec{Q})$.

Equivalence of (1) and (3) is proved similarly, using right exactness of $\mathbf{C}^-$. $\qquad\square$

Theorem 7.20. $\mathcal{R}(\vec{Q})$ *is an abelian category closed under extensions.*

Proof. First, let us prove that if $f\colon V \to W$ is a morphism of regular representations, then $\operatorname{Im} f$ is regular. Indeed, let $X = \operatorname{Im} f$, so that we can factor f as the composition $V \twoheadrightarrow X \hookrightarrow W$. By Lemma 7.19, $\partial(X) \leq 0$ since it is a subrepresentation of W; similarly, $\partial(X) \geq 0$ since it is a quotient of V. Thus, $\partial(X) = 0$. Since any subobject of X is also a subobject of W, X satisfies condition (2) of Lemma 7.19 and thus is regular.

Next we prove that if in a short exact sequence $0 \to A \to B \to C \to 0$ any two of the objects are regular, then so is the third one. Indeed, since $\partial(A) - \partial(B) + \partial(C) = 0$, if any two of them have zero defect, then so does the third one. Since every subobject of A is also a subobject of B, Lemma 7.19 shows that regularity of B, C implies regularity of A. Similarly, since every quotient of C is also a quotient of B, regularity of A, B would imply the regularity of C.

Finally, assume that A and C are regular. Then $\partial(B) = 0$, and for any subobject $B' \subset B$ we have a short exact sequence $0 \to A' \to B' \to C' \to 0$, with $A' = A \cap B' \subset A$, $C' = \pi(B') \subset C$. Thus, $\partial(B') = \partial(A') + \partial(C') \leq 0$, so B is regular. $\qquad\square$

Theorem 7.20 shows that in order to study regular representations, we should first classify simple objects in the category $\mathcal{R}$, in other words, regular representations of $\vec{Q}$ which have no nontrivial regular subrepresentations.

Note that for any simple regular representation X, its dimension $\mathbf{dim}\, X$ has a finite orbit under the action of the Coxeter element C, which follows from (7.5) and $\partial(X) = 0$.

Definition 7.21. We say that a simple regular representation X has period l if the C-orbit of $\mathbf{dim}\, X$ has size l.

7.6. Category of regular representations

Note that by Theorem 7.9, the period of X is 1 if and only if $\dim X \in \mathbb{Z}\delta$.

The following theorems list some basic properties of simple regular modules.

Theorem 7.22. *Let X, Y be simple regular representations such that X is of period $l > 1$. Then there are three possibilities:*

(1) $Y \simeq X$. *Then* $\operatorname{Hom}(X, X) = \mathbf{k}$, $\operatorname{Ext}^1(X, X) = 0$, *and* $\langle X, X \rangle = 1$, *where* $\langle, \rangle$ *is the Euler form* (1.15).

(2) $Y \simeq \mathbf{C}^+ X$. *Then* $\operatorname{Hom}(X, Y) = 0$, $\operatorname{Ext}^1(X, Y) = \mathbf{k}$, *and* $\langle X, Y \rangle = -1$.

(3) $Y \not\simeq X$, $Y \not\simeq \mathbf{C}^+ X$. *Then* $\operatorname{Hom}(X, Y) = \operatorname{Ext}^1(X, Y) = 0$ *and* $\langle X, Y \rangle = 0$.

Proof. Since the period of X is larger than 1, $\dim X \neq n\delta$, so $\langle X, X \rangle = \dim \operatorname{Hom}(X, X) - \dim \operatorname{Ext}^1(X, X) > 0$. Since $\operatorname{Hom}(X, X) = \mathbf{k}$ by Schur's lemma, we get $\operatorname{Ext}^1(X, X) = 0$.

Similarly, if $Y = \mathbf{C}^+ X$, then, since $Y \not\simeq X$, $\operatorname{Hom}(X, Y) = 0$. By Serre duality (Lemma 6.7), $\operatorname{Ext}^1(X, Y) = \operatorname{Hom}(Y, \mathbf{C}^+ X)^* = \mathbf{k}$.

In the last case, Schur's lemma gives $\operatorname{Hom}(X, Y) = 0$, and Serre duality gives $\operatorname{Ext}^1(X, Y) = \operatorname{Hom}(Y, \mathbf{C}^+ X)^* = 0$. □

Corollary 7.23. *If X is a simple regular representation of period $l > 1$, then it is the unique simple representation of dimension $\dim X$. In particular, $(\mathbf{C}^+)^l(X) \simeq X$.*

Indeed, since $\dim(\mathbf{C}^+)^l(X) = \dim X$, we get $\langle X, (\mathbf{C}^+)^l X \rangle = \langle X, X \rangle = 1$.

Similar statements hold in the case when X is of period 1.

Theorem 7.24. *Let X be a simple regular representation of period 1.*

(1) $\operatorname{Hom}(X, X) = \mathbf{k}$, $\operatorname{Ext}^1(X, X) = \mathbf{k}$, *and* $\langle X, X \rangle = 0$.

(2) $\mathbf{C}^+(X) \simeq X$.

(3) *If Y is a simple regular representation, $Y \not\simeq X$, then* $\operatorname{Hom}(X, Y) = \operatorname{Ext}^1(X, Y) = 0$ *and* $\langle X, Y \rangle = 0$.

Proof. By Schur's lemma, $\operatorname{Hom}(X, X) = \mathbf{k}$. Since by Theorem 7.9, $\dim X \in \mathbb{Z}\delta$, we have $\langle X, X \rangle = 0$, so $\operatorname{Ext}^1(X, X) = \mathbf{k}$.

For the second part, note that $\operatorname{Hom}(X, \mathbf{C}^+ X) = \operatorname{Ext}^1(X, X)^* \neq 0$; since both $X, \mathbf{C}^+ X$ are simple, they are isomorphic.

Part (3) is proved in the same way as the last part of Theorem 7.22. □

Let X now be a simple regular representation of period l. Denote by
$$\mathcal{O} = \{X, \mathbf{C}^+ X, \ldots, (\mathbf{C}^+)^{l-1} X\}$$
its orbit under the action of the Coxeter functor.

Definition 7.25. Let $\mathcal{O}$ be the $\mathbf{C}^+$-orbit of a simple regular representation. Let $\mathcal{R}_\mathcal{O}$ be the full subcategory in $\mathcal{R}(\vec{Q})$ consisting of objects $V \in \mathcal{R}(\vec{Q})$ such that all composition factors of V (in $\mathcal{R}(\vec{Q})$) are in $\mathcal{O}$. For a simple representation X, we will also use the notation $\mathcal{R}_X$ for $\mathcal{R}_\mathcal{O}$, where $\mathcal{O}$ is the orbit of X.

It follows from Theorem 7.22 and Theorem 7.24 that if $\mathcal{O} \neq \mathcal{O}'$, then for any $X \in \mathcal{R}_\mathcal{O}$, $X' \in \mathcal{R}_{\mathcal{O}'}$, $\mathrm{Hom}(X, X') = \mathrm{Ext}^1(X, X') = 0$. Thus, each $\mathcal{R}_\mathcal{O}$ is an abelian category, and

$$(7.6) \qquad \mathcal{R}(\vec{Q}) = \bigoplus_\mathcal{O} \mathcal{R}_\mathcal{O}(\vec{Q}).$$

In particular, each indecomposable object of $\mathcal{R}$ lies in exactly one of the categories $\mathcal{R}_\mathcal{O}$. The set of all indecomposables in $\mathcal{R}_\mathcal{O}$ is called a *tube*.

To analyze the category $\mathcal{R}_\mathcal{O}(\vec{Q})$, we need the following lemma.

Lemma 7.26. *Let X be a simple regular representation of period 1. Then for every $n \geq 1$, there exists a unique indecomposable object $X^{(n)}$ of length n in $\mathcal{R}_X$. Moreover,*

$$(7.7) \qquad \dim \mathrm{Hom}(X^{(n)}, X) = \dim \mathrm{Ext}^1(X^{(n)}, X) = 1.$$

Proof. The proof is by induction. For $n = 1$, uniqueness of the indecomposable is obvious, and (7.7) follows from Theorem 7.22.

Now, assume that the lemma is proven for some n. Let M be an indecomposable module of length $n + 1$. Then M has a simple submodule, isomorphic to X, and the quotient $Y = M/X$ has length n. Since M is indecomposable, it is easy to see that Y is also indecomposable; thus, by the induction assumption, $M/X \simeq X^{(n)}$, so we have a short exact sequence

$$0 \to X \to M \to X^{(n)} \to 0.$$

Since $\mathrm{Ext}^1(X^{(n)}, X)$ is 1-dimensional, this shows that M is unique up to isomorphism.

To prove equations (7.7) for $n+1$, note that we have a long exact sequence

$$0 \to \mathrm{Hom}(X^{(n)}, X) \to \mathrm{Hom}(M, X) \to \mathrm{Hom}(X, X)$$
$$\xrightarrow{\delta} \mathrm{Ext}^1(X^{(n)}, X) \to \mathrm{Ext}^1(M, X) \to \mathrm{Ext}^1(X, X) \to 0.$$

By the induction assumption, this becomes

$$0 \to \mathbf{k} \to \mathrm{Hom}(M, X) \to \mathbf{k} \xrightarrow{\delta} \mathbf{k} \to \mathrm{Ext}^1(M, X) \to \mathbf{k} \to 0.$$

Since M is a nonsplit extension, the connecting map δ is nonzero, which immediately gives $\mathrm{Hom}(M, X) = \mathbf{k}$, $\mathrm{Ext}^1(M, X) = \mathbf{k}$. $\square$

7.6. Category of regular representations

A similar statement, proof of which we omit, holds for orbits of period $l > 1$.

Lemma 7.27. *Let X be a simple regular representation of period $l > 1$. Denote $X_i = (\mathbf{C}^+)^i(X)$, $i \in \mathbb{Z}_l$. Then for every $i \in \mathbb{Z}_l$, $n \geq 1$, there exists a unique indecomposable object $X_i^{(n)}$ of length n in $\mathcal{R}_X$ which has X_i as a quotient. Moreover,*

$$\operatorname{Hom}(X_i^{(n)}, X_j) = \delta_{ij}\mathbf{k}, \qquad \operatorname{Ext}^1(X_i^{(n)}, X_j) = \delta_{i+n,j}\mathbf{k}.$$

The composition factors of $X_i^{(n)}$ are $X_i, X_{i+1}, \ldots, X_{i+n-1}$, necessarily in this order: for any composition series $X_i^{(n)} = M_0 \supset M_1 \supset \cdots \supset M_n = \{0\}$, we must have $M_k/M_{k+1} \simeq X_{i+k}$.

Remark 7.28. In fact, it can be shown (see [**Gab1973**]) that each of the categories $\mathcal{R}_X$ is equivalent to the category of nilpotent representations of the cyclic quiver C_l, with l vertices and all edges oriented the same way; we have studied its representations in Section 7.2. For $l = 1$, this gives the Jordan quiver ↻; in this case, the indecomposable objects $X^{(n)}$ correspond to the Jordan block J_n with zero eigenvalue.

The following is a summary of some properties of the regular indecomposable representations.

Lemma 7.29. *Let X be a simple regular representation of period l, and let $Y = X^{(n)}$ be the indecomposable regular representation of length n defined in Lemma 7.26 and Lemma 7.27.*

(1) *If n is a multiple of l (in particular, if $l = 1$), then $\dim Y \in \mathbb{Z}\delta$ and*
$$\operatorname{Hom}(Y, Y) = \operatorname{Ext}^1(Y, Y) = \mathbf{k}.$$

(2) *If n is not a multiple of l, then $\alpha = \dim Y$ is a positive real root with $\partial(\alpha) = 0$ and*
$$\operatorname{Hom}(Y, Y) = \mathbf{k}, \qquad \operatorname{Ext}^1(Y, Y) = 0.$$

Proof. Computation of $\operatorname{Hom}(Y, Y)$ and $\operatorname{Ext}^1(Y, Y)$ is easily done by induction, using the long exact sequence, Lemma 7.26, and Lemma 7.27. Since $\dim(Y) = (1 + C + \cdots + C^{n-1})\dim X$ is C-invariant if n is a multiple of l, we see that in this case, $\dim Y \in \mathbb{Z}\delta$. In the case when n is not a multiple of l, computation of $\operatorname{Hom}(Y, Y)$ and $\operatorname{Ext}^1(Y, Y)$ gives $\langle \alpha, \alpha \rangle = 1$; thus, α is a positive real root. Since by construction Y is regular, $\partial(\alpha) = 0$. □

A converse statement will be proved later (see Theorem 7.40).

7.7. Representations of the Kronecker quiver

Before proceeding with the general theory, we consider a special example: the Kronecker quiver $\bullet_1 \rightrightarrows \bullet_0$. This example is instructive; moreover, we will use the results of this section to prove the general theorem in the next section. Throughout this section, **k** is an algebraically closed field.

Representations of the Kronecker quiver were classified by Weierstrass and Kronecker in 1890 in relation to the problem of classification of (not necessarily symmetric) bilinear forms. Their results can be found, for example, in Gantmacher's book [**Gan1998**, Chapter XII]. However, here we use a different approach, based on the theory developed in the previous sections.

We begin by classifying preprojective and preinjective representations.

Theorem 7.30.

(1) *For any $n \geq 0$, the Kronecker quiver has a unique indecomposable representation of dimension $(\mathbf{v}_1, \mathbf{v}_0) = (n, n+1)$, shown in the figure below. This representation is preprojective, and all indecomposable preprojective representations are of this form.*

(7.8)
$$\mathbf{k}^n \bullet \underset{\begin{bmatrix} 0 \\ I_n \end{bmatrix}}{\overset{\begin{bmatrix} I_n \\ 0 \end{bmatrix}}{\rightrightarrows}} \bullet \mathbf{k}^{n+1}$$

Here I_n denotes the $n \times n$ identity matrix.

(2) *For any $n \geq 0$, the Kronecker quiver has a unique indecomposable representation of dimension $(\mathbf{v}_1, \mathbf{v}_0) = (n+1, n)$, shown in the figure below. This representation is preinjective, and all indecomposable preinjective representations are of this form.*

(7.9)
$$\mathbf{k}^{n+1} \bullet \underset{\begin{bmatrix} 0 & I_n \end{bmatrix}}{\overset{\begin{bmatrix} I_n & 0 \end{bmatrix}}{\rightrightarrows}} \bullet \mathbf{k}^n$$

Proof. From the explicit description of the root system and the Coxeter element given in Example 7.15, it is easy to see that the roots of the form $(n, n+1) = n\alpha_1 + (n+1)\alpha_0 = -\alpha_1 + (n+1)\delta$ are exactly the positive roots in the C-orbits of p_0, p_1. Thus, it follows from Theorem 7.16 that for each n, there is a unique indecomposable representation of dimension $(n, n+1)$,

namely
$$V_k = \begin{cases} (\mathbf{C}^-)^k P(0), & n = 2k, \\ (\mathbf{C}^-)^k P(1), & n = 2k+1, \end{cases}$$

and that these representations form a full list of indecomposable preprojective representations. Using induction, it is easy to show that V_k is given by (7.8).

The second part is proved similarly. □

We now need to classify indecomposable regular representations. Note that since $\partial(V) = 2(\dim V_1 - \dim V_0)$, any representation of zero defect must have dimension $(n, n) = n\delta$.

For any $(\lambda, \mu) \in \mathbf{k}^2$, consider the representation

$$S_{\lambda,\mu} = \mathbf{k} \bullet \underset{\mu}{\overset{\lambda}{\rightrightarrows}} \bullet \mathbf{k}.$$

Lemma 7.31.

(1) *If at least one of λ, μ is nonzero, then $S_{\lambda,\mu}$ is a simple regular representation. Up to isomorphism, it only depends on the ratio $[\lambda : \mu] \in \mathbb{P}^1$.*

Abusing the language, we will denote these representations by $S_t, t \in \mathbb{P}^1$.

(2) *Representations $S_t, t \in \mathbb{P}^1$, form a full list of simple regular representations of the Kronecker quiver.*

Proof. The first part is obvious. To prove the second part, let $V = (V_1, V_0; x_1, x_2)$ be a simple regular representation; note that $\dim V_0 = \dim V_1$. Choose a nonzero vector $v \in V_1$ such that $x_1(v)$ and $x_2(v)$ are linearly dependent. It is easy to see that such a v exists: if x_1 is invertible, we can take v to be an eigenvector of $x_1^{-1} x_2$; otherwise, we can take v to be any vector in the kernel of x_1. Let $W_1 = \mathbf{k}v$, and let W_2 be a one-dimensional subspace containing $x_1(v), x_2(v)$.

Then $W = (W_1, W_2)$ is a subrepresentation of V, of defect zero. Thus, if V was simple, we must have $W = V$. □

For future reference, we note that representations S_t can also be defined in a different way. Namely, let L be the irreducible preprojective representation of dimension $(1, 2)$ (see Theorem 7.30). Then one easily sees that

(7.10) $$S_t = L/L'(t),$$

where, for $t = [\lambda : \mu]$, the subrepresentation $L'(t)$ is given by $L'(t)_1 = 0$, $L'(t)_0 = \mathbf{k} \begin{bmatrix} \mu \\ -\lambda \end{bmatrix}$.

Thus, we have a full classification of simple regular representations. By (7.6), $\mathcal{R}(\vec{Q})$ can be written as a direct sum

$$\mathcal{R}(\vec{Q}) = \bigoplus_{t \in \mathbb{P}^1} \mathcal{R}_t(\vec{Q}),$$

where $\mathcal{R}_t(\vec{Q})$ is the full subcategory consisting of objects $X \in \mathcal{R}(\vec{Q})$ such that all composition factors of X (in $\mathcal{R}(\vec{Q})$) are copies of S_t. In particular, each indecomposable belongs to exactly one of the categories $\mathcal{R}_t$. Moreover, Lemma 7.26 shows that for every $n \geq 0, t \in \mathbb{P}^1$, there exists a unique indecomposable representation $S_t^{(n)}$ in $\mathcal{R}_t$. It is easy to construct explicitly that

(7.11)
$$S_t^{(n)} = \mathbf{k}^n \bullet \underset{J_{\mu,n}}{\overset{J_{\lambda,n}}{\rightrightarrows}} \bullet \mathbf{k}^n,$$

where $t = [\lambda : \mu] \in \mathbb{P}^1$ and $J_{\lambda,n}$ stands for the Jordan block of size n with eigenvalue λ. If $\mu \neq 0$, then this representation is equivalent to

$$S_t^{(n)} = \mathbf{k}^n \bullet \underset{I_n}{\overset{J_{t,n}}{\rightrightarrows}} \bullet \mathbf{k}^n.$$

Combining the results above, we get the following classification of indecomposable representations of the Kronecker quiver.

Theorem 7.32. *The following is the full list of all indecomposable representations of the Kronecker quiver.*

(1) *For every $n \geq 0$, the indecomposable preprojective representation of dimension $(n, n+1)$ given by (7.8).*

(2) *For every $n \geq 0$, the indecomposable preinjective representation of dimension $(n+1, n)$ given by (7.9).*

(3) *For every $t \in \mathbb{P}^1$, $n \geq 1$, the indecomposable regular representation $S_t^{(n)}$ of dimension (n, n) given by (7.11).*

We will give a geometric interpretation of these representations in terms of coherent sheaves on $\mathbb{P}^1$ in the next chapter (see Example 8.33).

7.8. Classification of regular representations

In this section, we give a complete classification of regular representations of Euclidean quivers. Our proof follows the works of Crawley-Boevey [**CB1992**] and Dlab and Ringel [**DR1976**], with some modifications. Our approach will be similar to the one used for the Kronecker quiver. Namely, we will construct a family $S_t^{\vec{Q}}$ of regular representations of $\vec{Q}$, parametrized by points of $\mathbb{P}^1$, and show that this family generates (in a suitable sense) all regular representations.

In this section, we keep the assumptions and notation of Section 7.4. As before, we also assume that the extending vertex i_0 is a sink. To simplify notation, we will frequently write α_0 for α_{i_0}, etc.

It is easy to see (for example, by analyzing the list of all Euclidean graphs in Figure 1.3) that in all cases, $\delta_0 = 1$ and that there are two possible cases:

(1) i_0 is connected to only one vertex i, by a single edge, and $\delta_i = 2$.

(2) i_0 is connected to two vertices i', i'', by a single edge to each one, and $\delta_{i'} = \delta_{i''} = 1$.

Together with the equality $(\delta, \delta) = 0$, this implies that $\delta = \alpha_0 + \theta$, where

$$(7.12) \qquad \theta = \sum_{I_f} \delta_i \alpha_i \in L_f$$

satisfies $(\theta, \theta) = 2$ and thus is a positive root of the finite root system R_f (in fact, it is known that θ is a maximal root of R_f). Let I_θ be the indecomposable representation of $\vec{Q}_f$ of dimension θ; by Gabriel's theorem, it exists and is unique up to isomorphism. It follows from the above that the space

$$U = \bigoplus_{i \to i_0} (I_\theta)_i$$

has dimension 2.

Let K_2 be the Kronecker quiver considered in Section 7.7. Define a functor

$$(7.13) \qquad T \colon \operatorname{Rep}(K_2) \to \operatorname{Rep}(\vec{Q})$$

as follows: for $V = (V_1, V_0; x_1, x_2) \in \operatorname{Rep}(K_2)$, we let $T(V)$ be defined by

$$T(V)|_{\vec{Q}_f} = V_1 \otimes I_\theta,$$
$$T(V)_{i_0} = V_0$$

and the operators assigned to edges connecting Q_f with i_0 are

- x_1, x_2, if there are two edges e_1, e_2, connecting $i', i'' \in I_f$ with i_0 (in which case, $T(V)_{i'} \simeq T(V)_{i''} \simeq V_1$),
- $(x_1 \oplus x_2)$ if there is one edge connecting i_0 with $i \in I_f$, in which case $T(V)_i = V_1 \otimes (I_\theta)_i \simeq V_1 \oplus V_1$.

This construction requires some choices: in the first case, ordering the vertices i', i''; in the second, choice of identification $(I_\theta)_i \simeq \mathbf{k}^2$. However, it is easy to show that all results of this section are independent of these choices.

Example 7.33. Let $\vec{Q} = \widehat{D}_4$. Then $T(V)$ is shown below.

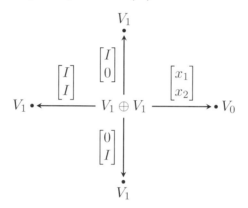

Theorem 7.34. *Let $\vec{Q}$ be a connected Euclidean quiver other than the Jordan quiver, and let $i_0 \in I$ be an extending vertex which is a sink. Then the functor $T \colon \mathrm{Rep}(K_2) \to \mathrm{Rep}(\vec{Q})$ is exact and fully faithful.*

Proof. Exactness is obvious. The fact that it is fully faithful immediately follows from the definition and the fact that $\mathrm{End}_{\vec{Q}_f}(I_\theta) = \mathbf{k}$ (see Theorem 3.25), so $\mathrm{Hom}_{\vec{Q}_f}(V_1 \otimes I_\theta, W_1 \otimes I_\theta) = \mathrm{Hom}_{\mathbf{k}}(V_1, W_1)$. $\square$

We will now use the functor T to construct simple regular representations of $\vec{Q}$. Namely, define, for every $t \in \mathbb{P}^1$, a representation $S_t^{\vec{Q}}$ of $\vec{Q}$ by

(7.14) $$S_t^{\vec{Q}} = T(S_t),$$

where S_t is the representation of the Kronecker quiver defined in Lemma 7.31. The following result immediately follows from Theorem 7.34.

Lemma 7.35. *Under the assumptions of Theorem 7.34, for any $t \in \mathbb{P}^1$, $S_t^{\vec{Q}}$ is an indecomposable regular representation of dimension δ. The representations $S_t^{\vec{Q}}$ are pairwise nonisomorphic, and $\mathrm{End}_{\vec{Q}}(S_t^{\vec{Q}}) = \mathbf{k}$, $\mathrm{Ext}^1(S_t^{\vec{Q}}, S_t^{\vec{Q}}) = \mathbf{k}$.*

It turns out that this family of representations contains indecomposable representatives from each subcategory $\mathcal{R}_{\mathcal{O}}$ defined in Definition 7.25.

7.8. Classification of regular representations

Lemma 7.36. *In the assumptions of Theorem 7.34, let X be a regular representation of $\vec{Q}$ such that $X_0 \neq 0$. Then for some $t \in \mathbb{P}^1$, there exists a nonzero homomorphism $S_t^{\vec{Q}} \to X$.*

Proof. We will use an alternative construction of the representations $S_t^{\vec{Q}}$. Recall that by (7.10), one can write the representation S_t of the Kronecker quiver as a quotient:

$$0 \to P \xrightarrow{f_t} L \to S_t \to 0,$$

where P is a representation of dimension $(0,1)$ and L is an indecomposable representation of dimension $(1,2)$.

Applying functor T to this short exact sequence gives a realization of $S_t^{\vec{Q}}$ as a quotient:

$$0 \to T(P) \xrightarrow{f_t} T(L) \to S_t^{\vec{Q}} \to 0.$$

Note that $T(P) = P(i_0)$ is a projective representation (since i_0 is a sink); its dimension is α_0. Since $\mathbf{dim}\, T(L) = \delta + \alpha_0$, we have $\partial(T(L)) = \partial(T(P)) < 0$ and thus $T(L)$ is also indecomposable preprojective by Corollary 7.18.

Let $\dim X_0 = m$. Then the space $\mathrm{Hom}_{\vec{Q}}(P(i_0), X) \simeq X_0$ has dimension m, which implies $\langle \alpha_0, \mathbf{dim}\, X \rangle = m$, and

$$\dim \mathrm{Hom}(T(L), X) \leq \langle \mathbf{dim}\, T(L), \mathbf{dim}\, X \rangle = \langle \alpha_0 + \delta, \mathbf{dim}\, X \rangle = m$$

(since X is regular, by Lemma 7.14, $\langle \delta, \mathbf{dim}\, X \rangle = 0$).

Since T is fully faithful, $\dim \mathrm{Hom}(T(P), T(L)) = \dim \mathrm{Hom}(P, L) = 2$. Let f_1, f_2 be a basis of this space; then they define $f_1^*, f_2^* \colon \mathrm{Hom}(T(L), X) \to \mathrm{Hom}(T(P), X)$.

It is a trivial exercise in linear algebra that if x_1, x_2 are two linear operators $V \to W$, and $\dim W \leq \dim V$, then one can find $v \in V, v \neq 0$ and $\lambda, \mu \in \mathbf{k}$, $(\lambda, \mu) \neq (0,0)$ so that $(\lambda x_1 + \mu x_2)(v) = 0$ (in the case when $V = W$, this was proved in the proof of Lemma 7.31). Applying it to the operators $f_1^*, f_2^* \colon \mathrm{Hom}(T(L), X) \to \mathrm{Hom}(T(P), X)$, we see that one can find $\varphi \in \mathrm{Hom}(T(L), X), \varphi \neq 0$, and $t \in \mathbb{P}^1$ so that $f_t^*(\varphi) = \varphi \circ f_t = 0$. Thus, φ descends to a morphism $T(L)/f_t T(P) \to X$. $\square$

This turns out to be enough to classify all simple regular representations of $\vec{Q}$ (and thus, by Lemma 7.26 and Lemma 7.27, all regular indecomposable representations).

Recall that for a simple regular representation X, we defined its period as the size of the orbit of $\mathbf{dim}\, X$ under the action of the Coxeter element (see Definition 7.21).

Theorem 7.37.

(1) *Let X be a simple regular representation of $\vec{Q}$ of period 1. Then $\dim X = \delta$, and $X \simeq S_t^{\vec{Q}}$ for some $t \in \mathbb{P}^1$.*

(2) *Let $\mathcal{O}$ be an orbit of a simple regular representation of period $l > 1$. Then*
$$\sum_{X \in \mathcal{O}} \dim X = \delta.$$
Thus, $\mathcal{O}$ contains a unique simple representation X satisfying $X_0 \neq 0$. Moreover, let $X^{(l)} \in \mathcal{R}_{\mathcal{O}}$ be the indecomposable representation of length l which has X as a quotient (see Lemma 7.27). Then $X^{(l)} \simeq S_t^{\vec{Q}}$ for some $t \in \mathbb{P}^1$.

Proof. If X is simple regular of period 1, then $\dim X \in \mathbb{Z}\delta$, so $X_0 \neq 0$. By Lemma 7.36, X is a quotient of $S_t^{\vec{Q}}$ for some t; thus, $\dim X = \delta$ and $X \simeq S_t^{\vec{Q}}$.

If $\mathcal{O}$ is an orbit of period $l > 1$, then $\sum_{\mathcal{O}} \dim X$ is C-invariant; thus, $\sum_{\mathcal{O}} \dim X \in \mathbb{Z}\delta$. This implies that $\mathcal{O}$ contains a simple regular representation X with $X_0 \neq 0$. By Lemma 7.36, X is a quotient of $S_t^{\vec{Q}}$ for some t; thus, $S_t^{\vec{Q}}$ is an indecomposable regular representation in the tube corresponding to $\mathcal{O}$, so this tube contains an indecomposable representation of dimension δ. Using Lemma 7.29, it is easy to see that $S_t^{\vec{Q}}$ must be of length l. $\square$

Corollary 7.38.

(1) *One has a bijection*
$$\mathbb{P}^1 \leftrightarrow \{\text{tubes of } \vec{Q}\}$$
given by $t \mapsto (\text{tube containing } S_t^{\vec{Q}})$.

(2) *The set*
$$D = \{t \in \mathbb{P}^1 \mid \text{tube containing } S_t^{\vec{Q}} \text{ has period} > 1\}$$
is finite. For all $t \notin D$, the corresponding tube has period 1 and S_t is simple.

Proof. By Theorem 7.37, every $\mathbf{C}^+$-orbit $\mathcal{O}$ of a simple regular representation contains a unique simple regular representation X with $X_0 \neq 0$, and the corresponding indecomposable representation $X^{(l)}$ of length l (where l is the period of X) must be isomorphic to $S_t^{\vec{Q}}$ for some Q. This shows that each tube contains a representative from the family $S_t^{\vec{Q}}$. Since representations $S_t^{\vec{Q}}$ are pairwise nonisomorphic (Lemma 7.35), such a representative is unique.

7.8. Classification of regular representations

Also, by Theorem 7.37, if $\mathcal{O}$ is an orbit of a simple regular representation of period $l > 1$, then for every $X \in \mathcal{O}$, $\alpha = \dim X$ is a positive real root satisfying $\alpha < \delta$. Since there are only finitely many such roots, the number of such orbits is finite. $\square$

Lemma 7.39. *Let α be a real positive root, $\alpha < \delta$, such that $\partial(\alpha) = 0$. Then there exists a regular indecomposable representation of dimension α.*

Proof. Consider the $GL(\alpha)$-orbit $\mathbb{O}$ of maximal dimension in the representation space $R(\alpha)$. Let X be the corresponding representation. By Theorem 2.13, we have $X \simeq \bigoplus I_i$, where I_i are indecomposable representations such that $\mathrm{Ext}^1(I_k, I_l) = 0$. Thus, $\langle I_k, I_l \rangle = \dim \mathrm{Hom}(I_k, I_l) \geq 0$, and therefore

$$\langle \alpha, \alpha \rangle \geq \sum \langle I_k, I_k \rangle.$$

Since $\dim I_k < \delta$, $\langle I_k, I_k \rangle \geq 1$. On the other hand, since α is a real root, $\langle \alpha, \alpha \rangle = 1$, which is only possible if the direct sum $X = \bigoplus I_k$ consists of one term. Since $\partial(X) = 0$, by Corollary 7.18 it must be regular. $\square$

Recall that by Lemma 7.29, for an indecomposable regular representation $Y = X_i^{(n)}$, where n is not a multiple of length of X_i, its dimension $\alpha = \dim Y$ is a positive real root. The following theorem provides a converse statement.

Theorem 7.40. *For every real positive root α with $\partial(\alpha) = 0$, there exists a unique indecomposable representation of dimension α. This representation has the form $Y = X_i^{(N)}$, where X_i is a simple regular representation of period $l > 1$ and N is not a multiple of l.*

Proof. Let α be a real positive root with $\partial(\alpha) = 0$. Write $\alpha = \beta + k\delta$, for some positive real root β with $\beta < \delta$. Obviously, $\partial(\beta) = \partial(\alpha) = 0$. By Lemma 7.39, there exists an indecomposable regular representation Y of dimension β. Using Lemma 7.26 and Lemma 7.27 we see that Y must be of the form $Y = X_i^{(n)}$ for some simple regular representation X_i of period $l > 1$, and n — not a multiple of l. Consider now the indecomposable representation $X_i^{(N)}$, $N = n + kl$. By Theorem 7.37, this representation has dimension $\beta + k\delta = \alpha$. This shows existence of an indecomposable representation of dimension α, which has the required form.

To show uniqueness, note that by Lemma 7.29, $\mathrm{Ext}^1(Y,Y) = 0$. By Corollary 2.6, this implies that the corresponding $GL(\alpha)$-orbit in $R(\alpha)$ is open. Since any other indecomposable representation I of dimension α must be regular and thus of the form $X_j^{(M)}$, it must also correspond to an open orbit. Since there can only be one open orbit, this shows the uniqueness of the indecomposable representation of dimension α. $\square$

7.9. Euclidean quivers are tame

In this section, we put together the results of the previous sections to give a complete description of the set of all indecomposable representations of a Euclidean quiver.

Throughout this section, $\vec{Q}$ is an arbitrary Euclidean quiver and the ground field **k** is algebraically closed.

Theorem 7.41. *Let $\vec{Q}$ be a connected Euclidean quiver other than the Jordan quiver and let R be the corresponding root system as defined in Section 7.3. Assume that the ground field **k** is algebraically closed. Then:*

(1) *An indecomposable representation of $\vec{Q}$ of dimension $\mathbf{v} \in L$ exists if and only if $\mathbf{v} \in R_+$.*

(2) *For any real positive root α, the indecomposable representation of $\vec{Q}$ of dimension α is unique up to an isomorphism.*

(3) *There exists a finite subset $D = \{p_1, \ldots, p_k\} \subset \mathbb{P}^1$ and a collection of positive integers $l_p > 1$, $p \in D$, such that for any positive imaginary root $\alpha = n\delta$, the set of isomorphism classes of indecomposable representations of $\vec{Q}$ of dimension α is in bijection with the set*

$$(\mathbb{P}^1 \setminus D) \cup \bigcup_{p \in D} \mathbb{Z}_{l_p}.$$

Proof. Assume first that $\vec{Q}$ is such that the extending vertex is a sink. Then $\vec{Q}$ has no oriented cycles, and by Theorem 7.16 and Theorem 7.17, the set of indecomposable representations of $\vec{Q}$ consists of

- preprojective representations (which are in bijection with real positive roots of defect $\partial(\alpha) < 0$),
- preinjective representations (which are in bijection with positive real roots with $\partial(\alpha) > 0$),
- regular representations: by Lemma 7.26 and Lemma 7.27, the regular indecomposables are of the form $X_i^{(n)}$, where X_i is a simple regular representation and $n \geq 1$; by Lemma 7.29 and Theorem 7.40, regular indecomposables of this form with n not divisible by the period l of X_i are in bijection with positive real roots with $\partial(\alpha) = 0$.

  Finally, regular indecomposables of the form $X_i^{(n)}$ for n a multiple of the period l (this includes all regular representations of period 1), by Corollary 7.38, are parametrized exactly by the set

$$(\mathbb{P}^1 \setminus D) \cup \bigcup_{p \in D} \mathbb{Z}_{l_p}.$$

To prove the general case (when i_0 is not necessarily a sink), note that if $\vec{Q}$ is the cyclic quiver C_n with all arrows oriented the same way, the theorem follows from Theorem 7.6; for any other orientation of the cyclic quiver, one can always find a vertex that is a sink. Since any vertex in a cyclic quiver is an extending vertex, the theorem is proved for a cyclic quiver.

For any Euclidean quiver which is not cyclic, the extending vertex is either a sink (in which case we have already proved the result) or a source. In the latter case, the result easily follows from the previous case by applying the reflection functor. $\square$

A full list of sets D for different Euclidean quivers can be found in [**DR1976**].

A geometric interpretation of these representations, using McKay correspondence between Euclidean graphs and finite subgroups in $SU(2)$, will be given in the next chapter (see Theorem 8.30). In particular, it will be shown that the set D corresponds to points in $\mathbb{P}^1$ which have nontrivial stabilizers in $G/\{\pm I\}$, where $G \subset SU(2)$ is the finite subgroup corresponding to Q.

Corollary 7.42. *Every Euclidean quiver is tame.*

7.10. Non-Euclidean quivers are wild

In this section, we prove one part of Theorem 7.5: if a quiver is neither Dynkin nor Euclidean, then it is wild (and thus not tame). Our exposition follows some ideas from [**Naz1973**] (with significant changes).

We begin with the following definition.

Definition 7.43. Let $\vec{Q}$ and $\vec{Q}'$ be quivers. We say that $\text{Rep}(Q) \subset \text{Rep}(\vec{Q}')$ if there exists a functor $T \colon \text{Rep}(Q) \to \text{Rep}(\vec{Q}')$ such that:

(1) T is fully faithful.

(2) For every representation $V = (V_i, x_h) \in \text{Rep}(Q)$, the representation $T(V)$ is given by
$$T(V)_j = \bigoplus_i V_i \otimes H^i_j, \qquad i \in I(Q), \; j \in I(Q'),$$
where H^i_j is some collection of finite-dimensional vector spaces, and the linear operators $x_h^{Q'}$ corresponding to edges of Q' have the form
$$x_h^{Q'} = \sum_p x_p^Q \otimes A_p,$$
where p is a path $i_1 \to i_2$ in Q and A_p is a linear operator $H^{i_1}_{j_1} \to H^{i_2}_{j_2}$.

We say that a quiver $\vec{Q}$ is *strongly wild* if $\operatorname{Rep}(L_2) \subset \operatorname{Rep}(\vec{Q})$, where L_2 is the double loop quiver from Example 7.3.

It is easy to see that the second condition on T is equivalent to requiring that $T(V) = M \otimes_{\mathbf{k}\vec{Q}} V$, where M is some $\vec{Q}' - \vec{Q}$ bimodule which is free of finite rank over the path algebra $\mathbf{k}\vec{Q}$.

It is obvious from the definition that every strongly wild quiver is also wild (as defined in Definition 7.4).

Lemma 7.44. *The tadpole quiver below (for any choice of orientation of the unmarked edge) is strongly wild*:

Proof. We give the proof in the case when the orientation of the "tail" edge is chosen so that the vertex is the sink. The proof for the other orientation is similar.

Define the embedding $T\colon \operatorname{Rep}(L_2) \subset \operatorname{Rep}(\vec{Q})$ by

$$T(V; x_1, x_2) = V \xleftarrow{A} V \oplus V \circlearrowright B, \qquad A = \begin{bmatrix} 0 & I \end{bmatrix}, \ B = \begin{bmatrix} x_1 & x_2 \\ I & 0 \end{bmatrix}.$$

It is obvious that the so-defined T is faithful; explicit computation shows that it is also full. $\square$

Lemma 7.45. *Let $\vec{Q}^E$ be a Euclidean quiver. Then there exists an embedding $\operatorname{Rep}(J) \subset \operatorname{Rep}(\vec{Q}^E)$, where J is the Jordan quiver.*

Proof. If $\vec{Q}^E$ is a cyclic quiver, with all edges oriented the same way, the embedding is obvious (take all edges but one to be the identity matrix).

In all other cases, we can choose an extending vertex which is either a sink or a source. If it is a sink, then we have an embedding $\operatorname{Rep}(K_2) \subset \operatorname{Rep}(\vec{Q}^E)$ defined by (7.13). Similar construction gives an embedding $\operatorname{Rep}(K_2) \subset \operatorname{Rep}(\vec{Q}^E)$ in the case when the extending vertex is a source.

Combining this with the embedding $\operatorname{Rep}(J) \subset \operatorname{Rep}(K_2)$ given by $(V; x) \mapsto (V, V; x, I)$, we get the statement of the lemma. $\square$

After these preliminary lemmas, we can prove the main result.

Theorem 7.46. *If $\vec{Q}$ is a connected quiver which is neither Dynkin nor Euclidean, then $\vec{Q}$ is strongly wild and thus wild.*

7.11. Kac's theorem

Proof. It follows from the proof of Theorem 1.28 that every such $\vec{Q}$ contains a Euclidean quiver $\vec{Q}^E$ as a proper subquiver. Clearly there are two possibilities.

(1) $\vec{Q}$ has a vertex $\imath$ which is not in $\vec{Q}^E$ and is connected by at least one edge to $\vec{Q}^E$.

(2) Vertices of $\vec{Q}$ are the same as vertices of $\vec{Q}^E$, and $\vec{Q}$ has an edge h which is not in $\vec{Q}^E$.

In the first case, consider the embedding $\text{Rep}(J) \subset \text{Rep}(\vec{Q}^E)$ constructed in Lemma 7.45. It is obvious that it can be extended to an embedding $\text{Rep}(T) \hookrightarrow \text{Rep}(\vec{Q})$, where T is the tadpole graph of Lemma 7.44 (with a suitable orientation of the "tail" edge). Since the tadpole graph is strongly wild, it proves the theorem in this case.

In the second case, the embedding $\text{Rep}(J) \subset \text{Rep}(\vec{Q}^E)$ can be extended to an embedding $\text{Rep}(L_2) \subset \text{Rep}(\vec{Q})$. Thus, the theorem is again proved. $\square$

We will discuss another proof of this theorem in the next section.

Combining Theorem 7.46 with Gabriel's theorem (Theorem 3.3) and Corollary 7.42, we get the main result of this chapter.

Theorem 7.47. *Let $\vec{Q}$ be a connected quiver.*

(1) *If $\vec{Q}$ is Dynkin, then it is of finite type (and thus tame).*

(2) *If $\vec{Q}$ is Euclidean, then it is tame but not of finite type.*

(3) *If $\vec{Q}$ is neither Dynkin nor Euclidean, then it is wild (and thus not tame).*

7.11. Kac's theorem

In this section, we give the statement of Kac's theorem, which gives partial information about indecomposable representations for arbitrary quivers, generalizing Theorem 7.41. Proofs of these results are quite complicated and will not be given here; instead, we refer the reader to the original papers.

Throughout this section, $\vec{Q}$ is an arbitrary connected quiver without edge loops.

As in Section 1.3, we denote by $L = \mathbb{Z}^I$ the root lattice of Q, by $(\,,\,)$, the symmetrized Euler form (1.17), and by W the Weyl group, generated by simple reflections s_i, $i \in I$.

As described in Appendix A, we can associate to the graph Q a generalized Cartan matrix C_Q and a Lie algebra $\mathfrak{g}(C_Q)$. We denote by $R \subset L$ the corresponding root system. An overview of the results on the root systems

is given in Section A.6. In particular, we have a decomposition of the set of positive roots into real and imaginary roots:
$$R_+ = R_+^{re} \sqcup R_+^{im}$$
and we have the following result, which is a reformulation of Theorem A.18.

Lemma 7.48.

If Q is Dynkin, then $R = R^{re} = \{\alpha \in L - \{0\} \mid (\alpha, \alpha) = 2\}$.

If Q is Euclidean, then $R = \{\alpha \in L - \{0\} \mid (\alpha, \alpha) \leq 2\}$.
In this case, the set $R^{im} = \{\alpha \in L - \{0\} \mid (\alpha, \alpha) = 0\}$ is nonempty.

If Q is neither Dynkin nor Euclidean, then there exists a root $\alpha \in R_+^{im}$ such that $(\alpha, \alpha) < 0$.

The following result was proved by Kac in [**Kac1980**], [**Kac1983**].

Theorem 7.49. *Let $\vec{Q}$ be an arbitrary connected quiver without edge loops. Assume that the ground field $\mathbf{k}$ is algebraically closed, and let $\alpha \in L$ be positive: $\alpha = \sum n_i \alpha_i, n_i \geq 0, \alpha \neq 0$. Then:*

(1) *An indecomposable representation I of $\vec{Q}$ with $\dim I = \alpha$ exists if and only if $\alpha \in R_+$.*

(2) *If $\alpha \in R_+^{re}$, then the indecomposable representation I_α with $\dim I_\alpha = \alpha$ is unique up to an isomorphism.*

(3) *If α is an imaginary root, then there exist infinitely many nonisomorphic indecomposable representations of dimension α.*

In fact, the last part can be refined. It can be shown, using Rosenlicht's theorem, that for any α, the set of isomorphism classes of indecomposable representations of dimension α is parametrized by a finite union of algebraic varieties $Z_1, \ldots, Z_N$ (details of this argument can be found in [**Kac1983**, §1.9]). We denote $\mu_\alpha = \max \dim Z_i$.

Theorem 7.50. *In the assumptions of Theorem 7.49, assume that $\alpha \in R_+$. Then*
$$\mu_\alpha = 1 - \tfrac{(\alpha,\alpha)}{2}.$$

The proof of this theorem is quite complicated; some steps require reduction to finite fields. We will not repeat the proof here, referring the interested reader to [**Kac1983**].

Note that this (together with Drozd's tame-wild theorem) gives another proof of Theorem 7.46: every quiver which is neither Dynkin nor Euclidean is wild. Indeed, by Lemma 7.48 there are imaginary roots with $(\alpha, \alpha) < 0$, so we would have $\mu_\alpha > 1$, which means that $\vec{Q}$ cannot be tame and thus must be wild.

7.11. Kac's theorem

Finally, it is also interesting to study the number of indecomposable representations of a quiver $\vec{Q}$ over a finite field. The following result was proved in [**Kac1983**]. We will call a representation *absolutely indecomposable* if it remains indecomposable over the algebraic closure of the ground field.

Theorem 7.51. *For any $q = p^r$, let $m(\vec{Q}, \alpha; q)$ be the number of isomorphism classes of absolutely indecomposable representations of $\vec{Q}$ with dimension α over the finite field $\mathbb{F}_q$. Then $m(\vec{Q}, \alpha; q)$ does not depend on the orientation of $\vec{Q}$ and is given by a polynomial in q with integer coefficients:*

$$m(\vec{Q}, \alpha; q) = q^N + a_1 q^{N-1} + \cdots + a_N, \qquad N = \mu_\alpha.$$

In [**Kac1983**], Kac also made two conjectures about the polynomial $m(\vec{Q}, \alpha; q)$. Both of them have been since proved.

Theorem 7.52. *Let $m(\vec{Q}, \alpha; q)$ be as defined in Theorem 7.51. Then:*

(1) *The constant term a_N is equal to the multiplicity of root α in $\mathfrak{g}(Q)$.*

(2) *All coefficients a_i are nonnegative.*

These two statements are known as the *Kac conjectures*. In the case when α is indivisible (i.e. not a multiple of another root), Kac's conjectures were proved in [**CBVdB2004**]. In full generality, the first conjecture was proved by Hausel [**Hau2010**]; the second conjecture was proved by Hausel, Letellier, and Rodriguez–Villegas in [**HLRV2013**]. Both proofs are well beyond the scope of this book.

Chapter 8

McKay Correspondence and Representations of Euclidean Quivers

In this chapter, we discuss a geometric approach to Euclidean quivers and their representations, based on finite subgroups in SU(2). This material will be revisited later, when discussing quiver varieties (see Chapter 12).

8.1. Finite subgroups in SU(2) and regular polyhedra

We begin by recalling some classic facts about finite subgroups in SU(2). They have been extensively studied, and for the most part, we omit the proofs, giving just the statement of the results. We refer the reader to [**Cox1974**] for a review.

Recall that one has a group homomorphism

$$\pi \colon \mathrm{SU}(2) \to \mathrm{SO}(3)$$

which is surjective; the kernel of π is $\mathrm{Ker}(\pi) = \{I, -I\} \simeq \mathbb{Z}_2$. Thus, every finite subgroup $\overline{G} \subset \mathrm{SO}(3)$ gives rise to a finite subgroup $G = \pi^{-1}(\overline{G}) \subset \mathrm{SU}(2)$. Therefore, classifying finite subgroups in SU(2) is almost the same as classifying finite subgroups in SO(3).

Example 8.1. For every $n > 1$, the cyclic group $G = \mathbb{Z}_n$ of order n can be embedded in SU(2) by

$$k \mapsto \begin{bmatrix} \zeta^k & 0 \\ 0 & \zeta^{-k} \end{bmatrix}, \qquad \zeta = e^{2\pi i/n}.$$

Let $\overline{G} = \pi(G) \subset \mathrm{SO}(3)$. If n is even, then $\overline{G} \simeq \mathbb{Z}_{n/2}$, $G = \pi^{-1}(\overline{G})$. If n is odd, then G does not contain $-I$, so it cannot be written in the form $\pi^{-1}(\overline{G})$; in this case, $\pi(G) = G = \mathbb{Z}_n$.

One way to construct finite subgroups in SO(3) (and thus in SU(2)) is by using regular polyhedra. Let X be a convex regular polyhedron in $\mathbb{R}^3$. Each such polyhedron can be described by a pair of numbers $\{p, q\}$, called its *Schläfli symbol*:

(8.1)
$$p = \text{number of sides of each face of } X,$$
$$q = \text{number of faces meeting at every vertex of } X.$$

For example, the cube has Schläfli symbol $\{4, 3\}$.

The classification of convex regular polyhedra in $\mathbb{R}^3$ has been known since antiquity. There are five regular polyhedra (Platonic solids): tetrahedron, octahedron, cube, icosahedron, and dodecahedron.

Projecting the vertices and edges of a regular polyhedron onto the sphere, we get a *spherical polyhedron*, i.e. a partition of the sphere into a union of spherical polygons. The spherical polyhedrons obtained by projecting regular polyhedra will be called *regular*. We will also include in the list of regular spherical polyhedrons the regular n-gon ($n \geq 2$); it partitions the sphere into two spherical n-gons, so its Schläfli symbol is $\{n, 2\}$. (In [**Cox1974**], regular spherical polyhedrons are called *regular spherical tessellations*.)

For each regular polyhedron X (or the corresponding spherical polyhedron), we can consider the group $\overline{G} = Sym(X) \subset \mathrm{SO}(3)$ of all rotational symmetries of X. Obviously it is a finite subgroup in SO(3); such groups are called *polyhedral groups*. One can also consider the corresponding subgroup in SU(2):

(8.2) $\qquad G = \pi^{-1}(Sym(X)), \qquad X$ a regular spherical polyhedron.

Finite subgroups in SU(2) of the form (8.2) are called *binary polyhedral groups*. Table 8.1 lists all regular spherical polyhedra and corresponding binary groups. It is obvious that dual polyhedra (for example, the cube and the octahedron) give the same binary polyhedral group; Table 8.1 shows that other than this, all binary polyhedral groups are pairwise nonisomorphic.

These groups have been extensively studied, starting with Klein's book [**Kle1884**]. We refer the reader to [**Cox1974**] and references there for a detailed discussion.

The following classical theorem, proof of which can be found in [**Cox1974**], gives a full classification of finite subgroups in SU(2).

Theorem 8.2. *Up to conjugation, each nontrivial finite subgroup in* SU(2) *is either a cyclic subgroup as in Example* 8.1, *or one of the binary polyhedral groups listed in Table* 8.1.

Table 8.1. Binary polyhedral groups.

Polyhedron	$\{p,q\}$	G	$|G|$
n-gon, $n \geq 2$	$\{n,2\}$	binary dihedral group BD_{4n}	$4n$
tetrahedron	$\{3,3\}$	binary tetrahedral	24
octahedron	$\{3,4\}$	binary octahedral	48
cube	$\{4,3\}$		
isocahedron	$\{3,5\}$	binary icosahedral	120
dodecahedron	$\{5,3\}$		

Corollary 8.3. *The only finite subgroups in* SU(2) *which are not of the form* $\pi^{-1}(\overline{G})$ *are cyclic groups of odd order.*

8.2. ADE classification of finite subgroups

It turns out that there is a direct connection between finite subgroups in SU(2) and Dynkin or Euclidean graphs. In fact, there are two such connections. We give one of them here; the other one will be described in the next section. As before, we skip most proofs, referring the reader to [**Cox1974**] for proofs and a detailed discussion.

Let G be a finite subgroup in SU(2), and let $\overline{G} = \pi(G)$ be the corresponding subgroup in SO(3). Let us consider the action of $\overline{G}$ (and thus of G) on the unit sphere S^2. From now on, we will identify $S^2 \simeq \mathbb{P}^1$ so that the action of G on S^2 is identified with the action of G on $\mathbb{P}^1$ given by

$$\begin{bmatrix} a & b \\ c & d \end{bmatrix}(t) = \frac{at+b}{ct+d} \tag{8.3}$$

(note that the element $-I \in$ SU(2) acts trivially).

Our first goal is to study the fundamental domain and points with nontrivial stabilizer for this action. Since the stabilizer of any point $x \in \mathbb{P}^1$ in SU(2) is isomorphic to U(1) = S^1, this implies that for every such x, the stabilizer of x in G (and thus in $\overline{G}$) is a cyclic group.

Let X be a regular spherical polyhedron, and let G be the corresponding binary polyhedral group. Let us consider the triangulation of the sphere obtained by the barycentric subdivision of X, i.e. by dividing each face of

X into spherical triangles by connecting the center of the face with each vertex and with each midpoint of an edge (thus, a face which is a p-gon is subdivided into $2p$ triangles). Each of the triangles of this triangulation has angles

(8.4) $$\alpha = \frac{\pi}{p}, \quad \beta = \frac{\pi}{q}, \quad \gamma = \frac{\pi}{2},$$

where $\{p,q\}$ is the Schläfli symbol of X.

It is easy to show that this triangulation admits a coloring using two colors, so that triangles having a common side have different colors. An example of such a barycentric triangulation for the binary icosahedral group is shown in Figure 8.1.

Figure 8.1. The triangulation of the sphere for the binary icosahedral group.

It is known that in this case, the group $\widetilde{G} \subset O(3)$ of all symmetries of X (rotations and reflections) is generated by reflections in the three great circles forming the sides of one of the spherical triangles of the barycentric subdivision; this group contains $\overline{G}$ as a subgroup of index 2.

Theorem 8.4. *Let X be a $\{p,q\}$ regular spherical polyhedron (including the regular n-gon, $n \geq 2$) and let $G \subset \mathrm{SU}(2)$ be the corresponding binary polyhedral group.*

(1) *Each of the spherical triangles of the barycentric subdivision is a fundamental domain for the action on S^2 of the group $\widetilde{G}$ of all symmetries of X (rotations and reflections). Any union of two adjacent triangles is a fundamental domain for the action of $\overline{G}$.*

(2) *There are exactly three $\overline{G}$-orbits in $\mathbb{P}^1 \simeq S^2$ which have nontrivial stabilizer:*
   - *Centers of faces of X. For each such point x, the stabilizer $\overline{G}_x$ of x in $\overline{G}$ is cyclic of order p, and the stabilizer in G is cyclic of order $2p$.*

8.2. ADE classification of finite subgroups

- Vertices of X. For each such point x, $\overline{G}_x \simeq \mathbb{Z}_q$ and $G_x \simeq \mathbb{Z}_{2q}$.
- Midpoints of edges of X. For each such point x, $\overline{G}_x \simeq \mathbb{Z}_2$ and $G_x \simeq \mathbb{Z}_4$.

(3) Let Δ be one of the spherical triangles of the barycentric subdivision, with vertices $(v_0, v_1, v_2) \in \mathbb{P}^1$, where
- v_0 is a vertex of X,
- v_1 is the midpoint of an edge incident to v_0,
- v_2 is the center of a face incident to the edge.

Let $A, B, C \in G$ be generators of the stabilizer groups $G_{v_2}, G_{v_0}, G_{v_1}$ respectively, chosen so that $\pi(A)$ is the counterclockwise rotation by angle $2\pi/p$ around v_2, and similarly for B, C. Then $A^p = B^q = C^2 = -I$, and elements A, B, C generate G.

Example 8.5. The binary dihedral group $BD_{4n} = \pi^{-1}(D_{2n})$ is generated by

$$A = \begin{bmatrix} \zeta & 0 \\ 0 & \zeta^{-1} \end{bmatrix}, \quad \zeta = e^{2\pi i/2n},$$

$$B = \begin{bmatrix} 0 & i \\ i & 0 \end{bmatrix},$$

and the orbits in $\mathbb{P}^1$ with nontrivial stabilizer are:

$\overline{G} \cdot 0 = \{0, \infty\}$ (for this two-point orbit, the stabilizer of each point in G is generated by A and is isomorphic to $\mathbb{Z}_{2n}$). These points correspond to the centers of the faces of X.

$\overline{G} \cdot 1$; this orbit has order n, and for each point, the stabilizer in G is isomorphic to $\mathbb{Z}_4$ (for $t = 1$, the stabilizer is generated by B). These points correspond to the vertices of X.

$\overline{G} \cdot \zeta$; this orbit has order n, and for each point, the stabilizer in G is isomorphic to $\mathbb{Z}_4$ (for $t = \zeta$, the stabilizer is generated by AB). These points correspond to the midpoints of the edges of X.

Similar results also hold for the cyclic group $G = \mathbb{Z}_n$. Obviously, this group has only two points with nontrivial stabilizer v_1, v_2 in $\mathbb{P}^1$ (the North and South poles); the stabilizer of each of them is the whole group. An analog of the barycentric subdivision for the cyclic group is given by partitioning the sphere into a union of n digons, with vertices at v_1, v_2 and with spherical angles $2\pi/n$. Such a spherical tessellation is sometimes called a hosohedron; an example is shown in Figure 8.2. Such a tessellation can be colored using two colors iff n is even.

138   8. McKay Correspondence and Representations of Euclidean Quivers

Figure 8.2. The spherical tessellation for the cyclic group $G = \mathbb{Z}_{10}$.

Note that each such digon can also be viewed as a spherical triangle with angles

(8.5) $$\alpha = \frac{2\pi}{n}, \quad \beta = \frac{2\pi}{n}, \quad \gamma = \pi.$$

Let A be the generator of $\mathbb{Z}_n$ chosen so that $\pi(A)$ is the counterclockwise rotation by angle $4\pi/n$ around v_1, and let $B = A^{-1}$ (thus, $\pi(B)$ is the counterclockwise rotation by angle $4\pi/n$ around v_2). Obviously, A, B generate G.

Further study shows that this can be used to give a presentation of the group G. Namely, we have the following theorem.

Theorem 8.6. *Let G be a nontrivial finite subgroup in* SU(2). *Let $p, q, r \in \mathbb{Z}_+$ be defined as follows*:

- *If G is a binary polyhedral group corresponding to a regular spherical polyhedron X, then $\{p, q\}$ is the Schläfli symbol of X and $r = 2$.*
- *If $G = \mathbb{Z}_n$, then p, q are arbitrary positive integers such that $p + q = n$ and $r = 1$.*

Then:

(1) *Group G is isomorphic to the group generated by three elements A, B, C with defining relations*

(8.6) $$A^p = B^q = C^r = ABC.$$

8.2. ADE classification of finite subgroups

(2) *The elements*

$$A^i, \quad 1 \leq i \leq p-1,$$
$$B^i, \quad 1 < i < q-1,$$
$$C^i, \quad 1 \leq i \leq r-1,$$
$$A^p = B^q = C^r$$

form a set of representatives of nonidentity conjugacy classes in G.

The proof of this theorem in the case of binary polyhedral groups can be found in [**Cox1974**, Section 11.7]; the proof for cyclic groups is left to the reader as a simple exercise.

In particular, it follows from this theorem that for $r = 2$ (i.e. in the case of binary polyhedral groups) relations (8.6) imply that the element $Z = ABC$ satisfies $Z^2 = 1$. However, this is not easy to prove.

Lemma 8.7. *In the assumptions of Theorem 8.6, we have*

$$(8.7) \qquad \frac{1}{p} + \frac{1}{q} + \frac{1}{r} > 1.$$

Proof. In the case of a cyclic group, $r = 1$, so the statement is obvious.

For a binary polyhedral group, consider a spherical triangle of the barycentric subdivision of X. By (8.4), it has angles π/p, π/q, π/r. Using the famous formula

$$S = \alpha + \beta + \gamma - \pi$$

for the area of the spherical triangle with angles α, β, γ on the sphere of unit radius, we see that

$$\frac{1}{p} + \frac{1}{q} + \frac{1}{r} - 1 = \frac{S}{\pi} > 0.$$

$\square$

Remark 8.8. Since for a binary polyhedral group, the fundamental domain for the $\overline{G}$ action on the sphere is the union of two spherical triangles, the area of each triangle is given by $S = 4\pi/(2|\overline{G}|) = 4\pi/|G|$, so

$$\frac{1}{p} + \frac{1}{q} + \frac{1}{r} - 1 = \frac{S}{\pi} = \frac{4}{|G|}.$$

Lemma 8.7 allows us to make a connection with the ADE classification. Namely, recall that in Section 1.6 we introduced the notation $\Gamma(p, q, r)$ for the "star" diagram consisting of three branches of lengths p, q, r meeting at the branch point (length includes the branch point). By Theorem 1.30, such a diagram is a Dynkin graph if and only if inequality (8.7) holds. Thus, every

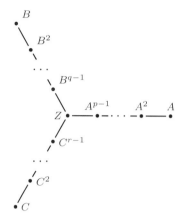

Figure 8.3. Representatives of conjugacy classes for $G \subset \mathrm{SU}(2)$. Here $Z = ABC = A^p = B^q = C^r$.

nontrivial finite subgroup $G \subset \mathrm{SU}(2)$ defines a Dynkin graph. Moreover, labeling vertices of this graph by the representatives of conjugacy classes described in Theorem 8.6 as shown in Figure 8.3, we see that vertices of the Dynkin graph Γ are in natural bijection with the nonidentity conjugacy classes in G. Comparing the classification of finite subgroups in $\mathrm{SU}(2)$ given in Theorem 8.2 with the classification of Dynkin graphs from Theorem 1.28, we get the following result.

Theorem 8.9. *We have a bijection between nontrivial finite subgroups $G \subset \mathrm{SU}(2)$ up to conjugation and Dynkin graphs. For every finite subgroup G, the vertices of the corresponding Dynkin graph are in bijection with nonidentity conjugacy classes in G.*

Table 8.2 lists all finite subgroups in $\mathrm{SU}(2)$ and the corresponding Dynkin graphs.

Table 8.2. Nontrivial finite subgroups in $\mathrm{SU}(2)$ and corresponding Dynkin graphs.

G	(p,q,r)	Γ
$\mathbb{Z}_n$, $n \geq 2$	$(p,q,1)$, $p+q=n$	A_{n-1}
binary dihedral group BD_{4n}, $n \geq 2$	$(n,2,2)$	D_{n+2}
binary tetrahedral	$(3,3,2)$	E_6
binary octahedral	$(4,3,2)$	E_7
binary icosahedral	$(5,3,2)$	E_8

For future use, we also note the following result.

Theorem 8.10. *Let $G \subset \mathrm{SU}(2)$ be a finite subgroup, $\overline{G} = \pi(G) \subset \mathrm{SO}(3)$. Then the quotient $\mathbb{P}^1/\overline{G}$ is isomorphic to $\mathbb{P}^1$: we have a holomorphic branched covering map*
$$\mathbb{P}^1 \to \mathbb{P}^1$$
whose fibers are the $\overline{G}$-orbits. It is branched at at most three points (which can be taken to be $0, 1, \infty$); the corresponding fibers are exactly the orbits of $\overline{G}$ action which have nontrivial stabilizer.

Proof. For a cyclic group, the proof is obvious. Let us consider the case when G is a binary polyhedral group, corresponding to the regular spherical polyhedron X. Consider the barycentric subdivision of $\mathbb{P}^1$ used in Theorem 8.4. Choosing one of the spherical triangles, we can holomorphically map it to the upper half-sphere in $\mathbb{P}^1$ so that the three vertices are mapped to points $0, 1, \infty$ respectively. Using the reflection principle, we can extend this map to an adjacent triangle, which will be mapped to the lower half-sphere. It is easy to see that the three vertices of the new triangle will again be mapped to $0, 1, \infty$. Continuing in this way, we get a holomorphic map $\mathbb{P}^1 \to \mathbb{P}^1$ which is G-invariant and branched over $0, 1, \infty$. $\square$

Remark 8.11. Covering maps $\Sigma \to \mathbb{P}^1$, where Σ is a Riemann surface, and which are branched over $0, 1, \infty$, are called Belyi maps. A deep theorem of Belyi says that such a map exists for any compact Riemann surface Σ defined over the field $\overline{\mathbb{Q}}$ of algebraic numbers. For more information about this result and related topics (such as Grothendieck's theory of *dessins d'enfants* and relation with the absolute Galois group $Gal(\overline{\mathbb{Q}}/\mathbb{Q}))$, see [**Sch1994**].

8.3. McKay correspondence

In this section, we give another way to relate finite subgroups in $\mathrm{SU}(2)$ and the ADE classification — this time, of Euclidean graphs. This correspondence was first observed by John McKay in [**McK1980**]; in his honor, it is usually called the McKay correspondence. We will describe how this relates with the construction of the previous section later (see Theorem 12.19).

Let G be a finite subgroup in $\mathrm{SU}(2)$. Let
$$I = \mathrm{Irr}(G)$$
be the set of isomorphism classes of irreducible complex representations of G; for every $i \in I$, we will denote by ρ_i the corresponding irreducible representation. In particular, we will use index 0 for the trivial representation: $\rho_0 = \mathbb{C}$. We denote by $K(G)$ the Grothendieck group of the category $\mathrm{Rep}(G)$; it is a free abelian group with basis $[\rho_i]$. It has a natural structure

of a commutative ring, given by tensor product of representations. It also has a natural bilinear form given by

$$([X],[Y])_0 = \dim \operatorname{Hom}_G(X,Y).$$

This form is symmetric and we have

$$([\rho_i],[\rho_j])_0 = \delta_{ij}.$$

Let ρ be the 2-dimensional representation given by the inclusion $G \subset \operatorname{SU}(2)$ (note that ρ is not necessarily irreducible). It is easy to see that $\rho^* \simeq \rho$.

We denote by A the operator $K(G) \to K(G)$ given by $x \mapsto [\rho] \cdot x$. In the basis ρ_i, it is given by a matrix A_{ij}:

$$(8.8) \qquad [\rho \otimes \rho_i] = \sum A_{ij}[\rho_j], \qquad A_{ij} = \dim \operatorname{Hom}_G(\rho_i, \rho \otimes \rho_j).$$

Note that $\rho^* \simeq \rho$ implies that A is symmetric with respect to $(\ ,\ )_0$: $A_{ij} = A_{ji}$.

Lemma 8.12. *Define the bilinear form on $K(G) \otimes_{\mathbb{Z}} \mathbb{R}$ by*

$$(8.9) \qquad\qquad (x,y) = (x,(2-A)y)_0$$

(thus, $(\rho_i, \rho_j) = 2\delta_{ij} - A_{ij}$). Then this form is symmetric and positive semidefinite, and the vector

$$(8.10) \qquad\qquad \delta = \sum \dim(\rho_i)[\rho_i] \in K(G)$$

is in the radical of this form.

Proof. First, note that for any representation V we have

$$(8.11) \qquad\qquad [V] \cdot \delta = \dim(V)\delta.$$

Indeed, for any ρ_i we have $([\rho_i], \delta)_0 = \delta_i$, where we denoted $\delta_i = \dim \rho_i$. Thus, for any W we have $([W], \delta)_0 = \dim(W)$. Applying it to $W = V^* \otimes \rho_i$ gives $([\rho_i], [V] \cdot \delta)_0 = \delta_i \dim(V)$, so $[V] \cdot \delta = \dim(V) \sum \delta_i[\rho_i] = \dim(V)\delta$.

Consider now the matrix A_V of multiplication by $[V]$ in $K(G)$. This matrix has nonnegative integer coefficients. By the Frobenius–Perron theorem (see [**Gan1998**, Chapter XIII, §2]), this matrix has a unique real eigenvalue $\lambda_{FP}(A_V)$ such that the corresponding eigenvector has real positive entries, and for any other eigenvalue we have $|\lambda| < \lambda_{FP}$. Comparing it with (8.11), we see that the Frobenius–Perron eigenvalue of A_V is equal to $\dim(V)$, so the matrix $\dim(V)I - A_V$ has nonnegative eigenvalues.

Applying this to $V = \rho$, we see that $2 - A$ is a symmetric nonnegative matrix, and that $(2 - A)\delta = 0$. $\square$

8.3. McKay correspondence

Theorem 8.13. *Let G be a nontrivial finite subgroup in $\mathrm{SU}(2)$. Let $Q(G)$ be the finite graph with set of vertices $I = \mathrm{Irr}(G)$ and with the number of edges connecting vertices i, j given by*

$$A_{ij} = \dim \mathrm{Hom}_G(\rho_i, \rho_j \otimes \rho). \tag{8.12}$$

Then $Q(G)$ is a connected Euclidean graph; the trivial representation ρ_0 corresponds to the extending vertex of Q, and the class of the regular representation in $K(G)$ is exactly the generator δ of the radical of the Euler form of Q.

Proof. For the so-defined $Q = Q(G)$, the root lattice is given by $L(Q) = K(G)$ and the symmetrized Euler form $(\ ,\ )_Q$ is exactly the form (8.9); thus, by Lemma 8.12, this form is positive semidefinite but not strictly positive.

To prove that Q is connected, note that for any $i, j \in I$, the number of paths of length n in Q connecting i with j is equal to multiplicity of ρ_i in $\rho_j \otimes \rho^{\otimes n}$. Thus, connectedness follows from the following lemma.

Lemma 8.14. *Every irreducible representation of G is a subrepresentation of $\rho^{\otimes n}$ for large enough n.*

Indeed, consider the space $S^n \rho$ of polynomials of degree n. Let $v \in \mathbb{C}^2$ be such that its stabilizer in G is trivial; then the orbit Gv is isomorphic to G, and the space $\mathbb{C}[Gv]$ of all functions on Gv is the regular representation of G. On the other hand, since polynomials separate points, for large enough n the restriction map $S^n \rho \to \mathbb{C}[Gv]$ is surjective; thus, for such n, the space $S^n \rho$ (and hence the space $\rho^{\otimes n}$) contains every irreducible representation with nonzero multiplicity. $\square$

We can now state McKay's theorem.

Theorem 8.15. *The construction of Theorem 8.13 gives a bijection between nontrivial finite subgroups $G \subset \mathrm{SU}(2)$ considered up to conjugation and connected Euclidean graphs other than the Jordan graph up to isomorphism.*

This correspondence is usually extended by letting the trivial subgroup $G = \{1\}$ correspond to the Jordan graph.

The most straightforward (but not very illuminating) proof of this theorem can be obtained by comparing the well-known classification of finite subgroups in $\mathrm{SU}(2)$ (see Theorem 8.2) with the list of Euclidean graphs in Section 1.6. This proof can be found in the original paper [**McK1980**].

Example 8.16. Let $G = \mathbb{Z}_n$ which is embedded in SU(2) by
$$k \mapsto \begin{bmatrix} \zeta^k & 0 \\ 0 & \zeta^{-k} \end{bmatrix}, \qquad \zeta = e^{2\pi i/n}$$
(see Example 8.1).

Then it has n irreducible representations ρ_i, $i \in \mathbb{Z}_n$, each of which is one-dimensional. The tensor product is given by $\rho_i \otimes \rho_j = \rho_{i+j}$, and $\rho \simeq \rho_1 \oplus \rho_{-1}$, so the matrix A is given by $A[\rho_i] = [\rho_{i+1}] + [\rho_{i-1}]$. Thus, the corresponding graph Q is the cyclic graph $\widehat{A}_{n-1}$ with n vertices.

The full list of finite subgroups and corresponding Euclidean graphs is given in Table 8.3.

Table 8.3. McKay correspondence. The table lists finite subgroups in SU(2) and the corresponding Euclidean graphs as defined in Theorem 8.13. J stands for the Jordan graph (1.1).

G	Q	$|I|$
$\{1\}$	J	1
$\mathbb{Z}_n$, $n \geq 2$	$\widehat{A}_{n-1}$	n
binary dihedral group BD_{4n}, $n \geq 2$	$\widehat{D}_{n+2}$	$n+3$
binary tetrahedral	$\widehat{E}_6$	7
binary octahedral	$\widehat{E}_7$	8
binary icosahedral	$\widehat{E}_8$	9

Note that since connected Euclidean graphs other than the Jordan graph are exactly the Dynkin diagrams of (untwisted) simply-laced affine Lie algebras (see [**Kac1990**]), the McKay theorem also provides a bijection between nontrivial finite subgroups $G \subset \mathrm{SU}(2)$ and simply-laced affine Lie algebras. Under this correspondence, the Cartan matrix of the affine Lie algebra is given by $C = 2I - A$, where A is defined by (8.8).

Comparing the McKay correspondence with the correspondence between finite subgroups and Dynkin graphs given in Theorem 8.9, we see that in all cases, the Dynkin graph Γ is the one obtained from the Euclidean graph $Q(G)$ by removing the extending vertex. Note however that the two constructions have very different natures: in the ADE classification of Theorem 8.9, branches of the diagram correspond to orbits in $\mathbb{P}^1$ with nontrivial stabilizer, and vertices correspond to nonidentity conjugacy classes of G; in the McKay correspondence, vertices correspond to irreducible representations of G. The relation between the two constructions will be clarified in Chapter 12 (see Theorem 12.19).

8.3. McKay correspondence

For future use, note that the McKay correspondence also gives a natural interpretation of the preprojective algebra of $Q(G)$ (over $\mathbb{C}$) in terms of G. As before, let

$$A_{ij} = \dim \operatorname{Hom}_G(\rho_i, \rho_j \otimes \rho)$$
$$= (\text{number of edges between } i, j \text{ in } Q)$$
$$= (\text{number of edges } i \to j \text{ in } Q^\sharp),$$

where $Q^\sharp$ is the double quiver of Q; see Definition 5.1.

Let us refine this. Namely, let

$$H_{ij} = \operatorname{Hom}_G(\rho_i, \rho_j \otimes \rho)$$

so that $\dim H_{ij} = A_{ij}$. Let us choose for each ordered pair (i,j) a basis $\{\varphi_h\}$ in H_{ij} indexed by edges $h \colon i \to j$ in $Q^\sharp$.

Then each path $p = h_l \ldots h_1$ from i to j in $Q^\sharp$ defines a G-morphism

$$(8.13) \qquad \varphi_p = \varphi_{h_l} \ldots \varphi_{h_1} \colon \rho_i \to \rho_j \otimes \rho^{\otimes l}.$$

Lemma 8.17. *Equation (8.13) defines an isomorphism*

$$(8.14) \qquad {}_j A_i^l \simeq \operatorname{Hom}_G(\rho_i, \rho_j \otimes \rho^{\otimes l}),$$

where $A = \mathbb{C} Q^\sharp$ and ${}_j A_i^l$ is the span of paths of length l from i to j.

We can now give a natural interpretation of the preprojective algebra. Namely, note that $\Lambda^2 \rho \simeq \mathbb{C}$ as a representation of $\operatorname{SU}(2)$ and thus as a representation of G. Let us fix a choice of such an isomorphism. This gives a bilinear pairing

$$\omega_{ij} \colon H_{ij} \otimes H_{ji} \to \mathbb{C}$$

defined by

$$\omega(\varphi, \psi) = \operatorname{tr}_{\rho_j}(\rho_j \xrightarrow{\psi} \rho_i \otimes \rho \xrightarrow{\varphi} \rho_j \otimes \rho^{\otimes 2} \to \rho_j \otimes \Lambda^2 \rho \simeq \rho_j).$$

It is easy to see from the definition that ω is skew-symmetric and nondegenerate.

Let us now assume that the bases φ_h were chosen so that $\omega(\varphi_h, \varphi_{h'}) = 0$ unless $h' = \bar{h}$. (If $A_{ij} = 1$, which is the most common case, this condition holds automatically.) In this case, $\omega(\varphi_h, \varphi_{\bar{h}}) \neq 0$, and we get a skew-symmetric function ε on the set of edges of $Q^\sharp$ defined by

$$\varepsilon(h) = \omega(\varphi_h, \varphi_{\bar{h}})^{-1}$$

which in turn defines a preprojective algebra Π.

Lemma 8.18. *The isomorphism (8.14) descends to an isomorphism*

$$_j\Pi_i^l \simeq \operatorname{Hom}_G(\rho_i, \rho_j \otimes S^l\rho),$$

where Π is the preprojective algebra of Q and $_j\Pi_i^l \subset \Pi$ is the span of paths of length l from i to j.

Indeed, under isomorphism (8.14) the elements θ_i exactly correspond to morphisms

$$\rho_i \to \rho_i \otimes \Lambda^2 \rho \subset \rho_i \otimes \rho^{\otimes 2}$$

and, by definition, $S^l\rho = \rho^{\otimes l}/J$, J is spanned by $\rho^{\otimes m} \otimes \Lambda^2\rho \otimes \rho^{\otimes n}$, $m + n = l - 2$.

8.4. Geometric construction of representations of Euclidean quivers

In this section, we give a geometric construction of representations of Euclidean quivers, using finite subgroups in SU(2). The results of this section are due to [**GL1987**] and, in a different language, to [**Kir2006**]. Our exposition follows [**Kir2006**]. The main result is given in Theorem 8.24, which establishes an equivalence between derived categories of representations of a Euclidean quiver and categories of equivariant sheaves on $\mathbb{P}^1$.

In this section, we will be using the language of derived categories. We assume that the reader is familiar with the basic notions of the theory of derived categories; a good reference is [**GM2003**].

Throughout this section, G is a nontrivial finite subgroup in SU(2) and $Q = Q(G)$ is the corresponding Euclidean graph as defined in Theorem 8.15. For simplicity, we assume that $G \supset \{\pm I\}$, so that $G = \pi^{-1}(\overline{G})$, for some subgroup $\overline{G} \subset \mathrm{SO}(3)$. This excludes the case $G = \mathbb{Z}_n$, n odd; see Corollary 8.3. This case could be included as well, but requires minor modifications.

Element $-I$ gives a $\mathbb{Z}_2$ grading of the category of representations of G:

(8.15)
$$\operatorname{Rep}(G) = \operatorname{Rep}_0(G) \oplus \operatorname{Rep}_1(G),$$
$$\operatorname{Rep}_i(G) = \{X \in \operatorname{Rep}(G) \mid (-I)|_X = (-1)^i\}.$$

We will write $p(X) = i$ if $X \in \operatorname{Rep}_i(G)$ and call $p(X)$ the parity of X. Note that the fundamental representation ρ has parity 1.

Let $\operatorname{Coh}_{\overline{G}}(\mathbb{P}^1)$ be the category of $\overline{G}$-equivariant coherent sheaves on $\mathbb{P}^1$. We refer the reader to [**CG1997**, Chapter 5] for basic facts about equivariant sheaves. Equivalently, this category can be described in terms of the group G: for any G-equivariant sheaf $\mathcal{F}$, the G-equivariant structure gives an isomorphism $(-I)^*: \mathcal{F} \to \mathcal{F}$. Then

$$\operatorname{Coh}_{\overline{G}}(\mathbb{P}^1) = \{\mathcal{F} \in \operatorname{Coh}_G(\mathbb{P}^1) \mid (-I)^*|_{\mathcal{F}} = \operatorname{id}\}.$$

8.4. Geometric construction of representations of Euclidean quivers

Example 8.19. For any $n \in \mathbb{Z}$, the sheaf $\mathcal{O}(n)$ has a canonical structure of the G-equivariant sheaf, and $(-I)^*|_{\mathcal{O}(n)} = (-1)^n$. Thus, $\mathcal{O}(n)$ is $\bar{G}$-equivariant iff n is even.

More generally, let X be a representation of G. Then

$$X(n) = X \otimes \mathcal{O}(n) \tag{8.16}$$

is a G-equivariant locally free sheaf; it is $\bar{G}$-equivariant iff $p(X) + n \equiv 0$ mod 2.

We list here some basic properties of the category $\mathrm{Coh}_{\bar{G}}(\mathbb{P}^1)$.

Theorem 8.20. *Let $\mathcal{C} = \mathrm{Coh}_{\bar{G}}(\mathbb{P}^1)$. Then:*

(1) *$\mathcal{C}$ is hereditary: for any $\mathcal{F}, \mathcal{G} \in \mathcal{C}$, we have $\mathrm{Ext}^i(\mathcal{F}, \mathcal{G}) = 0$ for $i > 1$.*

(2) *Serre duality: if $\mathcal{F}, \mathcal{G}$ are locally free, then we have an isomorphism*

$$\mathrm{Ext}^1_\mathcal{C}(\mathcal{F}, \mathcal{G}(-2)) = \mathrm{Ext}^1_\mathcal{C}(\mathcal{F}(2), \mathcal{G}) \simeq \mathrm{Hom}_\mathcal{C}(\mathcal{G}, \mathcal{F})^*.$$

(3) *For any locally free sheaf $\mathcal{F} \in \mathcal{C}$, we have a short exact sequence*

$$0 \to \mathcal{F} \to \rho \otimes \mathcal{F}(1) \to \Lambda^2 \rho \otimes \mathcal{F}(2) \simeq \mathcal{F}(2) \to 0, \tag{8.17}$$

where $\rho = \Gamma(\mathcal{O}(1)) \simeq \mathbb{C}^2$ is the standard two-dimensional representation of G. (Note that $\rho \simeq \rho^*$.)

(4) *Every G-equivariant coherent sheaf admits a resolution which consists of locally free G-equivariant sheaves. Every locally free G-equivariant sheaf is a direct sum of sheaves of the form (8.16).*

The proof of this theorem is straightforward. The interested reader can find details in [**Kir2006**].

We can now relate this category with the category of representations of Euclidean quivers. Let $Q = Q(G)$ be the Euclidean graph which corresponds to G under McKay correspondence as defined in Theorem 8.15. The set of vertices of this graph is the set $I = \mathrm{Irr}(G)$ of isomorphism classes of irreducible representations of G. The $\mathbb{Z}_2$ grading on $\mathrm{Rep}(G)$ given by (8.15) defines a bipartite structure on Q: $I = I_0 \sqcup I_1$.

Recall that in Section 6.3 we defined the translation quiver $\mathbb{Z}Q \subset Q \times \mathbb{Z}$. Let us assign to every vertex $q = (i, n)$ of $\mathbb{Z}Q$ a locally free $\bar{G}$-equivariant sheaf on $\mathbb{P}^1$ by

$$X_q = \rho_i \otimes \mathcal{O}(n), \quad q = (i, n), \ i \in I, \ n \in \mathbb{Z}, \tag{8.18}$$

where ρ_i is the irreducible representation of G corresponding to $i \in I$. Note that then

$$X_{\tau q} = X_{(i, n-2)} = X_q(-2).$$

Since the edges of Q correspond to morphisms $\rho_i \to \rho_j \otimes \rho$, we get, for every edge $h\colon i \to j$ in Q, a morphism

$$x_h \colon X_{(i,n)} = \rho_i \otimes \mathcal{O}(n) \to \rho_j \otimes \rho \otimes \mathcal{O}(n) \to \rho_j \otimes \mathcal{O}(n+1) = X_{(j,n+1)},$$

where the morphism $\rho \otimes \mathcal{O}(n) \to \mathcal{O}(n+1)$ is constructed using the isomorphism $\rho \simeq \Gamma(\mathcal{O}(1))$.

Thus, the collection of sheaves $X_q, q \in \mathbb{Z}Q$, forms a representation of $\mathbb{Z}Q$ (in $\mathcal{C}$).

Moreover, part (3) of Theorem 8.20 implies that for any $q = (i,n) \in \mathbb{Z}Q$, we have the following short exact sequence in $\mathcal{C}$:

$$(8.19) \qquad 0 \to X_{(i,n)} \to \bigoplus X_{(j,n+1)} \to X_{(i,n+2)} \to 0,$$

where the sum is over all edges $i \to j$ in Q. (Compare with Corollary 6.31, where a short exact sequence was constructed for preprojective representations. The relation between these two constructions will be described in Corollary 8.31.)

Let us now choose a slice $T \subset \mathbb{Z}Q$ (see Definition 6.12). Recall that every choice of a slice T determines an orientation of Q, giving a quiver $\vec{Q}_T$.

Lemma 8.21. *Let $T \subset \mathbb{Z}Q$ be a slice. Denote $\mathcal{C} = \operatorname{Coh}_{\overline{G}}(\mathbb{P}^1)$ and let $D_{\overline{G}}^b(\mathbb{P}^1)$ be the corresponding derived category: $D_{\overline{G}}^b(\mathbb{P}^1) = D^b(\mathcal{C})$.*

(1) *Sheaves $X_q, q \in T$, generate $D_{\overline{G}}^b(\mathbb{P}^1)$ as a triangulated category: the smallest triangulated subcategory in $D_{\overline{G}}^b(\mathbb{P}^1)$ containing all X_q is $D_{\overline{G}}^b(\mathbb{P}^1)$.*

(2) *If $q \in T$, $p \prec T$, then $\operatorname{Hom}_{\mathcal{C}}(X_q, X_p) = 0$. Similarly, if $p \succ T$, then $\operatorname{Ext}^1_{\mathcal{C}}(X_q, X_p) = 0$.*

(3) *If $p, q \in T$, then*

$$\operatorname{Hom}_{\mathcal{C}}(X_q, X_p) = \langle \text{paths in } T \text{ from } q \text{ to } p \rangle,$$
$$\operatorname{Ext}^1_{\mathcal{C}}(X_q, X_p) = 0.$$

Proof. To prove the first statement, recall that by Theorem 8.20, every coherent equivariant sheaf admits a resolution by locally free sheaves. Thus, it is easy to see that sheaves $X_q, q \in \mathbb{Z}Q$, generate $D_{\overline{G}}^b(\mathbb{P}^1)$. On the other hand, it follows from the short exact sequence (8.19) that if i is a sink for T, then the subcategory generated by $X_q, q \in T$, also contains all $X_q, q \in s_i^+ T$; similarly, if i is a source for T, then this subcategory also contains all $X_q, q \in s_i^- T$. Thus, this subcategory contains all $X_q, q \in \mathbb{Z}Q$.

8.4. Geometric construction of representations of Euclidean quivers

To compute homomorphisms between X_q, X_p, note that

$$\operatorname{Hom}_{\mathcal{C}}(X_{(i,n)}, X_{(j,m)}) = \operatorname{Hom}_G(\rho_i, \rho_j \otimes \Gamma(\mathcal{O}(m-n)))$$
$$= \operatorname{Hom}_G(\rho_i, \rho_j \otimes S^{m-n}\rho) = {}_j\Pi_i^{m-n},$$

where Π is the preprojective algebra of Q (see Lemma 8.18). Since for $p \prec T$, there are no paths in $\mathbb{Z}Q$ from q to p, which implies that ${}_p\Pi_q = 0$, we see that in this case there are no homomorphisms.

Since for $p, q \in T$, we have ${}_p\Pi_q = \langle \text{paths in } T \text{ from } q \text{ to } p \rangle$ (see Theorem 6.28), we get the statement of the second part.

The statement about Ext functors is easily obtained by using Serre duality, which gives $\operatorname{Ext}^1(X_q, X_p) = \operatorname{Hom}(X_p, X_q(-2))^* = \operatorname{Hom}(X_p, X_{\tau q})^*$. □

Definition 8.22. Let $T = \{(i, h_i)\} \subset \mathbb{Z}Q$ be a slice. We define the functor

$$\Psi_T \colon \operatorname{Coh}_{\bar{G}}(\mathbb{P}^1) \to \operatorname{Rep}(\vec{Q}_T)$$

by

$$\Psi_T(\mathcal{F}) = \bigoplus_{i \in I} \operatorname{Hom}_{\mathcal{C}}(X_{(i,h_i)}, \mathcal{F})$$

and the maps corresponding to edges of $\vec{Q}_T$ are given by

$$\operatorname{Hom}_{\mathcal{C}}(X_{(i,h_i)}, \mathcal{F}) \to \operatorname{Hom}_{\mathcal{C}}(X_{(j,h_j)}, \mathcal{F}), \quad h_i = h_j + 1,$$
$$f \mapsto f \circ x_{\tilde{e}},$$

where e is an edge between i and j in Q (and thus an edge $i \to j$ in $\vec{Q}_T$), $\tilde{e} \colon (j, h_j) \to (i, h_j + 1)$ is the corresponding edge in $\mathbb{Z}Q$.

Example 8.23. Let $\mathcal{F} = X_p$, $p = (i, h_i) \in T$. Then it follows from Lemma 8.21 that

$$\Psi_T(X_p) = P(i)$$

is the projective representation of $\vec{Q}_T$ defined by (1.9).

Theorem 8.24. Let $G \subset \operatorname{SU}(2)$ be a finite subgroup containing $-I$. Let Q be the corresponding Euclidean graph as in Theorem 8.15, and let $T \subset \mathbb{Z}Q$ be a slice.

(1) The functor $\Psi_T \colon \operatorname{Coh}_{\bar{G}}(\mathbb{P}^1) \to \operatorname{Rep}(\vec{Q}_T)$ is left exact.
(2) The derived functor

$$R\Psi_T \colon D^b_{\bar{G}}(\mathbb{P}^1) \to D^b(\vec{Q}_T)$$

is an equivalence of triangulated categories.

(3) Let T, T' be obtained from each by an elementary sink to source transformation: $T' = s_i^+(T)$ (see Definition 6.14). Then the following diagram is commutative:

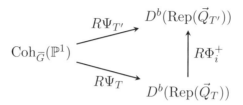

where $R\Phi_i^+$ is the derived reflection functor (see Theorem 3.12).

(4) $R\Psi_T$ identifies the derived Coxeter functor $RC^+ \colon D^b(\vec{Q}_T) \to D^b(\vec{Q}_T)$ with the twist functor
$$\mathcal{F} \mapsto \mathcal{F}(-2)$$
on $D_{\overline{G}}^b(\mathbb{P}^1)$.

Proof. The proof is based on the following general result of homological algebra.

Lemma 8.25. *Let $\mathcal{C}$ be an abelian category over a field $\mathbf{k}$ of finite homological dimension and such that every object has finite length and all spaces $\mathrm{Ext}^i(A, B)$ are finite-dimensional. Let $X = \bigoplus X_i \in \mathcal{C}$ be an object satisfying the following properties:*

(1) *X_i generate $D^b(\mathcal{C})$ as a triangulated category.*
(2) *$\mathrm{Ext}^i(X, X) = 0$ for all $i > 0$.*
(3) *The algebra $A = \mathrm{End}_\mathcal{C}(X)$ has finite homological dimension.*

Define the functor
$$\mathrm{Hom}_\mathcal{C}(X, -) \colon \mathcal{C} \to A^{op}\text{-mod},$$
where $A = \mathrm{End}_\mathcal{C}(X)$. Then this functor is left exact, and the corresponding derived functor
$$\mathrm{RHom}(X, -) \colon D^b(\mathcal{C}) \to D^b(A^{op}\text{-mod})$$
is an equivalence of triangulated categories.

This lemma is based on Beĭlinson's lemma in [**Beĭ1978**]; the formulation above can be found in [**GL1987**, Theorem 3.2].

By Lemma 8.21, the sheaf $X_T = \bigoplus_{q \in T} X_q$ satisfies the conditions of Lemma 8.25, and $\mathrm{End}_\mathcal{C}(X_T) = \mathbb{C}T = (\mathbb{C}\vec{Q}_T)^{opp}$. This proves the first two parts of the theorem.

8.4. Geometric construction of representations of Euclidean quivers

To prove the third part, we use the relation (8.19), which implies that if $q = (i, n) \in T$, i is a sink for $\vec{Q}_T$, then we have an exact triangle

$$\mathrm{RHom}(X_{(i,n+2)}, Y) \to \mathrm{RHom}(X', Y) \to \mathrm{RHom}(X_{(i,n)}, Y)$$
$$\to \mathrm{RHom}(X_{(i,n+2)}, Y)[1],$$

where $X' = \bigoplus X_{(j,n+1)}$, with sum over all edges $i \to j$ in Q. This is equivalent to

$$\mathrm{RHom}(X_{(i,n+2)}, Y) = Cone\Big(\mathrm{RHom}(X', Y) \to \mathrm{RHom}(X_{(i,n)}, Y)\Big)[-1].$$

But this is exactly the definition of the reflection functor in the derived category: $R\Psi_{T'}(Y) = R\Phi^+ R\Psi_T(Y)$. □

Note however that Ψ_T is not an equivalence of abelian categories: it only becomes an equivalence after we pass to derived categories.

Corollary 8.26. *Let $K_{\bar{G}}(\mathbb{P}^1)$ be the K-group of the category $\mathrm{Coh}_{\bar{G}}(\mathbb{P}^1)$ or, equivalently, of the category $D^b_{\bar{G}}(\mathbb{P}^1)$. Then a choice of a slice $T \in \mathbb{Z}Q$ gives an isomorphism $\psi_T \colon K_{\bar{G}}(\mathbb{P}^1) \to L$, where L is the root lattice of Q. This isomorphism has the following properties*:

(1) $\psi_T(\mathcal{F}(-2)) = C\psi_T(\mathcal{F})$, where C is the Coxeter element in W, adapted to the orientation $\vec{Q}_T$.
(2) *If $q = (i, h_i) \in T$, then $\psi_T(X_q) = [P(i)]$.*
(3) *We have $\langle \delta, \psi(\mathcal{F}) \rangle = \mathrm{rk}\, \mathcal{F}$, where rk is the rank of sheaf $\mathcal{F}$.*

Theorem 8.24 gives us a way to construct representations of $\vec{Q}_T$ geometrically. In particular, in the case when Q is a tree, any orientation of Q can be obtained from some slice T; thus, it gives us a way to construct representations of any Euclidean quiver which is a tree.

As an immediate corollary, we can easily get a description of all indecomposable representations, recovering results of Theorem 7.41. To do so, we need to classify indecomposable equivariant sheaves on $\mathbb{P}^1$.

Recall that a coherent sheaf $\mathcal{F}$ on a variety X is called a *torsion* sheaf if its stalk at a generic point is zero. For example, if X is defined over $\mathbb{C}$, then for any $x \in X$ we have the skyscraper sheaf $\mathbb{C}_x$ whose stalk at x is $\mathbb{C}$ and at all other points is zero. As a module over the structure sheaf $\mathcal{O}$, it can be defined as $\mathbb{C}_x = \mathcal{O}_X/m_x$, where m_x is the ideal sheaf consisting of functions vanishing at x. More generally, for any $x \in X$, $n \geq 1$, we can define the sheaf

(8.20) $$\mathbb{C}_{x,n} = \mathcal{O}_X/m_x^n.$$

If x is a nonsingular point on a curve and t is a local coordinate at x, then the stalk of $\mathbb{C}_{x,n}$ at x is isomorphic to $\mathbb{C}[t]/t^n$.

152   8. McKay Correspondence and Representations of Euclidean Quivers

These sheaves have equivariant analogs.

Lemma 8.27. *Let $x \in \mathbb{P}^1$ and let $\bar{G}_x \subset \bar{G}$ be the stabilizer of x. Let Y be a finite-dimensional representation of $\bar{G}_x$. Then for any $n \geq 1$, there is a unique $\bar{G}$-equivariant sheaf $Y_{\bar{G}x,n}$ with the following properties:*

(1) *The support of $Y_{\bar{G}x,n}$ is the $\bar{G}$-orbit of x.*

(2) *The stalk of $Y_{\bar{G}x,n}$ at x is $Y \otimes \mathcal{O}/m_x^n$ (as a representation of $\bar{G}_x$).*

For $n = 1$, we will use the shorter notation $Y_{\bar{G}x,1} = Y_{\bar{G}x}$.

The proof is left to the reader as an exercise.

Exercise 8.28.

(1) Show that for $n = 1$, $\Gamma(Y_{\bar{G}x})$ considered as a representation of $\bar{G}$ is the induced representation $\operatorname{Ind}_{\bar{G}_x}^{\bar{G}}(Y)$. In particular, if $x \in \mathbb{P}^1$ has a trivial stabilizer, then $\Gamma(\mathbb{C}_{\bar{G}x})$ is the regular representation of $\bar{G}$.

(2) Show that if $\bar{G}_x = \mathbb{Z}_p$, then $Y_{\bar{G}x,n} \otimes \mathcal{O}(-2) \simeq (Y \otimes \sigma)_{\bar{G}x,n}$, where σ is the one-dimensional representation of $\mathbb{Z}_p$ given by $k \mapsto \zeta^k$, $\zeta = e^{2\pi i/p}$. In particular, this implies that $Y_{\bar{G}x,n} \otimes \mathcal{O}(-2p) \simeq Y_{\bar{G}x,n}$.

The following theorem, the proof of which can be found in [**Kir2006**], gives a classification of all indecomposable objects in $D_{\bar{G}}^b(\mathbb{P}^1)$.

Theorem 8.29.

(1) *The following is a full list of nonzero indecomposable objects in $\operatorname{Coh}_{\bar{G}}(\mathbb{P}^1)$:*
   - *Locally free sheaves $\rho_i \otimes \mathcal{O}(n)$, $i \in \operatorname{Irr}(G)$, $n \in \mathbb{Z}$, $p(i) + n \equiv 0$ mod 2.*
   - *Torsion sheaves $\mathbb{C}_{\bar{G}x,n}$, where $n > 0$, $x \in \mathbb{P}^1$ is generic (i.e. has trivial stabilizer in $\bar{G}$).*
   - *Torsion sheaves $Y_{\bar{G}x,n}$, where $n > 0$, $x \in \mathbb{P}^1$ has nontrivial stabilizer $\bar{G}_x$ in $\bar{G}$, and Y is an irreducible representation of $\bar{G}_x$. (The pair (x,Y) is considered up to the action of $\bar{G}$.)*

(2) *Indecomposable objects of $D_{\bar{G}}^b(\mathbb{P}^1)$ are of the form $X[k]$, where X is an indecomposable object of $\operatorname{Coh}_{\bar{G}}(\mathbb{P}^1)$, $k \in \mathbb{Z}$.*

Combining this with Theorem 8.24, we see that indecomposable objects in $\operatorname{Rep}\vec{Q}$ must be of the form $R\Psi_T(X)[n]$, where X is an indecomposable object in $\mathcal{C}$ and n is chosen so that $R\Psi_T(X)[n] \in \operatorname{Rep}\vec{Q}$. By Lemma 8.21, for $p \succ T$, we have $R^1\Psi_T(X_p) = 0$, so $R\Psi_T(X_p) = \Psi(X_p) \in \operatorname{Rep}\vec{Q}$. Similarly, if $p \prec T$, then $\Psi_T(X_p) = 0$, and $R\Psi_T(X_p) = R^1\Psi_T(X_p)[-1] \in \operatorname{Rep}\vec{Q}[-1]$, so $R\Psi_T(X_p)[1] \in \operatorname{Rep}\vec{Q}$.

8.4. Geometric construction of representations of Euclidean quivers 153

For an indecomposable torsion sheaf X, it is easy to check using Serre duality that $\operatorname{Ext}^1(\mathcal{F}, X) = 0$ for any locally free sheaf $\mathcal{F}$, so $R^1\Psi_T(X) = 0$. Thus, $R\Psi_T(\mathcal{F}) = \Psi(\mathcal{F}) \in \operatorname{Rep}\vec{Q}$. Therefore, we get the following result.

Theorem 8.30. *In the assumptions of Theorem 8.24, the following is a full list of indecomposable objects in $\operatorname{Rep}\vec{Q}_T$:*

- $\Psi_T(\rho_i \otimes \mathcal{O}(n))$, $(i, n) \succcurlyeq T$. *These objects are preprojective.*
- $R^1\Psi_T(\rho_i \otimes \mathcal{O}(n)) = R\Psi_T(\rho_i \otimes \mathcal{O}(n))[1]$, $(i, n) \prec T$. *These objects are preinjective.*
- $\Psi_T(X)$, *where X is an indecomposable $\overline{G}$-equivariant torsion sheaf on $\mathbb{P}^1$ described in Theorem 8.29. These objects are regular.*

Corollary 8.31. *Let T, Ψ_T be as defined in Theorem 8.24.*

For $q \succcurlyeq T$, we have
$$\Psi_T(X_q) = I_q^T,$$
$$R^1\Psi_T(X_q) = 0,$$

where I_q^T is the indecomposable representation of $\vec{Q}_T$ constructed in Theorem 6.19.

Proof. Denote temporarily $\Psi_T(X_q) = \tilde{I}_q^T$. Then it follows from Example 8.23 and Theorem 8.24 that these representations have the following properties:

(1) If $q \prec T$, then $\tilde{I}_q^T = 0$.
(2) If $q = (i, n) \in T$ is a sink for $\vec{Q}_T$, then $\tilde{I}_q^T = S(i)$.
(3) If $T' = s_i^+ T$ and $q \succcurlyeq T'$, then $\tilde{I}_q^T = \Phi_i^- \tilde{I}_q^{T'}$, $\tilde{I}_q^{T'} = \Phi_i^+ \tilde{I}_q^T$.

These are exactly the defining properties of representations I_q^T (see Theorem 6.19). □

Exercise 8.32. In the assumptions of Theorem 8.24, prove that if $q = (i, h_i) \in T$ and $q' = \tau q = (i, h_i - 2)$, then $R^1\Psi_T(X_{q'}) = Q(i)$ is the indecomposable injective representation.

Theorem 8.30 gives a geometric interpretation of the construction of indecomposable regular representations of $\vec{Q}$ given in Theorem 7.41: the sheafs $X_i^{(n)}$ are exactly the images under Ψ_T of indecomposable $\overline{G}$-equivariant torsion sheaves on $\mathbb{P}^1$. They are parametrized by triples (x, n, Y), where $x \in \mathbb{P}^1/\overline{G} \simeq \mathbb{P}^1$ (see Theorem 8.10), $n \geq 1$, and $Y \in \operatorname{Irr}(\overline{G}_x) \simeq \mathbb{Z}_l$. The number l is the period of X, which is the same as the order of the stabilizer $\overline{G}_x$. The set D corresponding to tubes with period $l > 1$ (see Corollary 7.38) is exactly the set of $\overline{G}$-orbits with nontrivial stabilizer; by results of Section 8.2, this set has at most three elements.

Example 8.33. Let $G = \{\pm I\}$, $\overline{G} = \{1\}$. Then Theorem 8.24 shows that one has an equivalence of derived categories

$$D^b(\operatorname{Coh} \mathbb{P}^1) \simeq D^b(\operatorname{Rep} K),$$

where K is the Kronecker quiver:

$$K = \underset{1}{\bullet} \rightrightarrows \underset{0}{\bullet}$$

This is a special case of a result of Beĭlinson [**Beĭ1978**]. Explicitly, this equivalence can be described as follows.

For the Kronecker quiver, the corresponding translation quiver is shown in Figure 8.4.

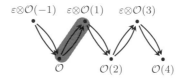

Figure 8.4. The translation quiver for the Kronecker graph. Next to each vertex q, we put the corresponding locally free sheaf X_q on $\mathbb{P}^1$ as defined by (8.18). ε stands for the sign representation of $\mathbb{Z}_2$.

Let us choose the slice T so that $X_T = \mathcal{O} \oplus (\varepsilon \otimes \mathcal{O}(1))$. Then the corresponding functor Ψ_T is given by

$$\Psi_T(\mathcal{F}) = \underset{1}{\overset{\Gamma(\mathcal{F}(-1))}{\bullet}} \rightrightarrows \underset{0}{\overset{\Gamma(\mathcal{F})}{\bullet}},$$

where the maps are given by multiplication by $z_1, z_2 \in \Gamma(\mathcal{O}(1))$.

Indecomposable objects of $\operatorname{Coh} \mathbb{P}^1$ are $\varepsilon^k \otimes \mathcal{O}(k)$, $k \in \mathbb{Z}$ (where ε is the sign representation of $\mathbb{Z}_2$), and the torsion sheaves $\mathbb{C}_{x,n}$, $x \in \mathbb{P}^1$.

The indecomposable representations of the Kronecker quiver are:

- Preprojective:

$$\Psi_T(\varepsilon^k \otimes \mathcal{O}(k)) = (S^{k-1}\rho, S^k\rho; z_{1,2}) = \underset{1}{\overset{\mathbb{C}^k}{\bullet}} \rightrightarrows \underset{0}{\overset{\mathbb{C}^{k+1}}{\bullet}} \qquad (k \geq 0).$$

- Preinjective:

$$R^1\Psi_T(\varepsilon^k \otimes \mathcal{O}(k)) = (\operatorname{Ext}^1(\mathcal{O}(k-1)), \operatorname{Ext}^1(\mathcal{O}(k))), \qquad k < 0.$$

Writing $k = -l - 1$, $l \geq 0$, and using Serre duality $\operatorname{Ext}^1(\mathcal{O}(n)) = \Gamma(\mathcal{O}(-n-2))^* \simeq \mathbb{C}^{-n-1}$, $n < 0$, we get

$$R^1\Psi_T(\varepsilon^k \otimes \mathcal{O}(k)) = \underset{1}{\overset{\mathbb{C}^{l+1}}{\bullet}} \rightrightarrows \underset{0}{\overset{\mathbb{C}^l}{\bullet}}, \qquad k = -l-1, l \geq 0.$$

- Regular: $\Psi_T(\mathbb{C}_{x,n})$.

8.4. Geometric construction of representations of Euclidean quivers   155

This matches the description of Theorem 7.32: namely, $\Psi_T(\varepsilon^k \otimes \mathcal{O}(k))$, $k \geq 0$, is exactly the preprojective representation of dimension $(k, k+1)$, and $R^1\Psi(\varepsilon^k \otimes \mathcal{O}(k))$, $k = -l - 1 < 0$, is the preinjective representation of dimension $(l+1, l)$.

For torsion sheaves, $\Psi_T(\mathbb{C}_{x,n}) = S_x^{(n)}$ is the indecomposable regular representation described in (7.11).

Part 3

Quiver Varieties

Chapter 9

Hamiltonian Reduction and Geometric Invariant Theory

In this chapter, we give an overview of several topics in differential and algebraic geometry which will be used in the subsequent chapters. Namely, we review the geometric invariant theory, symplectic geometry, Hamiltonian reduction, and hyperkähler quotients. Most of the results in this chapter are well known and we omit the proofs, referring the reader to the appropriate textbooks.

Throughout this chapter, $\mathbf{k}$ is an algebraically closed field of characteristic zero. In some sections we will restrict ourselves to the case $\mathbf{k} = \mathbb{C}$.

9.1. Quotient spaces in differential geometry

We begin with several well-known results from differential geometry (see, e.g., [**DK2000**]).

Let G be a real Lie group acting on a C^∞ manifold M. Recall that the action is called *proper* if the map

$$M \times G \to M \times M$$
$$(m, g) \mapsto (m, g.m)$$

is proper. This is equivalent to the requirement that for all compact subsets $K_1, K_2 \subset M$, the set

$$\{g \in G \mid gK_1 \cap K_2 \neq \varnothing\} \subset G$$

is compact. In particular, if G is compact, then its action is automatically proper.

Note that if the action is proper, then every G-orbit is closed.

An example of a nonproper action is the "irrational winding on the torus", i.e. the action of $\mathbb{R}$ on the two-dimensional torus $\mathbb{R}^2/\mathbb{Z}^2$ given by $t.(x,y) = (x+t, y+\alpha t)$, where $\alpha \in \mathbb{R}$ is irrational.

For proper group actions, one has a good local model.

Theorem 9.1. *Let M be a C^∞ manifold with a proper action of a Lie group G. For a point $x \in M$, let G_x be its stabilizer (which is necessarily a closed Lie subgroup in G) and let $\mathbb{O}_x$ be the orbit of x.*

Then there exists a (locally closed) submanifold S (slice) containing x, invariant under action of G_x, and a G-invariant open neighborhood $U \supset \mathbb{O}_x$ such that the map
$$G \times_{G_x} S \to U$$
$$(g, s) \mapsto gs$$
is an isomorphism.

In particular, this implies that $\mathbb{O}_x$ is a smooth closed submanifold in M, of dimension $\dim G - \dim G_x$.

This theorem is called the *slice theorem*. A proof can be found in [**DK2000**, Theorem 2.4.1].

As a corollary, we get the following result for free actions.

Theorem 9.2. *Let M be a C^∞ manifold with a free proper action of a Lie group G. Then the quotient space M/G is a manifold, and the projection $M \to M/G$ is a principal G-bundle.*

9.2. Overview of geometric invariant theory

In this and the next section, we give an overview of geometric invariant theory (GIT), which gives a way of constructing quotients by group actions in algebraic geometry. The standard reference for all of the results of these sections is Mumford's book [**MFK1994**]; however, reading it requires good knowledge of algebraic geometry. Easier to read references include books by Mukai [**Muk2003**] and Newstead [**New1978**], or a review paper [**New2009**]. For the reader's convenience, we list here the main results and outline some proofs.

Throughout this section, all algebraic varieties are defined over an algebraically closed field **k** of characteristic zero and are considered with Zariski topology.

9.2. Overview of geometric invariant theory

Let G be a linear algebraic group acting algebraically on an affine algebraic variety M. Recall that in this case, each orbit is a nonsingular (not necessarily closed) subvariety in M, and for every orbit $\mathbb{O}$, the boundary $\overline{\mathbb{O}} - \mathbb{O}$ is a union of orbits of lower dimension (see, for example, [**OV1990**, Section 3.1.5]).

Assume now additionally that G is reductive. Recall that an algebraic group G is called *reductive* if its radical is a torus; in this case, the Lie algebra $\mathfrak{g}$ of G is a direct sum of semisimple and commutative Lie algebras. In characteristic zero, an algebraic group is reductive if and only if every finite-dimensional algebraic representation of G is completely reducible. In particular, groups $\mathrm{GL}(n, \mathbf{k})$, $\mathrm{SL}(n, \mathbf{k})$ and their direct products are reductive.

Consider now an algebraic action of a reductive group G on an affine variety M. In this situation, we define the GIT quotient of M by the action of G by

$$(9.1) \qquad M /\!/ G = \mathrm{Spec}(\mathbf{k}[M]^G),$$

where $\mathbf{k}[M]^G$ is the algebra of G-invariant polynomial functions on M and Spec is the set of its maximal ideals. By a theorem due to Hilbert (see [**Muk2003**, Theorem 4.51]), this algebra is finitely generated, so $M /\!/ G$ is an affine algebraic variety.

One easily sees that every G-orbit $\mathbb{O} \subset M$ defines a maximal ideal $J_{\mathbb{O}} = \{f \mid f|_{\mathbb{O}} = 0\} \subset \mathbf{k}[M]^G$, so we have a natural map

$$(9.2) \qquad M/G \to M /\!/ G.$$

Note that M/G is just a topological space (usually non-Hausdorff), while $M /\!/ G$ is an affine algebraic variety.

Theorem 9.3. *If G is reductive and M is affine, then the map (9.2) is surjective. Moreover, two orbits $\mathbb{O}, \mathbb{O}' \subset M$ define the same point in $M /\!/ G$ iff*

$$\overline{\mathbb{O}} \cap \overline{\mathbb{O}'} \neq \varnothing$$

(here $\overline{\mathbb{O}}$ stands for the closure in Zariski topology).

The proof can be found in [**Muk2003**, Theorem 5.9].

Corollary 9.4. *As a topological space, $M /\!/ G = M/\sim$, where the equivalence relation is defined by*

$$(9.3) \qquad x \sim x' \text{ iff } \overline{\mathbb{O}}_x \cap \overline{\mathbb{O}}_{x'} \neq \varnothing,$$

where $\mathbb{O}_x$ is the G-orbit of x.

For an element $x \in M$, we will denote by $[x]$ the corresponding point in $M /\!/ G$.

In fact, the quotient $M/\!/G$ admits an even more explicit description.

Theorem 9.5. *In the assumptions of Theorem 9.3, we have an isomorphism*
$$M/\!/G = \{\text{closed orbits in } M\}$$
$$[x] \mapsto \text{unique closed orbit contained in } \overline{\mathbb{O}}_x$$
as a topological space.

Proof. It suffices to prove that the closure of any orbit contains a unique closed orbit. Existence is immediate by induction in dimension (recall that $\overline{\mathbb{O}} - \mathbb{O}$ is a union of orbits of lower dimension). Uniqueness easily follows from the following fundamental fact, the proof of which can be found in [**MFK1994**, Corollary 1.2].

Theorem 9.6 (Geometric Reductivity Principle). *If $X, Y \subset M$ are closed and G-invariant and $X \cap Y = \varnothing$, then there exists a G-invariant polynomial f such that $f|_X = 0$, $f|_Y = 1$.* $\square$

Example 9.7. Let $M = \mathbb{A}^1$ be a one-dimensional vector space over $\mathbf{k}$, considered as an algebraic variety, and let $G = \mathbf{k}^\times$ be the group of nonzero elements in $\mathbf{k}$, acting on M by $\lambda(x) = \lambda \cdot x$. (The group $\mathbf{k}^\times$ has an obvious structure of an algebraic group; in the theory of algebraic groups, this group is commonly denoted by $\mathbb{G}_m$.) Then there are exactly two orbits, $\mathbb{O}_1 = \{0\}$ and $\mathbb{O}_2 = M - \{0\} = \mathbf{k}^\times$, so the set-theoretic quotient consists of two points. However, only one of these orbits is closed: $\overline{\mathbb{O}}_1 = \mathbb{O}_1$, $\overline{\mathbb{O}}_2 = \mathbb{O}_2 \cup \mathbb{O}_1$, so $\mathbb{A}^1/\!/\mathbf{k}^\times = \{pt\}$. Indeed, $\mathbf{k}[\mathbb{A}^1]^G = \mathbf{k}[x]^{\mathbf{k}^\times} = \mathbf{k}$.

Corollary 9.8. *Let G be a reductive algebraic group acting on an affine variety M so that every orbit is closed. Then $M/\!/G$ coincides with the set-theoretic quotient M/G.*

An important special case is when the action is free; in this case, all orbits have the same dimension, so each orbit must be closed. Moreover, we can formulate an analog of Theorem 9.2.

Recall that a morphism $f: X \to Y$ of algebraic varieties over $\mathbf{k}$ of characteristic zero is called *étale* if for every $x \in X$, the natural map
$$f^*: \widehat{\mathcal{O}}_{f(x)} \to \widehat{\mathcal{O}}_x$$
is an isomorphism. Here $\widehat{\mathcal{O}}_x$ is the completed local ring at x (for nonsingular points, it is isomorphic to the ring of formal power series). In particular, if X is a nonsingular variety, and f is étale and surjective, then Y is also nonsingular, and for $\mathbf{k} = \mathbb{C}$, f is a local isomorphism (in the analytic topology) of the corresponding complex manifolds.

9.3. Relative invariants

Theorem 9.9. *Let M be a nonsingular affine variety over $\mathbf{k}$ with a free action of a reductive algebraic group G. Then $M /\!\!/ G = M/G$ is nonsingular, of dimension $\dim M/G = \dim M - \dim G$, and the projection $M \to M /\!\!/ G$ is an étale G-bundle: for every point $x \in M/G$, one can choose a neighborhood $U \subset M/G$, $x \in U$, and an étale cover $W \to U$ for some nonsingular W such that the pullback $W \times_U M$ is isomorphic to $G \times W$.*

The proof of this theorem can be obtained by using the algebro-geometric analog of the slice theorem, called the *Luna slice theorem*. A review can be found in [**MFK1994**, Appendix to Chapter 1, Section D].

Corollary 9.10. *In the assumptions of Theorem 9.9, let $\mathbf{k} = \mathbb{C}$. Let M^{an}, $(M/G)^{an}$ denote M, M/G considered as complex manifolds with analytic topology. Then $M^{an} \to (M/G)^{an}$ is a principal G^{an}-bundle; if the action is proper, the C^∞ structure on $(M/G)^{an}$ coincides with the one given in Theorem 9.2.*

Remark 9.11. As one can see, in the algebro-geometric situation it is not necessary to require properness of the action; instead, one requires that the group is reductive. It should be noted, however, that not every action of a reductive algebraic group is proper.

It should also be noted that the quotient can be nonsingular even if the action is not free, as is shown by the following example.

Example 9.12. Let $M = \text{End}(\mathbf{k}^n)$ be the space of $n \times n$ matrices. Since every matrix has a basis in which it is upper triangular, it is easy to show that any $\text{GL}(n)$-invariant polynomial function on $\text{End}(\mathbf{k}^n)$ is completely determined by its values on diagonal matrices, so
$$\mathbf{k}[M]^{\text{GL}(n)} = \mathbf{k}[\lambda_1, \ldots, \lambda_n]^{S_n}$$
and
$$M /\!\!/ \text{GL}(n) = \text{Spec}(\mathbf{k}[\lambda_1, \ldots, \lambda_n]^{S_n}) = \mathbf{k}^n/S_n.$$
The fundamental theorem about symmetric polynomials shows that the algebra $\mathbf{k}[\lambda_1, \ldots, \lambda_n]^{S_n}$ is in fact a polynomial algebra in elementary symmetric polynomials; thus, in this case $M /\!\!/ \text{GL}(n) = \mathbf{k}^n/S_n \simeq \mathbf{k}^n$ is nonsingular.

9.3. Relative invariants

The GIT quotient $M /\!\!/ G$ constructed in the previous section loses some information about the action, as it only depends on the set of closed orbits. It turns out that this definition can be generalized to recover some information about nonclosed orbits as well. To do that, we need to switch from affine varieties to projective varieties.

Recall that for a projective variety $X \subset \mathbb{P}^r$, there is a graded algebra

(9.4) $$A = \bigoplus_{n \geq 0} A_n,$$

A_n = homogeneous polynomials of degree n restricted to X.

It is well known that one can recover X from this graded algebra, namely

$$X = \mathrm{Proj}(A),$$

where

(9.5) $$\mathrm{Proj}(A) = \text{graded ideals } J \subset A, \text{ maximal among graded ideals} \\ \text{not containing } A_+ = \bigoplus_{n > 0} A_n.$$

More generally, if $A = \bigoplus_{n \geq 0} A_n$ is a finitely generated graded algebra without nilpotent elements, then one can define the variety $X = \mathrm{Proj}(A)$ which is a quasi-projective variety. There is also a natural morphism

$$\begin{array}{c} X = \mathrm{Proj}(A) \\ \pi \downarrow \\ X_0 = \mathrm{Spec}(A_0) \end{array}$$

induced by the embedding of algebras $A_0 \subset A$: for any $f \in \mathbf{k}[X_0] = A_0$, $x \in X$, we have $f(\pi(x)) = \pi^* f(x)$, where π^* is the embedding $A_0 \to A$. This morphism is projective; in particular, for every $x \in X_0$, the fiber $\pi^{-1}(x)$ is a projective variety. Proofs of these facts can be found in standard textbooks on algebraic geometry, such as [**Har1977**, Section II.7].

Note that if $A_0 = \mathbf{k}$, so that $X_0 = \{pt\}$, then X is a projective variety.

Let us now return to the situation when G is a reductive algebraic group acting on an affine algebraic variety M. Let χ be a character of G, i.e. a morphism of algebraic groups

$$\chi \colon G \to \mathbf{k}^\times.$$

Define

(9.6) $$\mathbf{k}[M]^{G,\chi} = \{f \in \mathbf{k}[M] \mid f(g.m) = \chi(g)f(m)\}.$$

Elements of $\mathbf{k}[M]^{G,\chi}$ are called *semi-invariants*, or *relative invariants*.

It is immediate from the definition that

$$A = \bigoplus_{n \geq 0} \mathbf{k}[M]^{G,\chi^n}$$

9.3. Relative invariants

is a graded algebra; a modification of Hilbert's arguments shows that it is finitely generated. Thus, we can define the corresponding quasi-projective variety, which will be denoted by $M/\!\!/_\chi G$:

$$(9.7) \qquad M/\!\!/_\chi G = \mathrm{Proj}\Big(\bigoplus_{n\geq 0} \mathbf{k}[M]^{G,\chi^n}\Big).$$

Note that $A_0 = \mathbf{k}[M]^{G,1} = \mathbf{k}[M]^G$, so as a special case of the general theory, we get a projective morphism

$$(9.8) \qquad \pi\colon M/\!\!/_\chi G \to M/\!\!/ G.$$

This "twisted" GIT quotient $M/\!\!/_\chi G$ also admits a description in terms of equivalence classes of orbits, similar to the description of $M/\!\!/ G$ given in Theorems 9.3 and 9.5.

Definition 9.13. Let χ be a character of G. Lift the action of G on M to an action on $M \times \mathbf{k}$ by letting $g(m,z) = (g(m), \chi^{-1}(g)z)$.

A point $x \in M$ is called χ-*semistable* if for any $z \in \mathbf{k}, z \neq 0$, the closure of the orbit of $\hat{x} = (x,z)$ in $M \times \mathbf{k}$ is disjoint from the zero section:

$$\overline{\mathbb{O}}_{\hat{x}} \cap (M \times \{0\}) = \varnothing.$$

The set of χ-semistable points in M is denoted M^{ss}_χ.

Similarly, we say that an orbit $\mathbb{O}$ is χ-semistable if $\mathbb{O} \subset M^{ss}_\chi$.

When there is no ambiguity about the choice of character χ, we will use the word semistable instead of χ-semistable, and use the notation M^{ss} rather than M^{ss}_χ.

Theorem 9.14. *An element $x \in M$ is χ-semistable iff there exists $f \in \mathbf{k}[M]^{G,\chi^n}$, $n \geq 1$, such that $f(x) \neq 0$.*

Proof. If $x \in M^{ss}$, then by the geometric reductivity principle (Theorem 9.6) applied to the action of G in $M \times \mathbf{k}$, there exists a function $\hat{f} \in \mathbf{k}[M \times \mathbf{k}]^G$ such that $\hat{f}_{M \times \{0\}} = 0$, $\hat{f}(x,1) \neq 0$. But G-invariance of $\hat{f}$ implies that $\hat{f}(x,z) = \sum_{n\geq 0} f_n(x) z^n$, $f_n \in \mathbf{k}[M]^{G,\chi^n}$. Since $\hat{f}(m,0) = 0$, we must have $f_0 = 0$; thus, there exists $f_n, n \geq 1$, such that $f_n(x) \neq 0$.

The proof in the opposite direction is similar. $\square$

Corollary 9.15.

(1) $M^{ss} \subset M$ is open and G-invariant (possibly empty).
(2) If $N \in \mathbb{Z}$, $N \geq 1$, then $x \in M$ is χ^N-semistable iff it is χ-semistable.

This corollary implies that the notion of χ-semistability can be defined for any $\chi \in X_{\mathbb{Q}}(G) = X(G) \otimes_{\mathbb{Z}} \mathbb{Q}$, where $X(G)$ is the abelian group of characters of G.

Example 9.16. Let $\chi = 1$. Then $M^{ss} = M$.

Theorem 9.14 shows that every $x \in M^{ss}$ defines an ideal $J_x = \{f \mid f(x) = 0\} \subset A = \bigoplus_n \mathbf{k}[M]^{G,\chi^n}$ which is maximal and does not contain A_+ and thus defines a point in $\mathrm{Proj}(A) = M/\!\!/_\chi G$. Therefore, we have a natural map

$$(9.9) \qquad M^{ss}/G \to M/\!\!/_\chi G.$$

The following theorem is an analog of Theorem 9.3.

Theorem 9.17. *Let G be a reductive group acting on an affine variety M, and let χ be a character of G. Then the map (9.9) is surjective; moreover, two semistable orbits $\mathbb{O}, \mathbb{O}' \subset M^{ss}$ define the same point in $M/\!\!/_\chi G$ iff*

$$\overline{\mathbb{O}} \cap \overline{\mathbb{O}'} \cap M^{ss} \neq \varnothing$$

(here $\overline{\mathbb{O}}$ is the closure of orbit $\mathbb{O}$).

Corollary 9.18. *As a topological space, $M/\!\!/_\chi G = M^{ss}/\sim$, where the equivalence relation is defined by*

$$(9.10) \qquad x \sim x' \text{ iff } \overline{\mathbb{O}}_x \cap \overline{\mathbb{O}}_{x'} \cap M^{ss} \neq \varnothing.$$

As before, for $x \in M^{ss}$ we denote by $[x]$ the corresponding point in $M/\!\!/_\chi G$.

We refer the reader to [**Muk2003**, Section 6.1] and [**New2009**, Theorem 3.4] for proofs.

One also has an easy description of the projective map $M/\!\!/_\chi G \to M/\!\!/ G$ in terms of orbits. Recall that by Theorem 9.5, the closure of every orbit contains a unique closed orbit.

Theorem 9.19. *The map $\pi \colon M/\!\!/_\chi G \to M/\!\!/ G$ is given by*

$$(9.11) \qquad \pi([x]) = \text{the unique closed orbit in } M \text{ contained in } \overline{\mathbb{O}}_x,$$

where $x \in M^{ss}$ and $\overline{\mathbb{O}}_x$ is the closure of $\mathbb{O}_x$ in Zariski topology on M.

Proof. Let $\mathbb{O}_1$ be the unique closed orbit in the closure of $\mathbb{O}_x$. It suffices to check that for any $f \in \mathbf{k}[M]^G$, we have $f(\mathbb{O}_1) = \pi^* f(\mathbb{O})$, which is immediate from the definition. $\square$

As before, the twisted GIT quotient can also be described in terms of closed orbits.

9.3. Relative invariants

Theorem 9.20. *Under the assumptions of Theorem 9.17, we have*
$$M/\!/_\chi G = \{\text{closed orbits in } M^{ss}\}$$
as a topological space. (Note that we are talking about orbits that are closed in M^{ss}, which is not the same as being closed in M.)

The proof is again parallel to the proof of Theorem 9.5 and is left to the reader.

Finally, there is a special class of semistable elements we will need.

Theorem 9.21. *Under the assumptions of Theorem 9.17, let $x \in M^{ss}$ be such that the stabilizer G_x of x in G is finite. Then the following conditions are equivalent:*

(1) *The orbit $\mathbb{O}_x$ is closed in M^{ss}.*
(2) *For any nonzero $z \in \mathbf{k}$, the orbit of $\hat{x} = (x, z) \in M \times \mathbf{k}$ is closed in $M \times \mathbf{k}$.*
(3) *There exists $f \in \mathbf{k}[M]^{G,\chi^n}$, $n \geq 1$, such that $f(x) \neq 0$ and the action of G in $A_f = \{x \in M^{ss} | f(x) \neq 0\}$ is closed (i.e. all orbits are closed in A_f).*

We skip the proof of this theorem, referring the reader to [**MFK1994**, Amplification 1.11] and [**New2009**, Theorem 3.4].

Definition 9.22. Under the assumptions of Theorem 9.17, an element $x \in M^{ss}$ is called *stable* if G_x is finite and it satisfies one (and hence all) of the conditions of Theorem 9.21.

We denote by M^s the set of all stable elements in M; we will write M^s_χ when we need to stress dependence on χ.

(In [**MFK1994**], such elements are called *properly stable*; we follow conventions of King and Nakajima, who call them simply stable.)

Note that by definition, if $x, x' \in M^{ss}$ are stable, then $\overline{\mathbb{O}_x}$ intersects $\overline{\mathbb{O}_{x'}}$ in M^{ss} iff $\mathbb{O}_x = \mathbb{O}_{x'}$; thus, by Theorem 9.17, any two stable orbits define different points in $M/\!/_\chi G$, so we have

(9.12) $$M^s/G \subset M/\!/_\chi G.$$

Theorem 9.23. *In the assumptions of Theorem 9.17, assume additionally that M^s is nonempty. Then:*

(1) *M^s is open in M^{ss} (and thus in M).*
(2) *If M is irreducible, then M^s is dense in M^{ss} and M^s/G is dense in $M/\!/_\chi G$.*

(3) If M is nonsingular and, for every $x \in M^s$, the stabilizer G_x is trivial, then M^s/G is a nonsingular variety of dimension
$$\dim M^s/G = \dim M - \dim G.$$

Proof. The fact that M^s is open is immediate if we use the definition of M^s given in part (3) of Theorem 9.21. Since a nonempty open subset of an irreducible variety is itself irreducible and dense, irreducibility of M implies that any nonempty open subset of M^s is dense, and hence M^s/G is dense in M^{ss}/G and thus in $M/\!/_\chi G$.

The last part follows from Theorem 9.9. □

Note that in particular, this shows that if M is smooth and irreducible and M^s is nonempty, then $M^s/G \subset M/\!/_\chi G$ is open, dense, and nonsingular.

Example 9.24. Let $M = \mathbb{A}^2$ be a two-dimensional vector space over $\mathbf{k}$, and let $G = \mathbf{k}^\times$ acting on M by multiplication. Then $M/G = \{0\} \cup \mathbb{P}^1(\mathbf{k})$ and $M/\!/G = \{pt\}$. If we let $\chi(\lambda) = \lambda$, then $M^{ss} = \mathbb{A}^2 - \{0\} = M^s$ and $M/\!/_\chi G = \mathbb{P}^1(\mathbf{k})$. It is easy to see that the same holds for $\chi(\lambda) = \lambda^n$, $n > 0$. On the other hand, if $\chi(\lambda) = \lambda^n$, $n < 0$, then $M^{ss} = \varnothing$, so $M/\!/_\chi G = \varnothing$.

For future use, we also give here a stability criterion, due to Mumford. Let $\lambda \colon \mathbf{k}^\times \to G$ be a one-parameter subgroup of G. For any such subgroup and a character χ of G, define $\langle \lambda, \chi \rangle \in \mathbb{Z}$ by
$$\chi(\lambda(t)) = t^{\langle \lambda, \chi \rangle}.$$

Theorem 9.25. *Under the assumptions of Theorem 9.17, an element $x \in M$ is semistable (respectively, stable) iff for any one-parameter subgroup λ such that $\lim_{t \to 0} \lambda(t).x$ exists, we have $\langle \lambda, \chi \rangle \geq 0$ (respectively, $\langle \lambda, \chi \rangle > 0$ for any nontrivial one-parameter subgroup λ).*

The proof of this theorem can be found in [**Kin1994**, Proposition 2.5]. It is obtained by applying Lemma 2.9 to the action of G in $M \times \mathbf{k}^\times$.

9.4. Regular points and resolution of singularities

As in the previous section, let M be an affine algebraic variety over $\mathbf{k}$ with an algebraic action of a reductive algebraic group G. Throughout this section, we assume that M is nonsingular.

Definition 9.26. A point $x \in M$ is called *regular* if the orbit $\mathbb{O}_x$ is closed and the stabilizer G_x of x in G is trivial.

We denote the set of regular points in M by M^{reg}.

Remark 9.27. Our use of the word regular is different from the use in [**MFK1994**].

9.4. Regular points and resolution of singularities

Since the set of regular points is obviously G-invariant, we can define the subset $(M/\!/G)^{reg} \subset (M/\!/G)$; points of this subset are in bijection with regular orbits in M.

Lemma 9.28. *The set M^{reg} is open in M (possibly empty), and $(M/\!/G)^{reg}$ is open in $(M/\!/G)$.*

This lemma immediately follows from Luna's slice theorem.

Let us now consider twisted GIT quotients.

Theorem 9.29.

(1) *Let $x \in M$ be regular. Then for any character χ, x is χ-stable.*

(2) *Let $\pi\colon M/\!/_\chi G \to M/\!/G$ be the projective morphism (9.8). Define*
$$(M/\!/_\chi G)^{reg} = \pi^{-1}(M/\!/G)^{reg}.$$
Then $(M/\!/_\chi G)^{reg}$, $(M/\!/G)^{reg}$ are nonsingular, and the restriction

(9.13) $$\pi\colon (M/\!/_\chi G)^{reg} \to (M/\!/G)^{reg}$$

is an isomorphism.

Proof. Since the orbit of x in M is closed and the stabilizer is trivial, we see that the orbit of (x, z) in $M \times \mathbf{k}$ is closed; thus, x is χ-stable.

To prove that the morphism (9.13) is an isomorphism, we note that orbits of $x \in M^{reg}$ have maximal dimension, so an orbit $\mathbb{O}_x$ cannot be properly contained in a closure of another orbit. Thus, morphism (9.13) is a bijection. $\square$

Note that in general, it is not true that every nonsingular point in $M/\!/G$ is regular, as is shown by the example below (suggested to the author by Radu Laza).

Example 9.30. Consider an action of $\mathbf{k}^\times$ on $M = \mathbf{k}^2$ given by $t.(x_1, x_2) = (tx_1, t^{-1}x_2)$. In this case, one easily sees that $\mathbf{k}[M]^G = \mathbf{k}[x]$, where $x = x_1 x_2$. Thus, in this case $M/\!/G \simeq \mathbf{k}$ is nonsingular. On the other hand, $(M/\!/G)^{reg} = \mathbf{k} - \{0\}$.

This theorem can be used to construct resolutions of singularities. Recall that a morphism $\pi\colon X \to Y$ is called a *resolution of singularities* if X is nonsingular and π is proper and birational. The last condition is equivalent to the requirement that there is an open dense subset $Y_0 \subset Y$ such that $\pi^{-1}(Y_0)$ is dense in X and the restriction $\pi\colon \pi^{-1}(Y_0) \to Y_0$ is an isomorphism.

Corollary 9.31. *Assume that M^{reg} is nonempty and that $M/\!/_\chi G$ is nonsingular and connected. Then $\pi\colon M/\!/_\chi G \to M/\!/G$ is a resolution of singularities.*

Proof. By Theorem 9.29, the morphism $(M/\!/_\chi G)^{reg} \to (M/\!/G)^{reg}$ is an isomorphism. Since M^{reg} is open in M, we see that $(M/\!/_\chi G)^{reg}$ is open in $M/\!/_\chi G$ and nonempty. Since $M/\!/_\chi G$ is connected, $(M/\!/_\chi G)^{reg}$ must be dense in $M/\!/_\chi G$. $\square$

Remark 9.32. There are different versions of the definition of resolution of singularities. In addition to the definition considered above, many people use a stronger definition, in which it is additionally required that $\pi\colon \pi^{-1}(Y^{ns}) \to Y^{ns}$ is an isomorphism, where Y^{ns} is the subvariety of nonsingular points in Y.

Most resolutions of singularities constructed by the GIT method as in Corollary 9.31 are also resolutions in this stronger sense. However, it is not always true. For example, if we consider the action of $\mathbf{k}^\times$ on $M = \mathbf{k}^3$ given by $t.(x_1, x_2, x_3) = (tx_1, tx_2, t^{-1}x_3)$, then $M/\!/\mathbf{k}^\times = \mathbf{k}^2$ (compare with Example 9.30), but $M/\!/_\chi \mathbf{k}^\times$ is the blowup of $\mathbf{k}^2$ at the origin.

Example 9.33. Consider the variety

$$M = \{(i,j) \mid j\colon \mathbf{k} \to \mathbf{k}^2,\ i\colon \mathbf{k}^2 \to \mathbf{k}, ij = 0\},$$

$$\mathbf{k} \underset{i}{\overset{j}{\rightleftarrows}} \mathbf{k}^2$$

(a motivation for this definition will be given later, in Example 9.56).

This variety has a natural action of the group $\mathbf{k}^\times$: $\lambda.(i,j) = (\lambda i, \lambda^{-1}j)$. Let us study the corresponding GIT quotients.

It is obvious that matrix elements of the operator $A = ji\colon \mathbf{k}^2 \to \mathbf{k}^2$ are $\mathbf{k}^\times$-invariant and thus we have the morphism

(9.14)
$$M/\!/\mathbf{k}^\times \to \mathrm{Mat}_{2\times 2}(\mathbf{k})$$
$$(i,j) \mapsto ji.$$

We leave it to the reader to check that the image of this morphism is the variety

$$Q = \{A \in \mathrm{Mat}_{2\times 2}(\mathbf{k}) \mid \mathrm{tr}\, A = 0, \det A = 0\} = \left\{\begin{pmatrix} a & b \\ c & -a \end{pmatrix} \mid a^2 + bc = 0\right\}$$

and, moreover, that this morphism is an isomorphism. Thus, in this case $M/\!/\mathbf{k}^\times \simeq Q$ is a quadric in $\mathbf{k}^3$. It obviously has a singularity at the origin.

Points with trivial stabilizer in M are those where at least one of i, j is nonzero; such a point has a closed orbit iff both i, j are nonzero, which is equivalent to the condition $ji \ne 0$. Therefore, in this case $(M/\!/\mathbf{k}^\times)^{reg} = Q - \{0\}$; as one easily sees, this variety is nonsingular.

Now let $\chi\colon \mathbf{k}^\times \to \mathbf{k}^\times$ be the identity morphism. Then one easily sees that $(i,j) \in M$ is χ-semistable iff $j \neq 0$, and every semistable point is stable. Therefore,
$$M/\!\!/_\chi \mathbf{k}^\times = \{(V,i) \mid V \subset \mathbf{k}^2,\ \dim V = 1,\ i\colon \mathbf{k}^2 \to V, i|_V = 0\}.$$

This variety admits an alternative description. Namely, we have an obvious projection $M/\!\!/_\chi \mathbf{k}^\times \to \mathbb{P}^1(\mathbf{k})$, given by $(V,i) \mapsto V$, and the fibers of this projection are one-dimensional vector spaces, so $M/\!\!/_\chi \mathbf{k}^\times$ is a line bundle over $\mathbb{P}^1(\mathbf{k})$. We claim that this is the cotangent bundle: $M \simeq T^*\mathbb{P}^1(\mathbf{k})$. Indeed, for a one-dimensional subspace $V \subset \mathbf{k}^2$, a choice of nonzero vector $v \in V$ gives an isomorphism $T_V(\mathbb{P}^1) = \mathbf{k}^2/V$, and thus an identification of the cotangent space $T_V^*(\mathbb{P}^1) = \{i\colon \mathbf{k}^2 \to \mathbf{k} \mid i|_V = 0\}$; combining it with identification $\mathbf{k} \to V : z \mapsto zv$ gives $T_V^*(\mathbb{P}^1) = \{i\colon \mathbf{k}^2 \to V \mid i|_V = 0\}$, which is easily shown to be independent of the choice of v.

It is obvious that $T^*\mathbb{P}^1$ is nonsingular and connected; thus, by Corollary 9.31, we get a resolution of singularities

(9.15) $$T^*\mathbb{P}^1 \to Q.$$

Of course, in this case it is also easy to check directly that this is a resolution of singularities.

Example 9.33 is a simplest special case of resolutions which will be considered later: Springer resolution (see Section 9.8) and quiver varieties, which will be studied in detail in the next chapter (see Example 10.45).

9.5. Basic definitions of symplectic geometry

In this and the following sections, we review some basic notions of symplectic geometry and Hamiltonian reduction, both in C^∞ and algebro-geometric settings. These results are well known and, for the most part, we omit the proofs. We refer the reader to textbooks such as Arnol'd [**Arn1989**] and Abraham–Marsden [**AM1978**] (in the C^∞ case) or Chriss–Ginzburg [**CG1997**] (in the algebraic case) for details and proofs. We are using conventions of [**Arn1989**]; when comparing with results of other authors, sometimes you need to introduce a minus sign in the formulas.

Throughout this section, the word "manifold" means either a real C^∞ manifold (in which case, the ground field $\mathbf{k} = \mathbb{R}$), or a nonsingular algebraic variety over the ground field $\mathbf{k}$, which is always assumed to be algebraically closed and of characteristic zero. We denote by $\mathcal{O}_X$ the structure sheaf of the manifold X; in the C^∞ case, this is the sheaf of germs of C^∞ functions on X. We will write $f \in \mathcal{O}_X$ if f is a local section of $\mathcal{O}_X$.

Definition 9.34. A Poisson structure on a manifold X is a **k**-bilinear morphism
$$\{\,,\,\}\colon \mathcal{O}_X \times \mathcal{O}_X \to \mathcal{O}_X$$
(called the Poisson bracket) which satisfies the following properties:

(1) Skew-symmetry: $\{f,g\} = -\{g,f\}$.
(2) Jacobi identity: $\{f,\{g,h\}\} = \{\{f,g\},h\} + \{g,\{f,h\}\}$.
(3) Leibniz property: $\{f,gh\} = \{f,g\}h + g\{f,h\}$.

A manifold equipped with a Poisson structure is called a Poisson manifold. A morphism $\varphi\colon M \to N$ of Poisson manifolds is called a *Poisson morphism* if it commutes with the Poisson bracket:
$$\varphi^*\{f,g\} = \{\varphi^*f, \varphi^*g\}, \qquad f,g \in \mathcal{O}_N.$$

The first two properties of the Poisson bracket say that for any open subset $U \subset X$, the space of sections $\mathcal{O}(U)$ is a Lie algebra over **k**. The last condition means that $\{f,-\}$ is a derivation of the commutative algebra $\mathcal{O}(U)$.

Example 9.35. Let $\mathfrak{g}$ be a finite-dimensional Lie algebra over **k**. Then the dual space $\mathfrak{g}^*$, considered as a manifold over **k**, has a natural Poisson structure, which is uniquely determined by the condition that for any $x, y \in \mathfrak{g}$ (considered as functions on $\mathfrak{g}^*$), one has $\{x,y\} = [x,y]$.

A large class of Poisson manifolds is given by symplectic manifolds.

Definition 9.36. A symplectic manifold is a manifold M together with a 2-form $\omega \in \Omega^2(M)$ which is closed ($d\omega = 0$) and nondegenerate: for every $x \in M$, the pairing $T_x M \otimes T_x M \to \mathbf{k}$ given by ω is nondegenerate.

Nondegeneracy of the form shows that it gives an isomorphism

(9.16)
$$T_x M \to T_x^* M$$
$$\xi \mapsto \omega(-,\xi).$$

In particular, for every function $f \in \mathcal{O}_M$, there is a unique vector field X_f such that

(9.17) $$\omega(-, X_f) = df$$

or equivalently

(9.18) $$\omega(v, X_f) = \langle v, df \rangle = \partial_v f.$$

The vector field X_f is sometimes called the skew-gradient of f.

9.5. Basic definitions of symplectic geometry

The following lemma is well known.

Lemma 9.37. *Let M be a symplectic manifold. Define, for every $f, g \in \mathcal{O}_M$,*
$$\{f, g\} = \partial_{X_g}(f) = -\partial_{X_f}(g) = \omega(X_g, X_f).$$
Then the so-defined bracket is a Poisson bracket on M and $X_{\{f,g\}} = [X_f, X_g]$, where $[\,,\,]$ is the commutator of vector fields.

Note that we are using conventions of [**Arn1989**]: the commutator of vector fields X, Y is defined by the formula
$$\partial_{[X,Y]} = \partial_Y \partial_X - \partial_X \partial_Y$$
(note the order — it is not a misprint). This convention is motivated by the following fact. Let G be a real Lie group with the Lie algebra $\mathfrak{g}$. Then action of G on M defines, for every $a \in \mathfrak{g}$, a vector field ξ_a, and we have the relation $[\xi_a, \xi_b] = \xi_{[a,b]}$.

Thus, every symplectic manifold is a Poisson manifold. We will call the Poisson bracket nondegenerate if it comes from a symplectic form.

Example 9.38. Let G be a real Lie group with Lie algebra $\mathfrak{g}$. Then G acts naturally on $\mathfrak{g}$ and thus, on the dual space $\mathfrak{g}^*$. Then every G-orbit $M \subset \mathfrak{g}^*$ has a canonical symplectic structure (Kirillov–Kostant–Souriau form); the corresponding Poisson bracket coincides with the restriction to M of the bracket on $\mathfrak{g}^*$ defined in Example 9.35. Similar results hold in the algebraic situation if we replace a real Lie group by a linear algebraic group over **k**.

Another (and more common) class of examples of symplectic manifolds are the cotangent bundles. Let X be a manifold of dimension n and let T^*X be its cotangent bundle: $T^*X = \{(x, \lambda)\}$, $x \in X, \lambda \in T_x^*X$, which is a manifold of dimension $2n$. The cotangent bundle has a canonical 1-form α defined by

(9.19) $$\langle \alpha, v \rangle = \langle \lambda, \pi_* v \rangle, \qquad v \in T_{(x,\lambda)}(T^*X),$$

where $\pi \colon T^*X \to X$ is the canonical projection, so $\pi_* v \in T_x X$. It is easy to see that for any isomorphism $\varphi \colon X \to X$, one has $\tilde{\varphi}^* \alpha = \alpha$, where $\tilde{\varphi} \colon T^*X \to T^*X$ is the lifting of φ.

Theorem 9.39. *For any manifold X, the cotangent bundle T^*X has a canonical symplectic structure given by the 2-form $\omega = d\alpha$, where α is given by (9.19).*

It is well known that if q^i are local coordinates on X and p_i are corresponding coordinates on T_x^*X, so that the $2n$-tuple p_i, q^i are local coordinates on T^*X, then locally the symplectic form ω is given by
$$\omega = \sum dp_i \wedge dq^i$$

and the Poisson bracket is given by
$$\{f,g\} = \sum_i \left(\frac{\partial g}{\partial p_i}\frac{\partial f}{\partial q^i} - \frac{\partial f}{\partial p_i}\frac{\partial g}{\partial q^i}\right).$$

Example 9.40. Let V be a finite-dimensional real vector space. Then
$$T^*V = V \oplus V^*$$
and the symplectic form on T^*V is given by
$$\omega((v_1, \lambda_1); (v_2, \lambda_2)) = \langle \lambda_1, v_2 \rangle - \langle \lambda_2, v_1 \rangle, \quad v_i \in V, \ \lambda_i \in V^*.$$

Definition 9.41. Let V be a vector space with a nondegenerate skew-symmetric bilinear form ω. A subspace $L \subset V$ is called *isotropic* if $\omega(a,b) = 0$ for all $a, b \in L$; it is called Lagrangian if it is isotropic and has dimension $\frac{1}{2} \dim V$ (which is the maximal possible dimension for an isotropic subspace).

A subvariety $L \subset M$ of a symplectic manifold M is called isotropic (respectively, Lagrangian) if for any nonsingular point $x \in L$, the subspace $T_x L \subset T_x M$ is isotropic (respectively, Lagrangian).

Example 9.42. Let X be a manifold and let $Y \subset X$ be a submanifold. Let N^*Y be the conormal bundle to N: $N_x^*Y = (T_x X / T_x Y)^* \subset T_x^* X$.

Then $N^*Y \subset T^*X$ is a Lagrangian subvariety, which can be easily shown by a direct computation.

9.6. Hamiltonian actions and moment map

Let M be a symplectic manifold and let G be a group acting on it: either a real Lie group (in the C^∞ case) or a linear algebraic group (in the algebraic case). We denote by $\mathfrak{g}$ the Lie algebra of G. In this case, every element $a \in \mathfrak{g}$ defines a vector field ξ_a on M (sometimes called the infinitesimal action of $\mathfrak{g}$).

Assume that the action of G is symplectic, i.e. preserves the symplectic form ω. In this case, every vector field ξ_a, $a \in \mathfrak{g}$, also preserves the form: $L_{\xi_a}(\omega) = 0$, where L_ξ is the Lie derivative. Such vector fields are called *symplectic*. In the C^∞ case, it is easy to show that locally, every symplectic vector field is generated by some function f: $\xi = X_f$ (see (9.17)). Thus, locally we can choose for any $a \in \mathfrak{g}$ a function $H_a(x)$ (the Hamiltonian) such that
$$\xi_a = X_{H_a}.$$
It is easy to show that for any choice of H_a, there is a relation between the commutator in $\mathfrak{g}$ and the Poisson bracket on M:
$$\{H_a, H_b\} = H_{[a,b]} + c(a,b),$$
where $c(a,b)$ is some locally constant function.

9.6. Hamiltonian actions and moment map

Note that the Hamiltonians H_a are defined only up to addition of a locally constant function. Thus, it is reasonable to ask whether they can be chosen so that $\{H_a, H_b\} = H_{[a,b]}$. Moreover, one would like $H_a(x)$ to be linear in a. This motivates the following definition, which can be used in both the C^∞ and the algebraic setting.

Definition 9.43. A symplectic action of G on M is called Hamiltonian if there exists a map
$$\mu \colon M \to \mathfrak{g}^*$$
such that:

(1) For any $a \in \mathfrak{g}$, the function $H_a(x) = \langle \mu(x), a \rangle$ is the Hamiltonian for the vector field ξ_a:
$$\xi_a = X_{H_a}.$$

(2) For any $a, b \in \mathfrak{g}$,
$$\{H_a, H_b\} = H_{[a,b]}.$$

(3) μ is G-equivariant.

In such a situation, μ is called the *moment map*.

It can be shown that if G is connected, then equivariance of μ follows automatically. The moment map is not unique; however, whenever we talk of Hamiltonian actions, we assume that we have chosen a moment map.

For future use, we rewrite condition (1) as follows:

$$(9.20) \quad \partial_v \langle \mu(x), a \rangle = \partial_v H_a(x) = \omega(v, \xi_a), \qquad a \in \mathfrak{g}, v \text{ -- a vector field.}$$

Theorem 9.44. *Let M be a symplectic manifold with a Hamiltonian action of a group G, and let $\mu \colon M \to \mathfrak{g}^*$ be the moment map. Then μ is a Poisson morphism (where Poisson structure on $\mathfrak{g}^*$ is defined by Example 9.35).*

Below we give some examples of Hamiltonian actions which will be needed in the future.

Example 9.45. Let V be a symplectic vector space, i.e. a $2n$-dimensional vector with a nondegenerate skew-symmetric bilinear form $\omega \in \Lambda^2 V^*$. Since we have a canonical identification $T_x V = V$, this shows that V has a canonical structure of a symplectic manifold.

Assume now that in addition we have a linear action of the group G on V such that ω is G-invariant. Then this action is Hamiltonian: for any $a \in \mathfrak{g}$, the corresponding Hamiltonian is given by
$$H_a(x) = \tfrac{1}{2}\omega(x, a.x),$$
where $a.x$ is the linear action of $a \in \mathfrak{g}$ on V.

Indeed, since ω is G-invariant, we have $\omega(a.x, w) + \omega(x, a.w) = 0$, so for $x \in V, v \in T_x V = V$ we have

$$\partial_v H_a(x) = \tfrac{1}{2}(\omega(v, a.x) + \omega(x, a.v)) = \omega(v, a.x) = \omega(v, \xi_a(x))$$

since the vector field generated by a is given by $\xi_a(x) = a.x$. Therefore, $X_{H_a} = \xi_a$.

Thus, in this case the moment map is given by

$$\langle \mu(x), a \rangle = \tfrac{1}{2}\omega(x, a.x).$$

It is easy to see that it is G-equivariant.

Example 9.46. Let X be a manifold with an action of G. Then the corresponding action of G on T^*X is Hamiltonian, and the moment map is given by

$$\langle \mu(x, \lambda), a \rangle = \langle \lambda, \xi_a(x) \rangle, \qquad x \in X, \lambda \in T_x^* X, a \in \mathfrak{g}.$$

Indeed, in this case it is easy to see that the canonical one-form α on T^*X defined by (9.19) is G-invariant, so for any $a \in \mathfrak{g}$, $L_{\tilde{\xi}_a}(\alpha) = 0$, where $\tilde{\xi}_a$ is the infinitesimal action of $a \in \mathfrak{g}$ on T^*X and L_ξ is the Lie derivative. By Cartan's formula, we have $L_\xi = i_\xi d + d i_\xi$, so

$$L_{\tilde{\xi}_a}(\alpha) = 0 = \omega(\tilde{\xi}_a, -) + d(\langle \alpha, \tilde{\xi}_a \rangle).$$

Using the definition of α, we see that $\omega(-, \tilde{\xi}_a) = dH_a$, where $H_a \in \mathcal{O}(T^*M)$ is given by $H(x, \lambda) = \langle \alpha, \tilde{\xi}_a \rangle(x, \lambda) = \langle \lambda, \xi_a(x) \rangle$.

In particular, if $X = V$ is a vector space, so that $T^*V = V \oplus V^*$ (see Example 9.40) and $G = \mathrm{GL}(V)$, then

$$\langle \mu(v, \lambda), a \rangle = \langle \lambda, a.v \rangle, \qquad v \in V, \ \lambda \in V^*.$$

This agrees with the previous example, since

$$\langle \lambda, a.v \rangle = \tfrac{1}{2}(\langle \lambda, a.v \rangle - \langle a.\lambda, v \rangle) = \tfrac{1}{2}\omega((v, \lambda), (a.v, a.\lambda)).$$

Example 9.47. Let $X = G/P$, where G is a Lie group (respectively, a linear algebraic group) and P is a closed Lie subgroup. Then one has a canonical isomorphism

$$T_x X = \mathfrak{g}/\mathrm{Ad}_x .\mathfrak{p},$$

where $\mathfrak{p} = \mathrm{Lie}(P)$ and Ad_x stands for the adjoint action of G on $\mathfrak{g}$. Thus,

$$T^*X = \{(x, \lambda) \mid x \in X, \lambda \in \mathrm{Ad}_x^* .\mathfrak{p}^\perp\}, \qquad \mathfrak{p}^\perp = \{\lambda \in \mathfrak{g}^* \mid \lambda|_\mathfrak{p} = 0\},$$

where Ad_x^* is the coadjoint action of G on $\mathfrak{g}^*$.

Consider now the action of G on G/P by left multiplication. By results of Example 9.46, the corresponding action of G on the $T^*(G/P)$ action is

Hamiltonian, and the moment map is given by $\langle \mu(x,\lambda), a \rangle = \langle \lambda, \xi_a(x) \rangle = \langle \lambda, a \rangle$. Thus, in this case the moment map is exactly the canonical projection

$$\mu \colon T^*X \to \mathfrak{g}^*$$
$$(x, \lambda) \mapsto \lambda.$$
(9.21)

Example 9.48. Let V, W be vector spaces, and let $L = \operatorname{Hom}(V, W)$. Consider $M = T^*L = L \oplus L^*$. Let us identify $L^* \simeq \operatorname{Hom}(W, V)$ using the bilinear form $(u, v) = \operatorname{tr}(uv)$. Let $G = \operatorname{GL}(V)$ act on L in the obvious way: $g.u = ug^{-1}$, $u \in L$. Then the induced action of G on T^*L is Hamiltonian, and the moment map is given by

$$\mu \colon L \oplus L^* \to \mathfrak{gl}(V) \colon (u, v) \mapsto -vu.$$

Indeed, by Example 9.46, we have

$$\langle \mu(u, v), a \rangle = \operatorname{tr}(v \cdot (a.u)) = \operatorname{tr}(-vua).$$

Similarly, if we consider the adjoint action of $\operatorname{GL}(V)$ on $L = \mathfrak{gl}(V)$, then the action on $L \oplus L^*$ is Hamiltonian and is given by

$$\mu \colon L \oplus L^* \to \mathfrak{gl}(V) \colon (u, v) \mapsto [u, v]$$

(identifying $\mathfrak{gl}(V) \simeq \mathfrak{gl}(V)^*$ using the bilinear form $(u, v) = \operatorname{tr}(uv)$).

9.7. Hamiltonian reduction

Let us again consider a Hamiltonian action of a group G (algebraic or a Lie group) on a symplectic manifold M. Note that in this case, even if the quotient M/G is smooth, in general it has no symplectic structure. In this section, we will discuss a construction that allows one to produce symplectic quotients $M /\!/ G$.

Theorem 9.49. *Let M be a C^∞ symplectic manifold with a proper Hamiltonian action of a real Lie group G; let $\mu \colon M \to \mathfrak{g}^*$ be the corresponding moment map. Let $p \in \mathfrak{g}^*$ be such that the following conditions hold:*

(1) *p is a regular value of μ, so $\mu^{-1}(p) \subset M$ is a submanifold.*
(2) *The stabilizer $G_p \subset G$ of p acts freely on $\mu^{-1}(p)$, so that $\mu^{-1}(p)/G_p$ is a smooth manifold.*

Then $\mu^{-1}(p)/G_p = \mu^{-1}(\mathbb{O}_p)/G$ has a canonical structure of a symplectic manifold, inherited from M.

This theorem is due to Marsden and Weinstein; we refer the reader to [**AM1978**, Theorem 4.3.1] for a proof.

Quotient spaces $\mu^{-1}(p)/G_p$ are usually called *Marsden–Weinstein quotients*. It is also common to say that $\mu^{-1}(p)/G_p$ is obtained from M by Hamiltonian reduction.

Corollary 9.50. *Let M be a C^∞ symplectic manifold with a proper Hamiltonian action of a real Lie group G and let $\mu\colon M \to \mathfrak{g}^*$ be the corresponding moment map.*

If $p \in \mathfrak{g}^*$ is G-invariant and G acts freely on $\mu^{-1}(p)$, then $\mu^{-1}(p)/G$ has a canonical structure of a symplectic manifold, inherited from M. In particular, if G acts freely on M, then $\mu^{-1}(0)/G$ has a canonical structure of a symplectic manifold.

Indeed, in this case, for every $x \in \mu^{-1}(p)$ the map $\mathfrak{g} \to T_x M\colon a \mapsto \xi_a(x)$ is injective, which implies that p is a regular value.

We will be mostly interested in the special case when X is a C^∞ manifold with an action of a group G and $M = T^*X$. In this case, this action is Hamiltonian; the moment map was given in Example 9.46. Assume for simplicity that the action of G is free.

Theorem 9.51. *Let X be a C^∞ symplectic manifold with a free proper action of a Lie group G; let $\mu\colon T^*X \to \mathfrak{g}^*$ be the corresponding moment map. By Corollary 9.50, $\mu^{-1}(0)/G$ has a canonical structure of a symplectic manifold, inherited from T^*X. Then we have a symplectomorphism*

$$T^*(X/G) \simeq \mu^{-1}(0)/G.$$

A similar result can also be formulated in the algebro-geometric setting. Namely, let X be a nonsingular affine algebraic variety over $\mathbf{k}$, and let G be a reductive algebraic group acting on X. Then the cotangent bundle T^*X is also an affine algebraic variety, with an algebraic symplectic form and a Hamiltonian action of G. Thus, we have an (algebraic) moment map $\mu\colon T^*X \to \mathfrak{g}^*$, and the zero level of moment map $\mu^{-1}(0)$ is again an affine algebraic variety with an action of G.

Theorem 9.52. *Let X be a nonsingular affine algebraic variety over $\mathbf{k}$ with a free action of a reductive algebraic group G so that the quotient $X/G = X/\!/G$ is a nonsingular affine algebraic variety (see Corollary 9.8). Then:*

(1) *For any G-invariant $p \in \mathfrak{g}^*$, the space $\mu^{-1}(p)/G$ has a natural structure of a symplectic manifold.*

(2) *One has a canonical symplectomorphism*

$$T^*(X/G) \simeq \mu^{-1}(0)/G.$$

Let us study more carefully what happens when the action is not free, so that the quotient $\mu^{-1}(0)/G$ is not smooth. In the remainder of this section, we work in the algebraic setting, assuming that X is a nonsingular algebraic variety over $\mathbf{k}$, with an algebraic action of a reductive algebraic group G. Since in this case $\mu^{-1}(0)$ is again an affine algebraic variety, we can then

9.7. Hamiltonian reduction

define the GIT quotient

(9.22) $$\mathcal{M}_0 = (\mu^{-1}(0))/\!\!/G, \qquad \mu \colon T^*X \to \mathfrak{g}^*.$$

More generally, for any character $\chi \colon G \to \mathbf{k}^\times$, we can define the twisted GIT quotient

(9.23) $$\mathcal{M}_\chi = (\mu^{-1}(0))/\!\!/_\chi G$$

together with a projective map $\pi \colon \mathcal{M}_\chi \to \mathcal{M}_0$ (see Theorem 9.19).

Theorem 9.53. *Let X be a nonsingular affine algebraic variety over $\mathbf{k}$, and let G be a reductive algebraic group acting on X. Let $\mathcal{M}_\chi$ be defined by (9.23). Then:*

(1) *For any χ, $\mathcal{M}_\chi$ has a Poisson structure.*
(2) *The morphism $\pi \colon \mathcal{M}_\chi \to \mathcal{M}_0$ is Poisson.*
(3) *Let $X^s \subset X$ be the subvariety of χ-stable points, so that $X^s/\!\!/G \subset X/\!\!/_\chi G$ is a smooth subvariety; similarly, let $(\mu^{-1}(0))^s \subset \mu^{-1}(0)$ be the subvariety of χ-stable points, so that $\mathcal{M}_\chi^s = (\mu^{-1}(0))^s /\!\!/ G \subset \mathcal{M}_\chi$ is smooth. Then the restriction of the Poisson bracket to $\mathcal{M}_\chi^s$ is nondegenerate, i.e. comes from a symplectic form on $\mathcal{M}_\chi^s$, and $\mathcal{M}_\chi^s$ contains $T^*(X^s/\!\!/G)$ as an open (possibly empty) subset.*

Proof. We first describe the Poisson structure on $\mathcal{M}_0$. By definition, $\mathbf{k}[\mu^{-1}(0)] = \mathbf{k}[T^*X]/I$, where I is the ideal generated by Hamiltonians H_a, $a \in \mathfrak{g}$. Thus, any $f \in \mathbf{k}[\mu^{-1}(0)]$ has a lifting to a polynomial $\tilde{f} \in \mathbf{k}[T^*X]$.

Let $f_1, f_2 \in \mathbf{k}[\mathcal{M}_0] = (\mathbf{k}[\mu^{-1}(0)])^G$. Define their Poisson bracket by

(9.24) $$\{f_1, f_2\} = \{\tilde{f}_1, \tilde{f}_2\} \mod I.$$

To check that it is well defined, we need to show that $\{\tilde{f}, I\} \subset I$ for any $f \in (\mathbf{k}[\mu^{-1}(0)])^G$. But by definition, G-invariance of f implies that for any $a \in \mathfrak{g}$, we have $\xi_a(\tilde{f}) \in I$ (where ξ_a is the vector field on T^*X corresponding to the infinitesimal action of $\mathfrak{g}$ on T^*X). Thus, $\{H_a, f\} = \xi_a(f) \in I$.

Since

$$g\{\tilde{f}_1, \tilde{f}_2\} = \{g\tilde{f}_1, g\tilde{f}_2\} = \{\tilde{f}_1 + i_1, \tilde{f}_2 + i_2\} = \{\tilde{f}_1, \tilde{f}_2\} + i_3, \qquad i_1, i_2, i_3 \in I,$$

we see that if f_1, f_2 are G-invariant, then so is $\{f_1, f_2\}$. Thus, (9.24) defines a bilinear operation on $\mathbf{k}[\mathcal{M}_0]$. It is obvious that it satisfies the axioms of Poisson bracket.

For the twisted GIT quotient, the proof is similar: we note that locally, functions on $\mathcal{M}_\chi$ have the form f/g, $f, g \in (\mu^{-1}(0))^{G, \chi^n}$. Since any such function is G-invariant, we can use the same argument as before.

The fact that $\pi \colon \mathcal{M}_\chi \to \mathcal{M}_0$ is Poisson is immediate from the definition. $\square$

9.8. Symplectic resolution of singularities and Springer resolution

In Section 9.4, we showed that if M is an affine algebraic variety with an action of a reductive group G, then under suitable conditions, the projective morphism $\pi\colon M/\!/_\chi G \to M/\!/G$ is a resolution of singularities (see Corollary 9.31).

Let us now apply it to the situation discussed in Section 9.7, when X is a nonsingular variety and $M = \mu^{-1}(0) \subset T^*X$.

Definition 9.54. Let $\mathcal{M}$ be an affine algebraic variety over $\mathbf{k}$ with a Poisson structure. A symplectic resolution of singularities is a nonsingular algebraic variety $\tilde{\mathcal{M}}$ together with an algebraic symplectic form on $\tilde{\mathcal{M}}$ and a morphism $\pi\colon \tilde{\mathcal{M}} \to \mathcal{M}$ such that:

(1) $\pi\colon \tilde{\mathcal{M}} \to \mathcal{M}$ is a resolution of singularities.

(2) π is a Poisson morphism.

Combining the results of Theorem 9.53 and Corollary 9.31, we get the following result.

Theorem 9.55. *Let X be a nonsingular affine algebraic variety over $\mathbf{k}$, and let G be a reductive algebraic group acting on X. Let $\mathcal{M}_\chi$ be defined by (9.23). Assume additionally that $\mathcal{M}_\chi$ is connected and nonsingular, and that the subset of regular points $\mathcal{M}_0^{reg} \subset \mathcal{M}_0$ is nonempty. Then $\pi\colon \mathcal{M}_\chi \to \mathcal{M}_0$ is a symplectic resolution of singularities.*

Example 9.56. Let $X = \mathbf{k}^2$ with the action of $G = \mathbf{k}^\times$ by rescaling. Then T^*X can be identified with the set $\{(i,j) \mid j\colon \mathbf{k} \to \mathbf{k}^2, i\colon \mathbf{k}^2 \to \mathbf{k}\}$. It follows from Example 9.46 that under this identification the moment map is given by $\mu(i,j) = ij$. Thus, in this case

$$\mu^{-1}(0) = \{(i,j) \mid ij = 0\}$$

which is exactly the variety considered in Example 9.33. Therefore, we see that in this case the resolution $T^*\mathbb{P}^1 \to Q$ constructed in Example 9.33 is a symplectic resolution.

The symplectic resolution constructed in Example 9.56 is the simplest case of the famous Springer resolution. For future reference, we recall the construction of the Springer resolution below. For simplicity, we give it here for the group $G = \mathrm{SL}(n, \mathbf{k})$; however, the same construction works for any semisimple algebraic group (see, for example, [**CG1997**, Chapter 3]).

9.8. Symplectic resolution of singularities and Springer resolution

Let

(9.25) $$\mathcal{N} = \{a \in \mathfrak{sl}(n, \mathbf{k}) \mid a \text{ is nilpotent}\}.$$

Since a is nilpotent iff $a^n = 0$, it is obvious that $\mathcal{N}$ is an affine algebraic variety which is stable under the action of $\mathbf{k}^\times$ acting on $\mathfrak{sl}(n, \mathbf{k})$ by $a \mapsto \lambda a$. The variety $\mathcal{N}$ is called the *nilpotent cone*. It is singular: for example, for $\mathfrak{sl}(2)$, the nilpotent cone is given by

(9.26) $$\mathcal{N} = \left\{ \begin{pmatrix} a & b \\ c & -a \end{pmatrix} \mid a^2 + bc = 0 \right\}$$

so it is a quadric in $\mathbf{k}^3$ (the same quadric was constructed as a GIT quotient in Example 9.33).

Variety $\mathcal{N}$ has a remarkable resolution of singularities. Namely, recall that the flag variety $\mathcal{F}$ for the group $\mathrm{SL}(n, \mathbf{k})$ is the set of all flags

(9.27) $$F = (0 \subset F_1 \subset \cdots \subset F_{n-1} \subset F_n = \mathbf{k}^n), \quad \dim F_i = i.$$

It can also be described as the variety of all Borel subgroups in $\mathrm{SL}(n, \mathbf{k})$, or as the quotient space $\mathrm{SL}(n, \mathbf{k})/B$, where B is the subgroup of all upper-triangular matrices with determinant 1.

Consider now the set

$$\widetilde{\mathcal{N}} = \{(F, y) \mid F \in \mathcal{F}, \ y \in \mathfrak{sl}(n, \mathbf{k}), \ yF_i \subset F_{i-1}\}.$$

Theorem 9.57.

(1) *One has an isomorphism $\widetilde{\mathcal{N}} \simeq T^*\mathcal{F}$; in particular, $\widetilde{\mathcal{N}}$ is a symplectic manifold.*

(2) *The action of $\mathrm{SL}(n, \mathbf{k})$ on $\widetilde{\mathcal{N}}$ is Hamiltonian, and the moment map is given by*

$$\mu \colon \widetilde{\mathcal{N}} \to \mathfrak{sl}(n, \mathbf{k})$$
$$(F, y) \mapsto y.$$

(We identify $\mathfrak{sl}(n) \simeq \mathfrak{sl}(n)^*$ using the bilinear form $(a, b) = \mathrm{tr}(ab)$.)

(3) *The image of the moment map coincides with $\mathcal{N} \subset \mathfrak{sl}(n, \mathbf{k})$, and the restriction*

$$\mu \colon \widetilde{\mathcal{N}} \to \mathcal{N}$$

is a symplectic resolution of singularities. It is called the Springer resolution.

Proof. Note that $\mathcal{F} = \mathrm{SL}(n, \mathbf{k})/B$, where B is the subgroup of all upper triangular matrices or, equivalently,
$$B = \{g \in \mathrm{SL}(n) \mid gF_i^0 \subset F_i^0\}, \qquad F_i^0 = \langle e_1, e_2, \ldots, e_i\rangle.$$
Thus, we can use results of Example 9.47. In particular, identifying $\mathfrak{sl}(n) \simeq \mathfrak{sl}(n)^*$ using the bilinear form $(a, b) = \mathrm{tr}(ab)$ we get an identification
$$T_F^*(\mathcal{F}) = \{y \in \mathfrak{sl}(n, \mathbf{k}) \mid \mathrm{tr}(ay) = 0 \text{ for any } a \text{ s.t. } aF_i \subset F_i\}.$$
An elementary linear algebra exercise shows that this condition is equivalent to $yF_i \subset F_{i-1}$. This gives part (1); formula (9.21) from Example 9.47 immediately gives part (2).

To prove that μ is a resolution of singularities, consider the subset $\mathcal{N}^{reg} \subset \mathcal{N}$ consisting of regular nilpotent elements, i.e. elements y such that $y^{n-1} \neq 0$ (this is equivalent to the condition that the Jordan canonical form of y consists of a single Jordan block of size n). It is obvious that $\mathcal{N}^{reg}$ is open and dense in $\mathcal{N}$, and that for any regular nilpotent y, there exists a unique flag F such that $yF_i \subset F_{i-1}$ (namely, $F_i = \mathrm{Ker}(y^i)$). Therefore, the restriction $\mu \colon \widetilde{\mathcal{N}}^{reg} \to \mathcal{N}^{reg}$, where $\widetilde{\mathcal{N}}^{reg} = \mu^{-1}(\mathcal{N}^{reg})$, is an isomorphism. This together with the fact that μ is proper, which follows from the compactness of the flag variety, implies that $\widetilde{\mathcal{N}}$ is a resolution of singularities of $\mathcal{N}$.

Since the moment map preserves the Poisson structure (Theorem 9.44), we see that $\mu \colon \widetilde{\mathcal{N}} \to \mathcal{N}$ is a symplectic resolution. $\square$

Example 9.58. Consider the Springer resolution for $\mathrm{SL}(2, \mathbf{k})$. In this case, by (9.26), the nilpotent cone is a quadric $Q \subset \mathbf{k}^3$ and the flag variety is $\mathcal{F} = \mathbb{P}^1$. Thus, in this case the Springer resolution gives a resolution
$$\mu \colon T^*\mathbb{P}^1 \to Q.$$
In particular, in this case the exceptional fiber is given by $\mu^{-1}(0) = \mathbb{P}^1$. This resolution can be shown to coincide with the one constructed by GIT methods in Example 9.56.

In the next chapter, we will show that in fact for any n, the Springer resolution for the group $\mathrm{SL}(n, \mathbf{k})$ can be constructed by GIT methods, as the quiver variety for a quiver of type A (see Section 10.7).

9.9. Kähler quotients

In this section, we show that in the special case of complex algebraic group actions on a Kähler vector space, the GIT quotient is closely related to the Marsden–Weinstein reduction for the compact form of the group.

9.9. Kähler quotients

We recall the following well-known definition.

Definition 9.59. A Kähler manifold is a C^∞ manifold M together with a Riemannian metric g and a complex structure I such that:

(1) For any tangent vectors $v, w \in T_x M$, we have

(9.28) $$g(Iv, Iw) = g(v, w).$$

(2) The 2-form ω defined by
$$\omega(v, w) = g(v, Iw)$$
is closed.

In this case, we can define the Hermitian form $(\,,\,)$ on M (considered as a complex manifold with complex structure I) by the formula

(9.29) $$(v, w) = g(v, w) + i\omega(v, w)$$

or, equivalently, $\omega(v, w) = \operatorname{Im}(v, w)$, $g(v, w) = \operatorname{Re}(v, w)$.

As is well known, the projective space $\mathbb{CP}^n$ is Kähler; since every complex submanifold of a Kähler manifold is Kähler, this implies that every smooth projective variety is Kähler.

Another example is given by vector spaces with a Kähler structure. Namely, let $V_\mathbb{R}$ be a finite-dimensional vector space over $\mathbb{R}$ together with a complex structure I and a positive-definite $\mathbb{R}$-bilinear form $g(\,,\,)$ satisfying (9.28). Then $V_\mathbb{R}$ is a Kähler manifold. We will denote by $V = V_I$ the space $V_\mathbb{R}$ considered as a vector space over $\mathbb{C}$ with the complex structure I. As usual, we denote by $\operatorname{GL}(V)$ the group of $\mathbb{C}$-linear invertible operators in V.

Recall that for a complex algebraic group G, a real algebraic subgroup $K \subset G$ is called a *real form* of G if the canonical map $K(\mathbb{C}) \to G$ is an isomorphism (see, for example, [**OV1990**, Section 3.1.2]). In this situation, we will also call G a complexification of K. In particular, this implies that one has an isomorphism of complex Lie algebras

$$\mathfrak{g} = \mathfrak{k} \otimes_\mathbb{R} \mathbb{C}, \quad \mathfrak{g} = \operatorname{Lie}(G), \quad \mathfrak{k} = \operatorname{Lie}(K).$$

For example, the group $\operatorname{U}(n)$ is a real form of the complex group $\operatorname{GL}(n, \mathbb{C})$.

We will need the following result establishing a correspondence between compact real groups and complex reductive groups.

Theorem 9.60. *Let V be a finite-dimensional Hermitian vector space and let $K \subset \operatorname{U}(V)$ be a closed subgroup. Then:*

(1) *K is a real algebraic subgroup in $\operatorname{U}(V)$.*

(2) *There is a unique complex algebraic subgroup $G \subset \operatorname{GL}(V)$ such that $K \subset G$ is a real form. Moreover, G is reductive and $K = G \cap \operatorname{U}(V)$.*

Conversely, if V is a finite-dimensional complex vector space and $G \subset \mathrm{GL}(V)$ is a complex reductive algebraic group, then it has a compact real form $K \subset G$. One can choose an inner product in V so that $K = G \cap \mathrm{U}(V)$.

The proof of this theorem can be found in [**OV1990**, Sections 3.4.4, 5.2.5].

Let V be a Kähler vector space and let $K \subset \mathrm{U}(V)$ be a compact group, preserving both the complex structure I and the Hermitian form (and thus, also the real inner product form $g(v,w) = \mathrm{Re}(v,w)$). We denote by $G \subset \mathrm{GL}(V_I)$ the complexification of K, and by $\mathfrak{g} = \mathfrak{k} \otimes \mathbb{C}$ the corresponding Lie algebra.

It is obvious from the definition that the action of K preserves the Kähler form ω and thus, by Example 9.45, is Hamiltonian. We denote by
$$\mu_\mathbb{R} \colon V_\mathbb{R} \to \mathfrak{k}^*, \qquad \mathfrak{k} = \mathrm{Lie}(K),$$
the corresponding moment map. By Example 9.45, the moment map is given by $\langle \mu_\mathbb{R}(x), a \rangle = \frac{1}{2}\omega(x, a.x)$ for $a \in \mathfrak{k}$. Since the action of K preserves the real inner product g, we have $g(x, a.x) = 0$; thus, the moment map can also be written as

(9.30) $$\langle \mu_\mathbb{R}(x), a \rangle = \tfrac{-i}{2}(x, a.x) = \tfrac{i}{2}(a.x, x),$$

where $(\,,\,)$ is the Hermitian form (9.29) on V.

Theorem 9.61. *Let V be a Kähler vector space, let $K \subset \mathrm{U}(V)$ be a compact group, and let $G \subset \mathrm{GL}(V)$ be the complexification of K. Then, for every $x \in \mu_\mathbb{R}^{-1}(0)$, its G-orbit is closed in V, and the map*

(9.31) $$\mu_\mathbb{R}^{-1}(0)/K \to V/\!/G$$

is a bijection.

This theorem is due to Kempf and Ness [**KN1979**]; a proof can also be found in [**Nak1999**, Theorem 3.12].

Example 9.62. Consider the action of $G = \mathrm{GL}(n, \mathbb{C})$ on the space of matrices $V = \mathrm{Mat}_{n \times n}(\mathbb{C})$ by conjugation. Define the Hermitian form on V by $(x, y) = \mathrm{tr}_{\mathbb{C}^n}(xy^*)$, where $y^* = \overline{y}^t$; thus, the symplectic from is given by
$$\omega(x, y) = \frac{(x,y) - (y,x)}{2i} = \tfrac{-i}{2}\mathrm{tr}(xy^* - yx^*).$$

Let $K = \mathrm{U}(n)$. By (9.30), the action of K is Hamiltonian and the moment map is given by
$$\langle \mu_\mathbb{R}(x), a \rangle = \tfrac{i}{2}\mathrm{tr}([a,x]x^*) = \tfrac{i}{2}\mathrm{tr}(a[x,x^*]).$$

Thus,
$$\mu_\mathbb{R}(x) = \tfrac{i}{2}[x, x^*].$$

9.9. Kähler quotients

In this case, the closed G-orbits are those consisting of diagonalizable matrices (see Theorem 2.10). On the other hand,

$$\mu_{\mathbb{R}}^{-1}(0) = \{x \in \operatorname{Mat}_{n\times n}(\mathbb{C}) \mid [x, x^*] = 0\}$$

is the set of normal matrices; since every normal matrix can be diagonalized by a unitary change of basis, we see that

$$V/\!/\operatorname{GL}(n, \mathbb{C}) = \mu_{\mathbb{R}}^{-1}(0)/U(n) = \mathbb{C}^n/S_n.$$

As in the algebro-geometric situation (Section 9.4), we can define the notion of a regular point.

Definition 9.63. In the assumptions of Theorem 9.61, a point $x \in \mu_{\mathbb{R}}^{-1}(0)$ is called *regular* if $K_x = \{1\}$, where $K_x \subset K$ is the stabilizer of x.

It is easy to see that the set of regular points $(\mu_{\mathbb{R}}^{-1}(0))^{reg}$ is open and K-invariant, and the quotient $(\mu_{\mathbb{R}}^{-1}(0))^{reg}/K$ is smooth. Note however that $(\mu_{\mathbb{R}}^{-1}(0))^{reg}$ can be empty.

Theorem 9.64.

(1) The quotient $(\mu_{\mathbb{R}}^{-1}(0))^{reg}/K$ has a natural structure of a Kähler manifold.

(2) The restriction of the map (9.31) is an isomorphism of complex manifolds

(9.32) $$(\mu_{\mathbb{R}}^{-1}(0))^{reg}/K \simeq V^{reg}/\!/G$$

(considering $V^{reg}/\!/G$ as a complex manifold, with analytic topology).

Proof. First, it easily follows from the proof of [**Nak1999**, Proposition 3.9] that if $K_x = \{1\}$, then for any nonzero $\xi \in \mathfrak{k}$, the tangent vector $i\xi.x \in T_x V_{\mathbb{R}}$ is nonzero and transversal to the K-orbit of x. This implies that the Lie algebra of G_x is trivial, i.e. G_x is finite. By [**KN1979**, Theorem 4.1], in this case $G_x \subset K_x$, so $G_x = \{1\}$. This proves that (9.32) is a bijection.

A construction of the Kähler structure on $(\mu_{\mathbb{R}}^{-1}(0))^{reg}$ is discussed in [**Nak1999**]. The fact that this map is an isomorphism of complex manifolds is also stated there (without a proof). $\square$

Corollary 9.65. *Assume that the action of K on $\mu_{\mathbb{R}}^{-1}(0)$ is free. Then $V/\!/G \simeq \mu_{\mathbb{R}}^{-1}(0)/K$ is nonsingular and Kähler.*

Theorem 9.61 can be generalized to twisted GIT quotients. Let $\chi\colon G \to \mathbb{C}^\times$ be an algebraic character of G; then its restriction to K is a morphism $K \to U(1)$. Let $\chi_*\colon \mathfrak{k} \to \mathfrak{u}(1) = i\mathbb{R}$ be the corresponding morphism of Lie algebras. We can consider χ_* as an element in $i\mathfrak{k}^*$.

Theorem 9.66. *In the assumptions of Theorem 9.61, for every $x \in \mu_{\mathbb{R}}^{-1}(i\chi_*)$ the G-orbit Gx is χ-semistable and closed in the set of χ-semistable elements of V, and the map $x \mapsto [x]$ is a bijection*

(9.33) $$\mu_{\mathbb{R}}^{-1}(i\chi_*)/K \to V/\!\!/_\chi G.$$

If the action of K on $\mu_{\mathbb{R}}^{-1}(i\chi_*)$ is free, then the set $\mu_{\mathbb{R}}^{-1}(i\chi_*)/K$ has a natural structure of a Kähler manifold, $V/\!\!/_\chi G$ is a nonsingular algebraic variety, and the map (9.33) is an isomorphism of complex manifolds.

The proof of this theorem can be found in [**Kin1994**, Theorem 6.1] or [**Nak1999**, Corollary 3.22].

9.10. Hyperkähler quotients

The Kähler quotients considered in the previous section can be generalized to the hyperkähler case. We review here the main results; proofs can be found in [**Nak1999**, Section 3.2].

Definition 9.67. A hyperkähler manifold is a C^∞ manifold M of dimension $4n$ together with a Riemannian metric g and three complex structures I, J, K such that:

(1) For any tangent vectors $v, w \in T_x M$, we have

(9.34) $$g(Iv, Iw) = g(Jv, Jw) = g(Kv, Kw) = g(v, w).$$

(2) The complex structures I, J, K satisfy the relations of the quaternion algebra:
$$I^2 = J^2 = K^2 = IJK = -1$$
(which implies $IJ = -JI = K$).

(3) The 2-forms ω_I, ω_J, ω_K defined by
$$\omega_I(v,w) = g(v, Iw),$$
$$\omega_J(v,w) = g(v, Jw),$$
$$\omega_K(v,w) = g(v, Kw)$$
are closed.

It can be shown (see [**Nak1999**, Lemma 3.37]) that it suffices to require in the definition that I, J, K are almost complex structures: integrability would automatically follow.

Hyperkähler manifolds have much richer structure than Kähler manifolds; we will show some of this below. However, they are also much more difficult to construct.

9.10. Hyperkähler quotients

Note that each of the complex structures I, J, K defines on M a structure of a Kähler manifold. We will denote by M_I (respectively, M_J, M_K) the manifold M considered as a complex manifold with complex structure I (respectively, J, K). Moreover, we have the following lemma.

Lemma 9.68. *Let M be a hyperkähler manifold. Let*
$$\omega_{\mathbb{C}} = \omega_J + i\omega_K.$$
Then $\omega_{\mathbb{C}}$ is a holomorphic $(2,0)$ form with respect to the complex structure I, and $d\omega_{\mathbb{C}} = 0$.

The proof of this lemma is left to the reader as an exercise. Note also that $\omega_{\mathbb{C}}$ can be written as

(9.35) $$\omega_{\mathbb{C}}(x,y) = (x, Jy)_I,$$

where $(\,,\,)_I$ is the Hermitian form defined by g and complex structure I as in (9.29).

In particular, this shows that M_I is a complex symplectic manifold. However, not every complex symplectic manifold is obtained in this way.

Definition 9.69. Let M be a hyperkähler manifold, and let K be a compact real Lie group acting on M and preserving the Riemannian metric g and each of the complex structures I, J, K. Assume that for each of the symplectic forms ω_I, ω_J, ω_K, the action of K is Hamiltonian; let $\mu_I, \mu_J, \mu_K \colon M \to \mathfrak{k}^*$ be the corresponding moment maps. Then the map
$$\boldsymbol{\mu} = (\mu_I, \mu_J, \mu_K) \colon M \to \mathbb{R}^3 \otimes \mathfrak{k}^*$$
is called the hyperkähler moment map.

Consider an element $\boldsymbol{\zeta} = (\zeta^1, \zeta^2, \zeta^3) \in \mathbb{R}^3 \otimes \mathfrak{k}^*$ which is K-invariant. Since $\boldsymbol{\mu}$ is K-equivariant, the set $\boldsymbol{\mu}^{-1}(\boldsymbol{\zeta}) \subset M$ is K-invariant, so we can consider the quotient $\boldsymbol{\mu}^{-1}(\boldsymbol{\zeta})/K$.

The following theorem was proved in [**HKLR1987**]. We omit the proof, referring the reader to the original paper or to exposition in [**Nak1999**, Theorem 3.35].

Theorem 9.70. *Let $M, K, \boldsymbol{\mu}$ be as in Definition 9.69. Let $\boldsymbol{\zeta} \in \mathbb{R}^3 \otimes \mathfrak{k}^*$ be a K-invariant element such that the action of K on $\boldsymbol{\mu}^{-1}(\boldsymbol{\zeta})$ is free. Then the quotient $\boldsymbol{\mu}^{-1}(\boldsymbol{\zeta})/K$ has a natural structure of a hyperkähler manifold of dimension $\dim M - 4\dim K$.*

Let us now consider an important special case, namely of hyperkähler quotients of vector spaces.

Let $V_\mathbb{R}$ be a vector space over $\mathbb{R}$ equipped with a positive definite bilinear form g and three complex structures I, J, K satisfying the relations in Definition 9.67. Then $V_\mathbb{R}$ is a hyperkähler manifold. It can also be considered as a complex symplectic manifold, with complex structure I and holomorphic symplectic form $\omega_\mathbb{C}$ defined in Lemma 9.68. As before, we will use the notation V for $V_\mathbb{R}$ considered as a complex vector space with respect to I.

Now let $K \subset \mathrm{SO}(V_\mathbb{R})$ be a compact Lie group acting on $V_\mathbb{R}$ so that the action preserves the metric g and each of the complex structures I, J, K. By Example 9.45, the action is Hamiltonian with respect to each of $\omega_I, \omega_J, \omega_K$; let $\boldsymbol{\mu} \colon V_\mathbb{R} \to \mathbb{R}^3 \otimes \mathfrak{k}^*$ be the corresponding moment map (see Definition 9.69). By results of Example 9.45, $\boldsymbol{\mu}$ is K-equivariant.

Let $G \subset \mathrm{GL}(V)$ be the complexification of K; recall that G must be reductive. It is immediate from the definition that the action of G on V preserves the holomorphic symplectic form $\omega_\mathbb{C}$ and thus is Hamiltonian (see Example 9.45).

Theorem 9.71. *Let $V_\mathbb{R}, V, G, K$ be as above. Then:*

(1) *The moment map for the action of G on V (with respect to the holomorphic symplectic form $\omega_\mathbb{C}$) is given by*

$$\mu_\mathbb{C} = \mu_J + i\mu_K \colon V \to \mathfrak{k}^* \oplus i\mathfrak{k}^* = \mathfrak{g}^*.$$

(2) *Let $\chi \colon G \to \mathbb{C}^*$ be a character of G. As in Theorem 9.66, χ defines a linear map $\chi_* \colon \mathfrak{k} \to \mathfrak{u}(1) = i\mathbb{R}$; thus, we can consider χ_* as a vector in $i\mathfrak{k}^*$. Let $\zeta^2, \zeta^3 \in \mathfrak{k}^*$ be K-invariant; then $\boldsymbol{\zeta} = (i\chi_*, \zeta^2, \zeta^3) \in \mathbb{R}^3 \otimes \mathfrak{k}^*$ is K-invariant.*

Assume that the action of K on $\boldsymbol{\mu}^{-1}(\boldsymbol{\zeta})$ is free, so that $\boldsymbol{\mu}^{-1}(\boldsymbol{\zeta})/K$ is a hyperkähler manifold (see Theorem 9.70). Then the algebraic variety

$$\mu_\mathbb{C}^{-1}(\zeta_\mathbb{C}) /\!/_\chi G, \qquad \zeta_\mathbb{C} = \zeta^2 + i\zeta^3,$$

is smooth and one has an isomorphism of complex manifolds

$$\boldsymbol{\mu}^{-1}(\boldsymbol{\zeta})/K \simeq \mu_\mathbb{C}^{-1}(\zeta_\mathbb{C}) /\!/_\chi G$$

(compare with Theorem 9.66).

Proof. Part (1) is obvious. To prove part (2), note that one can write

$$\boldsymbol{\mu}^{-1}(\boldsymbol{\zeta})/K = \{x \in \mu_I^{-1}(i\chi_*)/K \mid \mu_\mathbb{C}(x) = \zeta^2 + i\zeta^3\}.$$

Similarly, one can write

$$\mu_\mathbb{C}^{-1}(\zeta_\mathbb{C}) /\!/_\chi G = \{[x] \in V /\!/_\chi G \mid \mu_\mathbb{C}(x) = \zeta^2 + i\zeta^3\},$$

where we interpret $V /\!/_\chi G$ as the set of closed G-orbits in the set of χ-semistable points in V.

9.10. Hyperkähler quotients

On the other hand, by Theorem 9.66, we have a bijection $\mu_I^{-1}(i\chi_*)/K \simeq V/\!/_\chi G$. Thus, we get a bijection $\mu^{-1}(\boldsymbol{\zeta})/K \simeq \mu_\mathbb{C}^{-1}(\zeta_\mathbb{C})/\!/_\chi G$. □

Corollary 9.72. *Let $V_\mathbb{R}$ be a hyperkähler vector space with a Hamiltonian action of a compact group K and let $\boldsymbol{\zeta} = (0, \zeta^2, \zeta^3) \in \mathbb{R}^3 \otimes \mathfrak{k}^*$ be K invariant and such that the action of K on $\boldsymbol{\mu}^{-1}(\boldsymbol{\zeta})$ is free. Then the quotient complex manifold $\boldsymbol{\mu}^{-1}(\boldsymbol{\zeta})/K$ is an affine algebraic variety with respect to complex structure I.*

Indeed, by Theorem 9.71, one has
$$\boldsymbol{\mu}^{-1}(\boldsymbol{\zeta})/K \simeq \mu_\mathbb{C}^{-1}(\zeta_\mathbb{C})/\!/G$$
which is an affine algebraic variety.

Chapter 10

Quiver Varieties

Recall that in our constructions of the Hall algebra, we used $\mathrm{GL}(\mathbf{v})$-invariant functions on the representation space $R(\mathbf{v})$ or on the variety $\Lambda(\mathbf{v}) \subset \mu_{\mathbf{v}}^{-1}(0)$. Equivalently, one can think of such functions as functions on the quotient space $R(\mathbf{v})/\mathrm{GL}(\mathbf{v})$ (respectively, $\Lambda(\mathbf{v})/\mathrm{GL}(\mathbf{v})$).

However, these quotients are rather "bad" as topological spaces; they are typically non-Hausdorff. In this chapter, we replace these quotients by the GIT quotients constructed in Chapter 9. These new moduli spaces of (framed) quiver representations are called the quiver varieties; they were introduced and studied in Nakajima's papers [**Nak1994, Nak1998**].

To simplify the exposition, throughout this chapter we consider the ground field $\mathbf{k} = \mathbb{C}$.

10.1. GIT quotients for quiver representations

In this section we apply the general constructions of GIT to the study of the moduli spaces of quiver representations. The results of this section are due to King [**Kin1994**].

Recall that for a quiver $\vec{Q}$ and a dimension vector $\mathbf{v} \in \mathbb{Z}_+^I$ we defined the representation space $R(\mathbf{v}) = \bigoplus \mathrm{Hom}(\mathbb{C}^{\mathbf{v}_i}, \mathbb{C}^{\mathbf{v}_j})$ and the group $\mathrm{GL}(\mathbf{v})$ acting on it by conjugation (see Section 2.1). In fact, since the subgroup of constants $\mathbb{C}^\times \subset \mathrm{GL}(\mathbf{v})$ acts trivially, we have an action of the group

(10.1) $$\mathrm{PGL}(\mathbf{v}) = \mathrm{GL}(\mathbf{v})/\mathbb{C}^\times.$$

By Theorem 2.1, orbits of this action (or, equivalently, points of the set-theoretic quotient space $R(\mathbf{v})/\mathrm{PGL}(\mathbf{v})$) are in bijection with the set of isomorphism classes of representations of $\vec{Q}$ of dimension $\mathbf{v}$.

Consider now the GIT quotient

(10.2) $$\mathcal{R}_0(\mathbf{v}) = R(\mathbf{v}) /\!/ \mathrm{PGL}(\mathbf{v}).$$

Theorem 10.1. *There is a bijection between $\mathcal{R}_0(\mathbf{v})$ and the set of isomorphism classes of semisimple representations of $\vec{Q}$.*

Proof. By Theorem 9.5, points of $\mathcal{R}_0(\mathbf{v})$ are in bijection with closed orbits in $R(\mathbf{v})$. On the other hand, by results of Section 2.3, closed orbits correspond to semisimple representations. $\square$

Corollary 10.2. *If $\vec{Q}$ is a quiver without oriented cycles, then $\mathcal{R}_0(\mathbf{v})$ is a point.*

Indeed, by Corollary 2.11, in this case there is a unique closed orbit, namely the orbit of $x = 0$.

Remark 10.3. One can also give an explicit description of the algebra A of $\mathrm{PGL}(\mathbf{v})$-invariant functions on $R(\mathbf{v})$; namely, this algebra is generated by elements $a_{p,n}(x) = \mathrm{tr}(x_p)^n$, where $n \geq 1$, p is an oriented cycle in $\vec{Q}$, and $x_p \in \mathrm{End}(\mathbb{C}^{\mathbf{v}_i})$ is the corresponding linear operator (see [**LBP1990**]). This gives another proof of the fact that if $\vec{Q}$ has no oriented cycles, then $\mathcal{R}_0$ is a point: indeed, in this case $A = \mathbb{C}$, so $\mathrm{Spec}(A) = \{pt\}$.

Now let us consider the twisted GIT quotient. Let $\theta \in \mathbb{Z}^I$; define a character

(10.3) $$\chi_\theta \colon \mathrm{GL}(\mathbf{v}) \to \mathbb{C}^\times$$
$$g \mapsto \prod \det(g_i)^{-\theta_i}$$

(it can be shown that any character of $\mathrm{GL}(\mathbf{v})$ is of this form). We will call θ the *stability parameter*. Note that in order for the character to be well defined on $\mathrm{PGL}(\mathbf{v})$, we must have

(10.4) $$\theta \cdot \mathbf{v} = \sum_{i \in I} \theta_i \mathbf{v}_i = 0.$$

Assume that θ satisfies (10.4). Define

(10.5) $$\mathcal{R}_\theta(\mathbf{v}) = R(\mathbf{v}) /\!/_{\chi_\theta} \mathrm{PGL}(\mathbf{v}).$$

By results of Section 9.3, $\mathcal{R}_\theta$ can be described in terms of semistable orbits, and we have a projective morphism

$$\pi \colon \mathcal{R}_\theta \to \mathcal{R}_0$$

which sends every semistable orbit $\mathbb{O}$ to the unique closed orbit contained in $\overline{\mathbb{O}}$ (see Theorem 9.19). By Theorem 2.10, this map can be written in terms of representations by

(10.6) $$\pi(\mathbb{O}_V) = \mathbb{O}_{V^{ss}},$$

where V^{ss} is the semisimplification of representation V.

In particular, if $\vec{Q}$ has no oriented cycles, then by Corollary 10.2, $\mathcal{R}_0(\mathbf{v})$ is a point, so in this case $\mathcal{R}_\theta(\mathbf{v})$ is a projective variety.

The following result, due to King [**Kin1994**], shows that for the moduli space of quiver representations, the stability condition can be rewritten in terms of representation theory. Recall that for $x \in R(\mathbf{v})$, we denote by V^x the corresponding representation of $\vec{Q}$.

Definition 10.4. Let $\theta \in \mathbb{R}^I$. A representation V of $\vec{Q}$ is called θ-semistable (respectively, θ-stable) if $\theta \cdot \dim V = 0$ and for any subrepresentation $V' \subset V$ we have $\theta \cdot \dim V' \leq 0$ (respectively, for every nonzero proper subrepresentation V' we have $\theta \cdot \dim V' < 0$).

Theorem 10.5. *Let $\mathbf{v} \in \mathbb{Z}_+^I$, $\theta \in \mathbb{Z}^I$, be such that $\theta \cdot \mathbf{v} = 0$. Then an element $x \in R(\mathbf{v})$ is χ_θ-semistable (respectively, stable) iff the corresponding representation V^x is θ-semistable (respectively, θ-stable).*

Proof. Assume that V^x is θ-semistable. We will use Mumford's criterion (Theorem 9.25) to prove that x is semistable. Let $\lambda \colon \mathbb{C}^\times \to \mathrm{PGL}(\mathbf{v})$ be a one-parameter subgroup such that the limit $\lim_{t \to 0} \lambda(t).x$ exists. As in the proof of Theorem 2.8, such a subgroup gives rise to a weight decomposition $V = \bigoplus_{n \in \mathbb{Z}} V^n$. Thus, we have a decreasing filtration

(10.7) $$V = V^{\geq a} \supset V^{\geq a+1} \supset \cdots \supset V^{\geq b} = \{0\}, \qquad V^{\geq n} = \bigoplus_{m \geq n} V^m.$$

The same argument as in the proof of Theorem 2.8 shows that this filtration is x-stable, so each $V^{\geq n}$ is a subrepresentation. Since

$$\chi(\lambda(t)) = \prod_{i,n} t^{-n \dim V_i^n \theta_i} = t^{-\sum n \theta \cdot \dim V^n},$$

we see that

$$\langle \chi, \lambda \rangle = -\theta \cdot \mathbf{w}, \qquad \mathbf{w} = \sum_{n \in \mathbb{Z}} n \dim V^n.$$

On the other hand, each V^n appears in $n - a + 1$ steps of filtration (10.7), so

$$\sum_{m=a}^{b} \dim V^{\geq m} = \sum_{n=a}^{b} (n - a + 1) \dim V^n = (1 - a) \dim V + \sum_{n \in \mathbb{Z}} n \dim V^n$$
$$= \mathbf{w} + (1 - a)\mathbf{v}.$$

Since $\theta \cdot \mathbf{v} = 0$, we have

$$\langle \chi, \lambda \rangle = -\theta \cdot \mathbf{w} = -\sum_{n=a}^{b} \theta \cdot \dim V^{\geq n}.$$

Since each $V^{\geq n}$ is a subrepresentation, assumption of θ-stability of V gives $\theta \cdot \dim V^{\geq n} \leq 0$; thus, $\langle \chi, \lambda \rangle \geq 0$ and by Mumford's criterion, x is χ_θ-semistable.

The proof in the opposite direction is similar: if x is semistable and $V' \subset V$ is a subrepresentation, we can construct a one-parameter subgroup by letting $\lambda(t) = t$ on V', $\lambda(t) = 1$ on V/V'; then condition $\langle \chi, \lambda \rangle \geq 0$ implies $\dim V' \cdot \theta \leq 0$.

Easy modification of this argument (which we leave to the reader) gives the statement about stable representation. $\square$

Note that in particular, for $\theta = 0$ any $x \in R(\mathbf{v})$ is semistable, and it is stable iff V^x is simple. More generally, we have the following result.

Lemma 10.6. *Let $\theta \in \mathbb{R}^I$. Then the category $\mathrm{Rep}^\theta(\vec{Q})$ of θ-semistable representations of $\vec{Q}$ is an abelian category, and simple objects of this category are exactly θ-stable representations.*

Proof. To prove that $\mathrm{Rep}^\theta(\vec{Q})$ is abelian, it suffices to show that if $f\colon A \to B$ is a morphism of θ-semistable representations of $\vec{Q}$, then $\mathrm{Ker}\, f$, $\mathrm{Im}\, f$ are also θ-semistable. Since it is immediate from the definition that any subrepresentation S of a θ-semistable representation which has $\theta \cdot \dim S = 0$ is itself θ-semistable, it suffices to prove that $\theta \cdot \dim \mathrm{Ker}\, f = \theta \cdot \dim \mathrm{Im}\, f = 0$.

Since $\mathrm{Ker}\, f \subset A$, we have $\theta \cdot \dim \mathrm{Ker}\, f \leq 0$. Similarly, since $\mathrm{Im}\, f \subset B$, we have $\theta \cdot \dim \mathrm{Im}\, f \leq 0$. Since $\theta \cdot (\dim \mathrm{Ker}\, f + \dim \mathrm{Im}\, f) = \theta \cdot \dim A = 0$, we see that $\theta \cdot \dim \mathrm{Ker}\, f = \theta \cdot \dim \mathrm{Im}\, f = 0$. $\square$

In particular, for any $V \in \mathrm{Rep}^\theta(\vec{Q})$ we can define its θ-semisimplification by

$$V^{\theta ss} = \bigoplus M_i,$$

where M_i are the factors in the composition series of V in the category $\mathrm{Rep}^\theta(\vec{Q})$ (compare with (2.6)).

Theorem 10.7. *Let $\theta \in \mathbb{Z}^I$ be such that $\theta \cdot \mathbf{v} = 0$.*

(1) *Let $x, x' \in R^{ss}(\mathbf{v})$, and let $\sim$ be the equivalence relation defined in Corollary 9.18. Then $x \sim x'$ iff the corresponding representations have the same θ-semisimplification: $(V^x)^{\theta ss} \simeq (V^{x'})^{\theta ss}$.*

(2) *An orbit of $x \in R^{ss}(\mathbf{v})$ is closed iff V^x is a direct sum of θ-stable representations.*

The proof of this theorem is parallel to the proof of Theorem 2.10 and will not be given here. Interested readers can find it in [**Kin1994**].

Let us now consider points in $\mathcal{R}_\theta$ that correspond to the stable points $x \in R(\mathbf{v})$. Denote
$$\mathcal{R}_\theta^s(\mathbf{v}) = R^s(\mathbf{v})/\mathrm{PGL}(\mathbf{v}) \subset \mathcal{R}_\theta(\mathbf{v})$$
(compare with (9.12)).

Theorem 10.8. $\mathcal{R}_\theta^s(\mathbf{v}) \subset \mathcal{R}_\theta(\mathbf{v})$ *is smooth; if nonempty, it is open and dense in $\mathcal{R}_\theta$ and has dimension*
$$\dim \mathcal{R}_\theta^s(\mathbf{v}) = 1 - \langle \mathbf{v}, \mathbf{v} \rangle,$$
where $\langle \mathbf{v}, \mathbf{v} \rangle$ is the Euler form of $\vec{Q}$ (see (1.15)).

Proof. By Theorem 2.3, all stabilizers are connected; thus, for every $x \in R^s$, we have $\mathrm{Stab}(x) = \{1\}$. Now the result follows from Theorem 9.23 and explicit computation of dimension:
$$\dim \mathcal{R}_\theta^s = \dim R^s - \dim \mathrm{PGL}(\mathbf{v}) = \dim R - \dim \mathrm{GL}(\mathbf{v}) + 1$$
$$= \sum_{h \in \Omega} \mathbf{v}_{s(h)} \mathbf{v}_{t(h)} - \sum \mathbf{v}_i^2 + 1 = 1 - \langle \mathbf{v}, \mathbf{v} \rangle.$$
$\square$

Corollary 10.9. *If $\vec{Q}$ is Dynkin and $\mathbf{v} \neq 0$ is not a root, then $\mathcal{R}_\theta^s(\mathbf{v}) = \varnothing$; if $\mathbf{v}$ is a root, then $\mathcal{R}_\theta^s(\mathbf{v})$ is either empty or a point.*

This corollary shows that for Dynkin quivers, varieties $\mathcal{R}_\theta^s$ do not actually give anything useful. We will discuss how this can be corrected in the next chapter.

In fact, it can be shown (see [**Kin1994**]) that for any quiver without oriented cycles,
$$(\exists \theta : \mathcal{R}_\theta^s(\mathbf{v}) \text{ is nonempty}) \iff (\mathrm{End}(V^x) = \mathbb{C}\,\mathrm{id}\,\text{for generic } x \in R(\mathbf{v})).$$
In the Dynkin case, the last condition holds iff $\mathbf{v}$ is a root.

10.2. GIT moduli spaces for double quivers

In this section, we study the moment map for the representation space of the double quiver, considered in Chapter 5. As in Section 5.2, let Q be a graph, let $Q^\sharp$ be the corresponding double quiver with the set of oriented edges H, and let $R(Q^\sharp, \mathbf{v})$ be the representation space defined by (5.1). Recall that for any choice of orientation Ω we have an isomorphism
$$R(Q^\sharp, \mathbf{v}) = R(\vec{Q}, \mathbf{v}) \oplus R(\vec{Q}, \mathbf{v})^* = T^*(R(\vec{Q}, \mathbf{v}))$$

(see (5.2)). By Theorem 9.39, this defines a symplectic form ω_Ω on $R(Q^\sharp, \mathbf{v})$; explicitly, it can be written
$$\omega_\Omega(x_1 + y_1, x_2 + y_2) = \langle y_1, x_2\rangle - \langle y_2, x_1\rangle, \quad x_i \in R(\vec{Q}, \mathbf{v}), \tag{10.8}$$
$$y_i \in R(\vec{Q}^{opp}, \mathbf{v}) = R(\vec{Q}, \mathbf{v})^*.$$

More generally, let $\varepsilon\colon H \to \mathbf{k}^\times$ be a skew-symmetric function on the set of edges of $Q^\sharp$ (see (5.3)). Define the symplectic form ω_ε by
$$\omega_\varepsilon(z,w) = \operatorname{tr}_V \sum_{h\in H} \varepsilon(h) z_{\bar h} w_h, \quad z, w \in R(Q^\sharp, \mathbf{v}). \tag{10.9}$$

One easily sees that in the case when $\varepsilon = \varepsilon_\Omega$ is defined by orientation Ω (see (5.4)), this definition agrees with the previous one.

From now on, we fix a choice of function ε and the corresponding symplectic form $\omega = \omega_\varepsilon$ on $R(Q^\sharp, \mathbf{v})$.

Consider now the action of $G = \operatorname{GL}(\mathbf{v})$ on $R(Q^\sharp, \mathbf{v})$. One easily sees that this action preserves the form ω. This is a special case of the situation considered in Example 9.45. Thus, we get the following result.

Theorem 10.10. *The action of $\operatorname{GL}(\mathbf{v})$ on $R(Q^\sharp, \mathbf{v})$ is Hamiltonian; the corresponding moment map*
$$\mu_\mathbf{v}\colon R(Q^\sharp, \mathbf{v}) \to \bigoplus_i \mathfrak{gl}(\mathbf{v}_i, \mathbb{C})$$
is given by
$$z \mapsto \bigoplus_i \sum_{t(h)=i} \varepsilon(h) z_h z_{\bar h}.$$
(We identify $\mathfrak{gl}(n,\mathbb{C}) \simeq \mathfrak{gl}(n,\mathbb{C})^*$ using the bilinear form $(x,y) = \operatorname{tr}(xy)$.)

Proof. By Example 9.45, we have
$$\langle \mu_\mathbf{v}(z), a\rangle = \tfrac{1}{2}\omega_\varepsilon(z, a.z) = \tfrac{1}{2}\sum_{h\in H} \varepsilon(h)\operatorname{tr}(z_{\bar h}[a, z_h]).$$

Using the identity $\operatorname{tr}(a[b,c]) = \operatorname{tr}(b[c,a])$, we can rewrite it as
$$\langle \mu_\mathbf{v}(z), a\rangle = \tfrac{1}{2}\sum_{h\in H} \varepsilon(h)\operatorname{tr}(a[z_h, z_{\bar h}]) = \sum_{h\in H} \varepsilon(h)\operatorname{tr}(a z_h z_{\bar h}).$$

Thus, identifying $\mathfrak{gl}(n,\mathbb{C}) \simeq \mathfrak{gl}(n,\mathbb{C})^*$ we get
$$\mu_\mathbf{v}(z) = \sum_{h\in H} \varepsilon(h) z_h z_{\bar h} = \bigoplus_i \sum_{t(h)=i} \varepsilon(h) z_h z_{\bar h}.$$

□

10.2. GIT moduli spaces for double quivers

In particular, if $\varepsilon = \varepsilon_\Omega$ is defined by an orientation Ω, then the moment map can be written as

$$\mu_{\mathbf{v}}(z) = \sum_{h \in \Omega} [z_h, z_{\bar{h}}]. \tag{10.10}$$

In Chapter 5, we had defined the space $\mu_{\mathbf{v}}^{-1}(0) = \{z \in R(Q^\sharp, \mathbf{v}) \mid \mu_{\mathbf{v}}(z) = 0\}$; then $\mathrm{GL}(\mathbf{v})$-orbits in $\mu_{\mathbf{v}}^{-1}(0)$ are in bijection with the isomorphism classes of $\mathbf{v}$-dimensional representations of the preprojective algebra Π. We later used the space of $\mathrm{GL}(\mathbf{v})$-invariant functions on $\mu_{\mathbf{v}}^{-1}(0)$ (or, rather on the variety $\Lambda(\mathbf{v}) \subset \mu_{\mathbf{v}}^{-1}(0)$ corresponding to nilpotent representations; see (5.9)) to construct the positive part of universal enveloping algebra $U\mathfrak{g}$.

Let us now study the GIT analog of the quotient space $\mu_{\mathbf{v}}^{-1}(0)/\mathrm{GL}(\mathbf{v})$.

Definition 10.11. Let Q be a graph and $\mathbf{v} \in \mathbb{Z}_+^I$. Let

$$\mu_{\mathbf{v}} \colon R(Q^\sharp, \mathbf{v}) \to \mathfrak{gl}(\mathbf{v})$$

be the moment map defined in Theorem 10.10. Then we define

$$\mathcal{M}_0(\mathbf{v}) = \mu_{\mathbf{v}}^{-1}(0) /\!/ \mathrm{PGL}(\mathbf{v}).$$

More generally, if $\theta \in \mathbb{Z}^I$ is such that $\theta \cdot \mathbf{v} = 0$, then we define

$$\mathcal{M}_\theta(\mathbf{v}) = \mu_{\mathbf{v}}^{-1}(0) /\!/_{\chi_\theta} \mathrm{PGL}(\mathbf{v})$$

and

$$\mathcal{M}_\theta^s(\mathbf{v}) = \{z \in R(Q^\sharp, \mathbf{v}) \mid \mu_{\mathbf{v}}(z) = 0, z \text{ is } \chi_\theta\text{-stable}\}/\mathrm{PGL}(\mathbf{v}) \subset \mathcal{M}_\theta(\mathbf{v})$$

to be the subvariety of stable points (compare with (9.12)).

By general theory developed in Section 9.3, we see that $\mathcal{M}_0$ is an affine algebraic variety, $\mathcal{M}_\theta$ is quasiprojective, and we have a projective morphism

$$\pi \colon \mathcal{M}_\theta(\mathbf{v}) \to \mathcal{M}_0(\mathbf{v}).$$

Moreover, by Theorem 9.53, the morphism π is a morphism of Poisson varieties.

Theorem 10.12.

(1) $\mathcal{M}_0(\mathbf{v})$ is the set of isomorphism classes of $\mathbf{v}$-dimensional semisimple representations of the preprojective algebra Π, and the map $\pi \colon \mathcal{M}_\theta \to \mathcal{M}_0$ is given by $[V] \mapsto [V^{ss}]$, where V^{ss} is the semisimplification of representation V.

(2) $\mathcal{M}_\theta^s(\mathbf{v}) \subset \mathcal{M}_\theta(\mathbf{v})$ is open; if nonempty, it is a nonsingular variety of dimension

$$\dim \mathcal{M}_\theta^s(\mathbf{v}) = 2 - 2\langle \mathbf{v}, \mathbf{v} \rangle.$$

(3) $\mathcal{M}_\theta^s(\mathbf{v})$ contains $T^* \mathcal{R}_\theta^s(\mathbf{v})$ as an open subset.

Proof. By Theorem 9.5, $\mathcal{M}_0$ is the set of closed orbits in $\mu_{\mathbf{v}}^{-1}(0)$; since $\mu_{\mathbf{v}}^{-1}(0) \subset R(Q^\sharp, \mathbf{v})$ is closed and G-invariant, this is the same as closed orbits in $R(Q^\sharp, \mathbf{v})$ which lie in $\mu_{\mathbf{v}}^{-1}(0)$. By Theorem 2.10, closed orbits in $R(Q^\sharp, \mathbf{v})$ correspond to semisimple representations of the path algebra $\mathbb{C}Q^\sharp$, and such an orbit is in $\mu_{\mathbf{v}}^{-1}(0)$ iff the corresponding representation descends to Π.

Part (2) follows from Theorem 9.23 together with the fact that stabilizers are connected (Theorem 2.3) and thus, for any stable x we have $G_x = \{1\}$.

Part (3) immediately follows from Theorem 9.53. $\square$

As before, the notion of stability can be rewritten in terms of representation theory.

Definition 10.13. Let $\theta \in \mathbb{R}^I$. A representation V of the preprojective algebra Π is called θ-*semistable* (respectively, θ-*stable*) if $\theta \cdot \dim V = 0$ and for any subrepresentation $V' \subset V$ we have $\theta \cdot \dim V' \leq 0$ (respectively, for every nonzero proper subrepresentation V' we have $\theta \cdot \dim V' < 0$).

Lemma 10.14. *Let $\theta \in \mathbb{R}^I$. Then the category of θ-semistable representations of the preprojective algebra Π is an abelian category, and simple objects of this category are exactly θ-stable representations.*

The proof of this lemma is identical to the proof of the similar statement for quiver $\vec{Q}$ (Lemma 10.6).

Theorem 10.15. *Let $\mathbf{v} \in \mathbb{Z}_+^I$, $\theta \in \mathbb{Z}^I$, be such that $\theta \cdot \mathbf{v} = 0$. Then an element $x \in \mu_{\mathbf{v}}^{-1}(0)$ is χ_θ-semistable (respectively, χ_θ-stable) iff the corresponding representation V^x of Π is θ-semistable (respectively, θ-stable).*

Proof. The proof is the same as for Theorem 10.5. $\square$

In the Dynkin case the situation is simplified.

Theorem 10.16. *If Q is a Dynkin graph, then $\mathcal{M}_0(\mathbf{v}) = \{pt\}$ for any $\mathbf{v}$, so $\mathcal{M}_\theta(\mathbf{v})$ is projective.*

Proof. By Theorem 10.12, points of $\mathcal{M}_0$ are in bijection with semisimple representations of the preprojective algebra. On the other hand, by Corollary 5.10, for a Dynkin graph, there is only one semisimple representation of Π of given dimension $\mathbf{v}$, namely $\bigoplus \mathbf{v}_i S(i)$. $\square$

Example 10.17. Let $Q = \bullet \!\!-\!\!\!-\!\! \bullet$, $\mathbf{v} = (1,1)$. Then $Q^\sharp = \bullet \rightleftarrows \bullet$ and
$$\mu_{\mathbf{v}}^{-1}(0) = \{(x,y) \in \mathbb{C}^2 \mid xy = yx = 0\}$$
and $\mathcal{M}_0(\mathbf{v}) = \{pt\}$ (namely, it is the orbit of $(0,0)$).

10.2. GIT moduli spaces for double quivers

If we choose $\theta = (1, -1)$, then $(x, y) \in \mu_{\mathbf{v}}^{-1}(0)$ is θ-semistable iff $y \neq 0$; any such pair is automatically stable, so
$$\mathcal{M}_\theta = \mathcal{M}_\theta^s = \{pt\}.$$

Similarly, if we choose $\theta = (-1, 1)$, then (x, y) is semistable iff $x \neq 0$, so again, $\mathcal{M}_\theta = \mathcal{M}_\theta^s = \{pt\}$.

Example 10.18. Let $\vec{Q} = \underset{\bullet}{\circlearrowright}$, $\mathbf{v} = n$. Note that in this case condition $\theta \cdot \mathbf{v} = 0$ implies $\theta = 0$.

In this case $R(\vec{Q}, \mathbf{v}) = \operatorname{End}(\mathbb{C}^n)$; as was shown in Example 9.12, in this case the GIT quotient is
$$\mathcal{R}_0(n) = R(n) /\!/ \operatorname{PGL}(n) = \operatorname{Spec}(\mathbb{C}[\lambda_1, \ldots, \lambda_n]^{S_n}) = \mathbb{C}^n/S_n \simeq \mathbb{C}^n.$$

Let us now consider the corresponding double quiver. Then
$$\mu_n^{-1}(0) = \{(x, y) \mid x, y \in \operatorname{End}(\mathbb{C}^n), \quad [x, y] = 0\}$$
is the variety of pairs of commuting matrices. Since any pair of commuting matrices can be brought simultaneously to an upper triangular form, any invariant polynomial function on $\mu^{-1}(0)$ is uniquely determined by its values on pairs of diagonal matrices; thus,
$$A = \mathbb{C}[\mu^{-1}(0)]^{\operatorname{PGL}(n)} = \mathbb{C}[\lambda_1, \ldots, \lambda_n, \mu_1, \ldots, \mu_n]^{S_n}$$
and
$$\mathcal{M}_0 = \operatorname{Spec}(A) = \mathbb{C}^{2n}/S_n.$$

This variety is no longer smooth. For example, for $n = 2$, the algebra A is given by
$$A = \mathbb{C}[\lambda_1, \lambda_2, \mu_1, \mu_2]^{\mathbb{Z}_2},$$
where $\mathbb{Z}_2$ acts on the variables by $\lambda_1 \leftrightarrow \lambda_2$, $\mu_1 \leftrightarrow \mu_2$. If we make a change of variables
$$x = \lambda_1 + \lambda_2,$$
$$y = \mu_1 + \mu_2,$$
$$s = \lambda_1 - \lambda_2,$$
$$t = \mu_1 - \mu_2$$
so that the action of $\mathbb{Z}_2$ is given by $s \to -s, t \to -t$, then
$$A = \mathbb{C}[x, y] \otimes \mathbb{C}[s, t]^{\mathbb{Z}_2} = \mathbb{C}[x, y, u, v, w]/(uv - w^2), \quad u = s^2, v = t^2, w = st.$$

Thus, in this case $\mathcal{M}_0 = Q \times \mathbb{C}^2$, where Q is a quadric in $\mathbb{C}^3$ defined by the equation $uv = w^2$, which is obviously not smooth.

10.3. Framed representations

We now introduce the final ingredient in the construction of the quiver varieties, called *framing*. As before, $\vec{Q}$ is an arbitrary quiver, with the set of vertices I and set of (oriented) edges Ω.

Definition 10.19. Let W be an I-graded vector space: $W = \bigoplus_{k \in I} W_k$. A W-*framed* representation of $\vec{Q}$ is a representation $V = (V_k, x_h)$ of $\vec{Q}$ together with a collection of linear maps
$$j_k \colon V_k \to W_k$$
for every $k \in I$.

A morphism $f \colon V \to V'$ of two W-framed representations (with the same W) is a morphism of representations of $\vec{Q}$ which also commutes with j_k: $j'_k \circ f_k = j_k \colon V_k \to W_k$.

An illustration of such a representation is given in Figure 10.1.

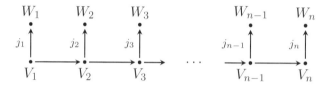

Figure 10.1. A framed representation.

Framed representations are closely related with representations of a new quiver obtained from $\vec{Q}$ by adding, for every $k \in I$, a new vertex k' and an edge $k \to k'$. The difference is in the definition of morphisms: the definition of framed representation requires the morphisms to be identity on W. Note in particular that because of this, morphisms of framed representations do not form an abelian group.

The importance of framing will be clarified later. For now we just mention that various constructions we will do later will turn out to be independent of the choice of the framing space W as long as it is large enough.

As in Section 2.1, it is easy to identify isomorphism classes of framed representations of given dimension with the orbits of a certain group action. Namely, let V, W be I-graded vector spaces. Define

$$(10.11) \quad R(V, W) = \left(\bigoplus_{k \in I} \mathrm{Hom}_{\mathbb{C}}(V_k, W_k) \right) \oplus \left(\bigoplus_{h \in \Omega} \mathrm{Hom}_{\mathbb{C}}(V_{s(h)}, V_{t(h)}) \right).$$

10.3. Framed representations

Similarly, for $\mathbf{v}, \mathbf{w} \in \mathbb{Z}_+^I$, define

(10.12) $$R(\mathbf{v}, \mathbf{w}) = R(\mathbb{C}^{\mathbf{v}}, \mathbb{C}^{\mathbf{w}}).$$

It is clear that we have isomorphisms $R(V, W) \simeq R(\dim V, \dim W)$.

The following theorem (the proof of which is left as an exercise) is an analog of Theorem 2.1.

Theorem 10.20. *Let W be an I-graded vector space. Then there is a natural bijection between the isomorphism classes of W-framed representations of $\vec{Q}$ of dimension $\mathbf{v}$ and the quotient space*

(10.13) $$R(\mathbf{v}, \mathbf{w})/\mathrm{GL}(\mathbf{v}).$$

Note that we are only taking the quotient by the action of $\mathrm{GL}(\mathbf{v})$ and not $\mathrm{GL}(\mathbf{w})$.

This quotient space is usually non-Hausdorff. We will instead consider the GIT quotient

(10.14) $$\mathcal{R}_0(\mathbf{v}, \mathbf{w}) = R(\mathbf{v}, \mathbf{w}) /\!/ \mathrm{GL}(\mathbf{v}).$$

Theorem 10.21. *For any quiver $\vec{Q}$, we have*

$$\mathcal{R}_0(\mathbf{v}, \mathbf{w}) = \mathcal{R}_0(\mathbf{v}).$$

In particular, if $\vec{Q}$ is Dynkin, then $\mathcal{R}_0(\mathbf{v}, \mathbf{w})$ is a point.

Proof. Since $\mathcal{R}_0$ is the set of closed orbits, it suffices to prove that if $(x, j) \in R(\mathbf{v}, \mathbf{w})$ is such that its orbit is closed, then $j_k = 0$ for all k. To prove this, consider a one-parameter subgroup $\lambda(t) = t^{-1} \cdot \mathrm{id} \subset \mathrm{GL}(\mathbf{v})$. Obviously, $\lambda(t).(x, j) = (x, tj)$, so $\lim_{t \to 0} \lambda(t)(x, j) = (x, 0)$. Therefore, if the orbit of (x, j) is closed, then $j = 0$. $\square$

Thus, in this case framing does not give any new quiver varieties.

We can also consider the twisted GIT quotients. Let $\theta \in \mathbb{Z}^I$, and let $\chi_\theta \colon \mathrm{GL}(\mathbf{v}) \to \mathbb{C}^\times$ be the corresponding character as in (10.3). Note that because of the framing, the subgroup of scalars acts nontrivially on the space $R(\mathbf{v}, \mathbf{w})$, so we no longer require that $\theta \cdot \mathbf{v} = 0$.

From now on, we will be mostly interested in the special choice of θ. Namely, for an element $\theta \in \mathbb{R}^I$, we write

(10.15) $$\theta > 0 \quad \text{if } \theta_i > 0 \text{ for all } i$$

and $\theta < 0$ if $-\theta > 0$.

Theorem 10.22. *Let $\vec{Q}$ be an arbitrary quiver. Let $\theta \in \mathbb{Z}^I$, $\theta > 0$, and let $\chi = \chi_\theta$ be the corresponding character of $\mathrm{GL}(V)$ as in (10.3). Then:*

(1) *Element $(x,j) \in R(\mathbf{v}, \mathbf{w})$ is χ-semistable iff the following condition holds:*

(10.16)
If $V' \subset V^x$ is a subrepresentation of $\vec{Q}$ and $V' \subset \mathrm{Ker}\, j$, then $V' = 0$.

(2) *Any χ-semistable element is automatically stable.*

Proof. The proof repeats with minor changes the proof of Theorem 10.5.

Assume that (10.16) holds. We need to show that then (x,j) is semistable. We will use Mumford's criterion (Theorem 9.25); thus, we need to show that if $\lambda(t)$ is a one-parameter subgroup such that $\lim_{t\to 0} \lambda(t).(x,j)$ exists, then $\langle \chi, \lambda \rangle \geq 0$.

Let $\lambda(t)$ be such a subgroup. As in the proof of Theorem 10.5, such a subgroup gives rise to decreasing filtration

$$(10.17) \quad V = V^{\geq a} \supset V^{\geq a+1} \supset \cdots \supset V^{\geq b} = \{0\}, \quad V^{\geq n} = \bigoplus_{m \geq n} V^m,$$

which is x-stable. Since $\lambda(t).(x,j) = (\lambda(t)x\lambda^{-1}(t), j\lambda(t)^{-1})$, existence of limit as $t \to 0$ implies that $j|_{V^k} = 0$ for $k > 0$, so $V^{\geq 1}$ is an x-stable subrepresentation contained in $\mathrm{Ker}\, j$. By assumption, this means $V^{\geq 1} = 0$, so $V = \bigoplus_{n \leq 0} V^n$. Then

$$(10.18) \quad \langle \chi, \lambda \rangle = -\theta \cdot \sum_{n \leq 0} n \dim V^n = -\sum_{n \leq 0, k} n\theta_k \dim V_k^n \geq 0.$$

Conversely, if (x,j) is χ-semistable and $V' \subset \mathrm{Ker}\, j$, then we can define a one-parameter subgroup $\lambda(t)$ such that $\lambda(t)|_{V'} = t$, $\lambda(t)|_{V/V'} = 1$. Then the limit $\lambda(t).(x,j)$ exists and, by Mumford's criterion, $\langle \chi, \lambda \rangle = -\theta \cdot \dim V' \geq 0$. Since $\theta > 0$, this implies $\dim V' = 0$.

To prove part (2), assume that (x,j) is semistable but not stable. Then there exists a nontrivial one-parameter subgroup $\lambda(t)$ such that the limit $\lim_{t\to 0} \lambda(t).(x,j)$ exists, yet $\langle \chi, \lambda \rangle = 0$. Writing again the eigenspace decomposition $V = \bigoplus V^n$, we get that $V^n = 0$ for $n > 0$ (from semistability) and $V^n = 0$ for $n < 0$ (from (10.18) and $\langle \chi, \lambda \rangle = 0$). Thus, $V = V^0$, i.e. $\lambda(t) = 1$, which contradicts the assumption of nontriviality of λ. $\square$

Note that this shows that if θ, θ' are two stability parameters which are both positive, $\theta > 0$, $\theta' > 0$, then they define the same stability condition, so $\mathcal{R}_\theta(\mathbf{v}, \mathbf{w}) = \mathcal{R}_{\theta'}(\mathbf{v}, \mathbf{w})$. We will prove a more general result later (see Corollary 10.39).

10.3. Framed representations

Corollary 10.23. *Let $\theta > 0$. Then, if $\mathcal{R}_\theta(\mathbf{v}, \mathbf{w})$ is nonempty, it is a nonsingular algebraic variety of dimension*
$$\dim \mathcal{R}_\theta(\mathbf{v}, \mathbf{w}) = \mathbf{v} \cdot \mathbf{w} - \langle \mathbf{v}, \mathbf{v} \rangle.$$

Proof. By Theorem 10.22, we have $\mathcal{R}_\theta = R^{ss}/\!/\mathrm{GL}(\mathbf{v}) = R^s/\mathrm{GL}(\mathbf{v})$. By Theorem 9.23, we see that $\mathcal{R}_\theta$ is nonsingular of dimension
$$\dim \mathcal{R}_\theta = \dim R(\mathbf{v}, \mathbf{w}) - \dim \mathrm{GL}(\mathbf{v})$$
$$= \sum_{i \in I} \mathbf{v}_i \mathbf{w}_i + \sum_{h \in \Omega} \mathbf{v}_{s(h)} \mathbf{v}_{t(h)} - \sum_{i \in I} \mathbf{v}_i^2$$
$$= \mathbf{v} \cdot \mathbf{w} - \langle \mathbf{v}, \mathbf{v} \rangle.$$
□

As before, we can also describe the map $\mathcal{R}_\theta \to \mathcal{R}_0$; namely, it is given by

(10.19) $$\pi([x, j]) = [x^{ss}, 0],$$

where x^{ss} is the semisimplification of representation V^x (compare with (10.6)).

Example 10.24. Let $\theta > 0$, $\vec{Q} = \bullet$; then representations of $\vec{Q}$ are just vector spaces and, for $\dim V = n$, $\dim W = r$, we get
$$\mathcal{R}_\theta(n, r) = \{j \colon \mathbb{C}^n \to \mathbb{C}^r \mid \operatorname{Ker} j = \{0\}\}/\mathrm{GL}(n, \mathbb{C})$$
$$= G(n, r)$$
is the Grassmannian of n-dimensional subspaces in $\mathbb{C}^r$.

Example 10.25. Let $\theta > 0$ and let $\vec{Q}$ be a quiver of type A:

$$\vec{Q} = \bullet \longrightarrow \bullet \longrightarrow \bullet \longrightarrow \quad \cdots \quad \longrightarrow \bullet \longrightarrow \bullet \qquad (l \text{ vertices})$$

and let $\mathbf{w} = (0, 0, \ldots, 0, r)$. Then
$$R^{ss}(\mathbf{v}, \mathbf{w}) = \{(x_1, \ldots, x_{l-1}, j)\},$$
where $x_i \colon V_i \to V_{i+1}$, $j \colon V_l \to \mathbb{C}^r$ are linear operators such that any composition $j \circ x_{l-1} \circ x_{l-2} \circ \cdots \circ x_a \colon V_a \to \mathbb{C}^r$ is injective:
$$V_1 \xrightarrow{x_1} V_2 \xrightarrow{x_2} V_3 \to \quad \cdots \quad \to V_{l-1} \xrightarrow{x_{l-1}} V_l \xrightarrow{j} \mathbb{C}^r.$$

From this, it is easy to deduce that any element $(x, j) \in R^{ss}$ defines a flag $V_1 \subset V_2 \subset \cdots \subset V_l \subset \mathbb{C}^r$ with $\dim V_i = \mathbf{v}_i$ and that
$$\mathcal{R}_\theta(\mathbf{v}, \mathbf{w}) = \mathcal{F}(\mathbf{v}_1, \mathbf{v}_2, \ldots, \mathbf{v}_l, r)$$
is the corresponding flag variety.

10.4. Framed representations of double quivers

Let us now combine the two ideas of the previous sections, doubling the quiver and framing. Namely, let Q be a graph with set of vertices I and let V, W be I-graded vector spaces. Define

(10.20) $\qquad R(Q^\sharp, V, W) = R(Q^\sharp, V) \oplus L(V, W) \oplus L(W, V),$

where
$$R(Q^\sharp, V) = \bigoplus_{h \in H} \operatorname{Hom}(V_{s(h)}, V_{t(h)}),$$
$$L(V, W) = \bigoplus_{k \in I} \operatorname{Hom}(V_k, W_k),$$

and H is the set of edges of the double quiver $Q^\sharp$ (compare with Section 5.1).

We will write elements of $R(Q^\sharp, V, W)$ as triples $m = (z, i, j)$, where

(10.21)
$$z = \bigoplus x_h, \quad x_h \colon V_{s(h)} \to V_{t(h)},$$
$$j = \bigoplus j_k, \quad j_k \colon V_k \to W_k,$$
$$i = \bigoplus i_k, \quad i_k \colon W_k \to V_k,$$

as illustrated in Figure 10.2.

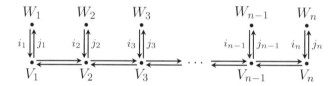

Figure 10.2. A framed representation of a double quiver.

Note that the definition of $R(Q^\sharp, V, W)$ is independent of the choice of orientation of Q; however, for every choice of orientation $\vec{Q} = (Q, \Omega)$ we have a canonical isomorphism

$$R(Q^\sharp, V, W) = T^* R(\vec{Q}, V, W)$$

(compare with (5.2)). In particular, every choice of orientation of Q gives rise to a symplectic structure on $R(Q^\sharp, V, W)$ defined in the same way as in (10.9). More generally, every choice of a skew-symmetric function $\varepsilon \colon H \to \mathbb{C}^\times$ (see (5.3)) gives rise to a symplectic form on $R(Q^\sharp, V, W)$:

(10.22) $\qquad \omega_\varepsilon(z, i, j; z', i', j') = \operatorname{tr}_V \left(\sum_{h \in H} \varepsilon(h) z_{\bar{h}} z'_h + \sum_{k \in I} i_k j'_k - i'_k j_k \right)$

(compare with (10.9)).

10.4. Framed representations of double quivers

We have a natural action of $\mathrm{GL}(V)$ on $R(Q^\sharp, V, W)$, given by $g(z, i, j) = (gzg^{-1}, gi, jg^{-1})$.

Lemma 10.26. *The action of $\mathrm{GL}(V)$ on $R(Q^\sharp, V, W)$ is Hamiltonian, and the moment map*
$$\mu_{V,W} \colon R(Q^\sharp, V, W) \to \mathfrak{gl}(V)$$
is given by

(10.23) $$\mu_{V,W}(z, i, j) = \sum_{h \in H} \varepsilon(h) z_h z_{\bar{h}} - \sum_k i_k j_k.$$

(We identify $\mathfrak{gl}(V) \simeq \mathfrak{gl}(V)^*$ using the bilinear pairing $(a, b) = \mathrm{tr}(ab)$ on $\mathfrak{gl}(V)$.)

This lemma is obtained by combining the arguments in the proof of Theorem 10.10 and Example 9.48; details are left to the reader.

We can now consider the corresponding GIT quotients. First, we can define the usual GIT quotient

(10.24) $$\mathcal{M}_0(V, W) = \mu_{V,W}^{-1}(0) /\!\!/ \mathrm{GL}(V)$$

(note that we are not taking the quotient by action of $\mathrm{GL}(W)$). This is an affine algebraic variety whose points are in bijection with closed orbits in $R(Q^\sharp, V, W)$ satisfying the extra condition $\mu(z, i, j) = 0$. Note that because of the framing, we can no longer interpret them as the semisimple representations of the preprojective algebra.

We can also define the twisted GIT quotients. Recall that any $\theta \in \mathbb{Z}^I$ defines a character χ_θ of the group $\mathrm{GL}(V)$ by $\chi_\theta(g) = \prod(\det g_i)^{-\theta_i}$ (see (10.3)).

Definition 10.27. Let Q be a graph, $\varepsilon \colon H \to \mathbb{C}^\times$ a skew-symmetric function (see (5.3)), and $\chi = \chi_\theta, \theta \in \mathbb{Z}^I$. Then the variety

(10.25) $$\mathcal{M}_\theta(Q; V, W) = \mu_{V,W}^{-1}(0) /\!\!/_\chi \mathrm{GL}(V)$$

is called the *quiver variety*.

These varieties were introduced by Nakajima in [**Nak1994**].

We will usually drop Q from the notation. As before, we will also frequently use the notation

(10.26)
$$\begin{aligned} R(Q^\sharp, \mathbf{v}, \mathbf{w}) &= R(Q^\sharp, \mathbb{C}^{\mathbf{v}}, \mathbb{C}^{\mathbf{w}}), \\ \mathcal{M}_0(\mathbf{v}, \mathbf{w}) &= \mathcal{M}_0(\mathbb{C}^{\mathbf{v}}, \mathbb{C}^{\mathbf{w}}), \\ \mathcal{M}_\theta(\mathbf{v}, \mathbf{w}) &= \mathcal{M}_\theta(\mathbb{C}^{\mathbf{v}}, \mathbb{C}^{\mathbf{w}}) \end{aligned}$$

for $\mathbf{v}, \mathbf{w} \in \mathbb{Z}_+^I$.

It follows from the general theory discussed in Chapter 9 that for any θ, the variety $\mathcal{M}_\theta(\mathbf{v},\mathbf{w})$ is quasiprojective, with a projective morphism $\pi\colon \mathcal{M}_\theta(\mathbf{v},\mathbf{w}) \to \mathcal{M}_0(\mathbf{v},\mathbf{w})$. Moreover, by Theorem 9.53 both $\mathcal{M}_\theta(\mathbf{v},\mathbf{w})$ and $\mathcal{M}_0(\mathbf{v},\mathbf{w})$ have natural Poisson structure, and $\pi\colon \mathcal{M}_\theta(\mathbf{v},\mathbf{w}) \to \mathcal{M}_0(\mathbf{v},\mathbf{w})$ is a morphism of Poisson varieties.

Remark 10.28. One can also consider more general varieties
$$\mathcal{M}_{\theta,\lambda}(V,W) = \mu_{V,W}^{-1}(\lambda)/\!\!/_{\chi_\theta} \mathrm{GL}(V),$$
where $\lambda = \sum_I \lambda_i \, \mathrm{id}_{V_i} \in \mathfrak{gl}(V)$. We will return to these varieties in Section 10.8.

Lemma 10.29. *One has an isomorphism of algebraic varieties*
$$\mathcal{M}_\theta(\mathbf{v},\mathbf{w}) \simeq \mathcal{M}_{-\theta}(\mathbf{v},\mathbf{w}).$$

Proof. Indeed, if $m = (z,i,j) \in R(Q^\sharp, V, W)$ is θ-stable, then $m^* = (z^*, j^*, i^*) \in R(Q^\sharp, V^*, W^*)$ is $(-\theta)$-stable. $\square$

In particular, we denote
(10.27) $\qquad \mathcal{M}(\mathbf{v},\mathbf{w}) = \mathcal{M}_\theta(\mathbf{v},\mathbf{w}) \simeq \mathcal{M}_{-\theta}(\mathbf{v},\mathbf{w}), \qquad \theta > 0$

(recall that by Theorem 10.22, all $\theta > 0$ give the same stability condition).

10.5. Stability conditions

We continue using the notation and conventions of the previous section. In this section, we show that the stability condition for $(z,i,j) \in R(Q^\sharp, \mathbf{v}, \mathbf{w})$ can be described in terms of quiver representations. A similar result for framed representations of $\vec{Q}$ (before doubling) was given in Theorem 10.22. Using this, we prove that for a generic value of θ, the variety $\mathcal{M}_\theta(\mathbf{v},\mathbf{w})$ is nonsingular.

Lemma 10.30. *Let $m = (z,i,j) \in R(Q^\sharp, V, W)$ and let $\lambda\colon \mathbb{C}^\times \to \mathrm{GL}(V)$ be a one-parameter subgroup such that the limit $m_0 = \lim_{t\to 0} \lambda(t).m$ exists. Let $V = \bigoplus_{n\in\mathbb{Z}} V^n$ be the eigenspace decomposition defined by λ, as in the proof of Lemma 2.9. Consider the filtration*

(10.28) $\qquad \dots V^{\geq -1} \supset V^{\geq 0} \supset V^{\geq 1} \dots, \qquad V^{\geq n} = \bigoplus_{m \geq n} V^m.$

Then this filtration has the following properties:

(1) *Each subspace $V^{\geq n}$ is z-stable.*
(2) $\mathrm{Im}(i) \subset V^{\geq 0}$.
(3) $V^{\geq 1} \subset \mathrm{Ker}(j)$.

10.5. Stability conditions

Moreover, in this case the framed representation V^{m_0} defined by $m_0 = \lim_{t\to 0} \lambda(t).m$ is isomorphic (as a W-framed representation) to $\operatorname{gr} V$, where $\operatorname{gr} V = \bigoplus (\operatorname{gr} V)^m$, $\operatorname{gr}(V)^m = V^{\geq m}/V^{\geq m+1}$, with the structure of W-framed representation defined by induced maps

$$\operatorname{gr} z \colon \operatorname{gr} V \to \operatorname{gr} V,$$
$$\operatorname{gr} j \colon (\operatorname{gr} V)^0 \to W,$$
$$\operatorname{gr} i \colon W \to (\operatorname{gr} V)^0.$$

The proof of this lemma is similar to the proof of Lemma 2.9; details are left to the reader.

The following definition is a framed version of Definition 10.13.

Definition 10.31. Let $\theta \in \mathbb{R}^I$. Then a W-framed representation V of $Q^\sharp$ is called θ-semistable if for any $Q^\sharp$ subrepresentation $V' \subset V$ we have

(10.29) $\qquad (V' \subset \operatorname{Ker} j) \implies \theta \cdot \mathbf{dim}\, V' \leq 0,$

(10.30) $\qquad (V' \supset \operatorname{Im} i) \implies \theta \cdot \mathbf{dim}\, V' \leq \theta \cdot \mathbf{dim}\, V.$

Similarly, V is called θ-stable if for any proper nonzero $Q^\sharp$ subrepresentation $V' \subset V$ we have

(10.31) $\qquad (V' \subset \operatorname{Ker} j) \implies \theta \cdot \mathbf{dim}\, V' < 0,$

(10.32) $\qquad (V' \supset \operatorname{Im} i) \implies \theta \cdot \mathbf{dim}\, V' < \theta \cdot \mathbf{dim}\, V.$

Note that (10.30) can be rewritten as follows:

(10.33) for any quotient representation $V'' = V/V'$ of $Q^\sharp$ such that the composition $W \xrightarrow{i} V \to V''$ is zero, one has $\theta \cdot \mathbf{dim}\, V'' \geq 0$,

and (10.32) can be rewritten as follows:

(10.34) for any proper quotient representation $V'' = V/V'$ of $Q^\sharp$ such that the composition $W \xrightarrow{i} V \to V''$ is zero, one has $\theta \cdot \mathbf{dim}\, V'' > 0$.

As before, this is related to the notion of stability as defined in GIT.

Theorem 10.32. *Let $\theta \in \mathbb{Z}^I$ and let χ_θ be the corresponding character of $\operatorname{GL}(\mathbf{v})$. Then an element $m = (z, i, j) \in R(Q^\sharp, \mathbf{v}, \mathbf{w})$ is χ_θ-semistable (respectively, θ-stable) iff the corresponding W-framed representation V is θ-semistable (respectively, θ-stable).*

Note that while the notion of χ_θ-stability is only defined for integer θ, Definition 10.31 can be used for any real vector θ.

Proof. The proof is similar to the proof of Theorem 10.22.

Assume that conditions (10.29), (10.33) hold. We will use Mumford's criterion to prove that m is semistable. Let $\lambda(t)$ be a one-parameter subgroup such that $\lim_{t\to 0}\lambda(t).m$ exists. By Lemma 10.30, it gives a filtration (10.28) on V, and each of $V^{\geq n}$, $n > 0$, is z-stable and is contained in $\text{Ker}(j)$, so by (10.29), $\theta \cdot \dim V^{\geq n} \leq 0$ for $n > 0$, which gives

$$\theta \cdot \sum_{n>0} n \dim V^n = \theta \cdot \sum_{n>0} \dim V^{\geq n} \leq 0.$$

Similarly, existence of limit implies that each of the quotient spaces $V^{\leq n} = V/V^{>n}$, $n < 0$, satisfies the condition of (10.33), so $\theta \cdot \dim V^{\leq n} \geq 0$. Thus,

$$\theta \cdot \sum_{n<0} n \dim V^n = -\theta \cdot \sum_{n<0} \dim V^{\leq n} \leq 0.$$

Combining the two inequalities, we get that $\langle \chi, \lambda \rangle = -\theta \sum_n n \dim V^n \geq 0$, so by Mumford's criterion, m is χ-semistable.

The proof in the opposite direction, as well as the part about stable representations, are obtained by easy modification of the proof in Theorem 10.22; details are left to the reader. $\square$

It turns out that for generic values of θ, every semistable representation is actually stable. Namely, let

(10.35) $$R'_+(\mathbf{v}) = \{\alpha \in \mathbb{Z}^I_+ \mid \alpha \neq 0,\ (\alpha, \alpha) \leq 2,\ \alpha \leqslant \mathbf{v}\},$$

where $(\,,\,)$ is the symmetrized Euler form. As was discussed in Section 7.11, for Dynkin and Euclidean graphs nonzero elements $\alpha \in \mathbb{Z}^I_+$ satisfying $(\alpha, \alpha) \leq 2$ are exactly the positive roots of the corresponding Kac–Moody algebra; in general, it is not so.

Definition 10.33. An element $\theta \in \mathbb{R}^I$ is called **v**-*generic* if for any $\alpha \in R'_+(\mathbf{v})$, $\theta \cdot \alpha \neq 0$.

The following theorem was proved in [**Nak1994**].

Theorem 10.34. *Let $\theta \in \mathbb{R}^I$ be **v**-generic. Then every θ-semistable element $m \in \mu_{\mathbf{v},\mathbf{w}}^{-1}(0)$ is θ-stable.*

Proof. Let $V = V^m$ be θ-semistable. We need to show that it is stable. We will argue by contradiction, assuming that $S \subset V$ is a proper subrepresentation such that $S \subset \text{Ker}\,j$, yet $\theta \cdot \dim S = 0$. Since $S \subset \text{Ker}\,j$, the ij term in the expression (10.23) for the moment map vanishes, so that $z|_S \in \mu_{\dim S}^{-1}(0)$, and S is a representation of the preprojective algebra Π; by Theorem 10.15, it is θ-semistable. It follows immediately from Lemma 10.14 that any θ-semistable representation of Π contains a nonzero θ-stable representation; thus, there exists θ-stable representation $V' \subset S$. Therefore,

10.5. Stability conditions

$\mathcal{M}_\theta(\mathbf{v}') \neq \varnothing$; since $\dim \mathcal{M}_\theta(\mathbf{v}') = 2 - (\mathbf{v}', \mathbf{v}')$ (Theorem 10.12), this implies that $(\mathbf{v}', \mathbf{v}') \leq 2$, so $\mathbf{v}' \in R'_+(\mathbf{v})$. Since $\mathbf{v}'$ is θ-stable, $\mathbf{v}' \cdot \theta = 0$, which contradicts the assumption that θ is $\mathbf{v}$-generic. $\square$

Combining this with Theorem 9.53, we get the following result.

Theorem 10.35. *Let $\theta \in \mathbb{Z}^I$ be $\mathbf{v}$-generic. Assume that the quiver variety $\mathcal{M}_\theta(\mathbf{v}, \mathbf{w})$ is nonempty. Then*

(1) $\mathcal{M}_\theta(\mathbf{v}, \mathbf{w})$ *is nonsingular of dimension*
$$\dim \mathcal{M}_\theta(\mathbf{v}, \mathbf{w}) = 2\mathbf{v} \cdot \mathbf{w} - 2\langle \mathbf{v}, \mathbf{v} \rangle.$$

(2) *The Poisson structure on $\mathcal{M}_\theta(\mathbf{v}, \mathbf{w})$ is nondegenerate, i.e. it comes from a symplectic form on $\mathcal{M}_\theta(\mathbf{v}, \mathbf{w})$.*

(3) *If the skew-symmetric function ε is defined by an orientation Ω of Q, so that $R(Q^\sharp, \mathbf{v}, \mathbf{w}) = T^*R(\vec{Q}, \mathbf{v}, \mathbf{w})$, and $\theta > 0$, then $\mathcal{M}_\theta$ contains $T^*\mathcal{R}_\theta(\mathbf{v}, \mathbf{w})$ as an open (possibly empty) subset.*

Example 10.36. Let $\theta > 0$. Then for any nonzero $\alpha \in \mathbb{Z}^I_+$, $\alpha \cdot \theta > 0$, so θ is $\mathbf{v}$-generic for any $\mathbf{v}$ and thus the quiver variety $\mathcal{M}(\mathbf{v}, \mathbf{w})$ is always smooth. Note also that in this case, the stability conditions (10.29), (10.30) are simplified: the first condition becomes

(10.36) $\qquad (V' \subset \operatorname{Ker} j, \quad V' \text{ is } z\text{-stable}) \implies V' = \{0\}$

(compare with the stability condition (10.16) for the quiver before doubling), and the second condition is automatic.

Similarly, if $\theta < 0$, then it is automatically $\mathbf{v}$-generic for any $\mathbf{v}$ and the stability conditions (10.29), (10.30) become

(10.37) $\qquad (V' \supset \operatorname{Im} i, \quad V' \text{ is } z\text{-stable}) \implies V' = V$.

Moreover, one has the following result.

Theorem 10.37. *Let $\theta \in \mathbb{Z}^I$ be $\mathbf{v}$-generic. Then the variety $\mathcal{M}_\theta(\mathbf{v}, \mathbf{w})$ is connected.*

The proof of this theorem is not given here. It can be found in [**CB2001**, pp. 261–262]. (Note that this result is only stated there for $\theta < 0$; however, the same proof works for any $\mathbf{v}$-generic θ.)

The set of $\mathbf{v}$-generic elements in $\mathbb{R}^I$ can be written as a complement to a collection of hyperplanes:

(10.38)
$$\mathbb{R}^I(\mathbf{v}) = \{\theta \in \mathbb{R}^I \mid \theta \text{ is } \mathbf{v}\text{-generic}\}$$
$$= \mathbb{R}^I - \bigcup_{\alpha \in R'_+(\mathbf{v})} D_\alpha, \qquad D_\alpha = \{\theta \mid \theta \cdot \alpha = 0\}.$$

We will refer to connected components of $\mathbb{R}^I(\mathbf{v})$ as *chambers* (this is a natural analog of Weyl chambers in the theory of root systems).

Theorem 10.38. *Let V be a W-framed representation of $Q^\sharp$ defined by an element $m \in \mu_{\mathbf{v},\mathbf{w}}^{-1}(0)$. Let $\theta, \theta' \in \mathbb{R}^I(\mathbf{v})$ be in the same chamber. Then V is θ-stable iff it is θ'-stable.*

Proof. It is immediate from the definitions that for a fixed representation V, the set $\{\theta \in \mathbb{R}^I(\mathbf{v}) \mid V$ is θ-stable$\}$ is given by a finite collection of inequalities $\theta \cdot \mathbf{c}^k > 0$, $\theta \cdot \mathbf{d}^l < 0$ and thus is open. Similarly, the set $\{\theta \in \mathbb{R}^I(\mathbf{v}) \mid V$ is θ-semistable$\}$ is closed in $\mathbb{R}^I(\mathbf{v})$. But by Theorem 10.34, these sets coincide. Thus, such a set is a union of connected components of $\mathbb{R}^I(\mathbf{v})$. $\square$

Corollary 10.39. *If $\theta, \theta' \in \mathbb{Z}^I$ are in the same connected component of $\mathbb{R}^I(\mathbf{v})$, then $\mathcal{M}_\theta(\mathbf{v}, \mathbf{w}) \simeq \mathcal{M}_{\theta'}(\mathbf{v}, \mathbf{w})$ as algebraic varieties.*

This corollary makes it possible to define $\mathcal{M}_\theta(\mathbf{v}, \mathbf{w})$ for any $\theta \in \mathbb{R}^I(\mathbf{v})$, by replacing θ by an arbitrary $\theta' \in \mathbb{Z}^I$ in the same chamber as θ.

10.6. Quiver varieties as symplectic resolutions

As in Section 10.4, let Q be a graph, let ε be a skew-symmetric function on the set of edges of $Q^\sharp$, and let $\mathcal{M}_\theta(\mathbf{v}, \mathbf{w})$ be the corresponding quiver variety.

As in Section 9.4, a point $m = (z, i, j) \in R(Q^\sharp, V, W)$ will be called *regular* if its orbit is closed and the stabilizer is trivial. We denote by $\mathcal{M}_0^{reg}(\mathbf{v}, \mathbf{w}) \subset \mathcal{M}_0(\mathbf{v}, \mathbf{w})$ the set of regular orbits. For convenience, we also let $\mathcal{M}_0^{reg}(0, \mathbf{w}) = \mathcal{M}_0(0, \mathbf{w}) = \{pt\}$.

Note that $\mathcal{M}_0^{reg}(\mathbf{v}, \mathbf{w})$ can be empty. We will give a criterion of nonemptiness of $\mathcal{M}_0^{reg}(\mathbf{v}, \mathbf{w})$ in Section 13.4.

Theorem 10.40. *Let $\theta \in \mathbb{Z}^I$ be $\mathbf{v}$-generic, and assume that $\mathcal{M}_0^{reg}(\mathbf{v}, \mathbf{w})$ is nonempty. Then $\pi \colon \mathcal{M}_\theta(\mathbf{v}, \mathbf{w}) \to \mathcal{M}_0(\mathbf{v}, \mathbf{w})$ is a symplectic resolution of singularities.*

Indeed, in this case Theorems 10.35 and 10.37 imply that $\mathcal{M}_\theta(\mathbf{v}, \mathbf{w})$ is smooth and irreducible. Now the result follows from Theorem 9.55.

Note that the conditions listed in the theorem are sufficient but not required for $\pi \colon \mathcal{M}_\theta(\mathbf{v}, \mathbf{w}) \to \mathcal{M}_0(\mathbf{v}, \mathbf{w})$ to be a resolution of singularities. For example, π could be a resolution even if $\mathcal{M}_0^{reg}$ is empty.

For future use, we note that the regularity condition can also be rewritten in terms of quiver representations as follows.

10.6. Quiver varieties as symplectic resolutions

Theorem 10.41. *A point $m = (z, i, j) \in R(Q^\sharp, V, W)$ is regular iff the following conditions are satisfied*:

(10.39)
for any z-invariant subspace $V' \subset V$ we have $V' \subset \operatorname{Ker}(j) \implies V' = \{0\}$,

(10.40)
for any z-invariant subspace $V'' \subset V$ we have $V'' \supset \operatorname{Im}(i) \implies V'' = V$.

Note that by results of Example 10.36, (10.39) is equivalent to the requirement that m is θ-stable for some $\theta > 0$, and (10.40) is equivalent to the requirement that m is θ-stable for some $\theta < 0$.

Proof. If m is regular, then by Theorem 9.29 it is θ-stable for any θ; thus, (10.39), (10.40) hold.

Conversely, assume that (10.39), (10.40) hold. Let us prove that the orbit $\mathbb{O}_m$ is closed. Indeed, if m_0 is in the closure of $\mathbb{O}_m$, then there exists a one-parameter subgroup $\lambda \colon \mathbb{C}^\times \to \operatorname{GL}(\mathbf{v})$ such that $\lim_{t \to 0} \lambda(t).m \in \mathbb{O}_{m_0}$. Consider the corresponding filtration on V as in Lemma 10.30. Then (10.40) implies $V^{\geq 0} = V$; similarly, (10.39) implies $V^{\geq 1} = 0$. Therefore, $\lambda(t).m = m$, so $m_0 \in \mathbb{O}_m$, and the orbit is closed.

To prove that the stabilizer is trivial, assume that $g \in \operatorname{GL}(V)$ satisfies $g.m = m$. Consider the subspace $V' = V^g = \{v \in V \mid gv = v\}$. This subspace is z-invariant and contains $\operatorname{Im}(i)$, so $V' = V$. Thus, $g = 1$. $\square$

For future use, we also give here the following result, which is a modification of [**VV1999**, Lemma 1]. Let $\mathbf{v}', \mathbf{v}'', \mathbf{w}', \mathbf{w}'' \in \mathbb{Z}_+^I$ and let $\mathbf{v} = \mathbf{v}' + \mathbf{v}''$, $\mathbf{w} = \mathbf{w}' + \mathbf{w}''$. Then we can write $\mathbb{C}^{\mathbf{v}} = \mathbb{C}^{\mathbf{v}'} \oplus \mathbb{C}^{\mathbf{v}''}$. It is easy to see that it gives rise to a morphism

(10.41)
$$\mathcal{M}_0(\mathbf{v}', \mathbf{w}') \times \mathcal{M}_0(\mathbf{v}'', \mathbf{w}'') \to \mathcal{M}_0(\mathbf{v}, \mathbf{w})$$
$$(z', i', j') \times (z'', i'', j'') \mapsto (z' \oplus z'', i' \oplus i'', j' \oplus j'')$$

which is independent of any choices.

Theorem 10.42. *We have an isomorphism*

$$\mathcal{M}_0(\mathbf{v}, \mathbf{w}) = \bigsqcup_{\mathbf{v}' \leq \mathbf{v}} \mathcal{M}_0^{reg}(\mathbf{v}', \mathbf{w}) \times \mathcal{M}_0(\mathbf{v} - \mathbf{v}', 0).$$

In particular, if $\vec{Q}$ is Dynkin, then we have

$$\mathcal{M}_0(\mathbf{v}, \mathbf{w}) = \bigsqcup_{\mathbf{v}' \leq \mathbf{v}} \mathcal{M}_0^{reg}(\mathbf{v}', \mathbf{w}).$$

Proof. Let $m = (z, i, j) \in \mu_{\mathbf{v},\mathbf{w}}^{-1}(0)$ be such that its orbit is closed. Let $G_m \subset \operatorname{GL}(\mathbf{v})$ be the stabilizer of m. Then $\mathbb{O}_m = \operatorname{GL}(\mathbf{v})/G_m$ is an affine

algebraic variety; by [**Ric1977**], this implies that G_m is reductive. Thus, V as a representation of G_m is completely reducible: we have a decomposition
$$V = \bigoplus_{\lambda \in \mathrm{Irr}(G_m)} V(\lambda),$$
where the sum is over all irreducible representations of G_m and $V(\lambda)$ is the isotypic component: $V(\lambda) \simeq n_\lambda \rho_\lambda$, where ρ_λ is the irreducible representation.

It is easy to see that each isotypic component $V(\lambda)$ is z-stable and we have $\mathrm{Im}\, i \subset V(1)$, $j|_{V(\lambda)} = 0$ for $\lambda \neq 1$. Thus, we can write $V = V' \oplus V''$, $V' = V(1)$, $V'' = \bigoplus_{\lambda \neq 1} V(\lambda)$. Under this decomposition, we get $(z, i, j) = (z', i, j) \oplus (z'', 0, 0)$. Since the orbit of m is closed, orbits of $m' = (z', i, j)$ and $m'' = (z'', 0, 0)$ must also be closed. Since the stabilizer G_m acts trivially on V', this implies that the stabilizer of m' in $\mathrm{GL}(\mathbf{v}')$ is trivial. Thus, we have the required decomposition. Uniqueness of such a decomposition follows from the fact that in this decomposition, we must have $V' = V^{G_m}$.

The statement for Dynkin quivers easily follows since by Theorem 10.16, $\mathcal{M}_0(\mathbf{v}'', 0)$ is a point. $\square$

10.7. Example: Type A quivers and flag varieties

In this section, we consider an explicit example of the quiver variety. As in Example 10.25, let $\theta > 0$ and let $\vec{Q}$ be a quiver of type A:

$$\vec{Q} = \bullet \longrightarrow \bullet \longrightarrow \bullet \longrightarrow \quad \cdots \quad \longrightarrow \bullet \longrightarrow \bullet \qquad (l \text{ vertices})$$

and let $\mathbf{w} = (0, 0, \ldots, 0, r)$.

Then
$$R^{ss}(Q^\sharp, \mathbf{v}, \mathbf{w}) = \{(x_1, \ldots, x_{l-1}, y_1, \ldots, y_{l-1}, i, j)\},$$
where $x_i \colon V_i \to V_{i+1}$, $y_i \colon V_{i+1} \to V_i$, $i \colon \mathbb{C}^r \to V_l$, $j \colon V_l \to \mathbb{C}^r$ are linear operators such that any composition $j \circ x_{l-1} \circ x_{l-2} \circ \cdots \circ x_a$ is injective.

$$V_1 \underset{y_1}{\overset{x_1}{\rightleftarrows}} V_2 \underset{y_2}{\overset{x_2}{\rightleftarrows}} V_3 \rightleftarrows \cdots \rightleftarrows V_{l-1} \underset{y_{l-1}}{\overset{x_{l-1}}{\rightleftarrows}} V_l \underset{i}{\overset{j}{\rightleftarrows}} \mathbb{C}^r$$

Injectivity of compositions implies that R^{ss} is empty unless $\mathbf{v}_1 \leq \mathbf{v}_2 \leq \cdots \leq r$. If for some i, we have $\mathbf{v}_i = \mathbf{v}_{i+1}$, one can easily see that removing one of these vertices does not change the quiver variety; thus, without loss of generality, we can assume that $\mathbf{v}$ satisfies the following inequality:

(10.42) $$\mathbf{v}_1 < \mathbf{v}_2 < \cdots < \mathbf{v}_l < r.$$

From now on we assume that condition (10.42) holds.

10.7. Example: Type A quivers and flag varieties

As in Example 10.25, any element $(x, j) \in R^{ss}$ defines a flag $V_1 \subset V_2 \subset \cdots \subset V_l \subset \mathbb{C}^r$, with $\dim V_i = v_i$. The equation $\mu(x, y, i, j) = 0$ gives $y_i|_{V_i} = y_{i-1}$, $i|_{V_l} = y_{l-1}$, which is equivalent to the condition $y_k = i|_{V_{k+1}}$. Therefore,

$$\mathcal{M}(\mathbf{v}, \mathbf{w}) = \{(F, y)\},$$
$$F = (0 \subset V_1 \subset V_2 \subset \cdots \subset V_l \subset \mathbb{C}^r) - \text{a partial flag in } \mathbb{C}^r, \dim V_i = \mathbf{v}_i,$$
$$y \colon \mathbb{C}^r \to \mathbb{C}^r, \ y(V_i) \subset V_{i-1}$$

(in the last formula, we let $V_0 = \{0\}, V_{l+1} = \mathbb{C}^r$).

Note that in the special case when $r = l + 1$, $\mathbf{v} = (1, 2, \ldots, l)$ this coincides with the definition of the Springer resolution $\widetilde{\mathcal{N}}$ of the nilpotent cone given in Section 9.8.

As in the case of Springer resolution, variety $\mathcal{M}(\mathbf{v}, \mathbf{w})$ can be identified with a cotangent bundle. Namely, the partial flag variety can be written as the quotient

$$\mathcal{F} = \operatorname{GL}(r, \mathbb{C})/P(\mathbf{v}),$$

where $P(\mathbf{v})$ is the stabilizer of the standard flag:

$$P(\mathbf{v}) = \{A \in \operatorname{GL}(r, \mathbb{C}) \mid AV_i \subset V_i\},$$
$$V_i = \langle e_1, \ldots, e_{\mathbf{v}_i} \rangle$$

(it is a parabolic subgroup in $\operatorname{GL}(r, \mathbb{C})$).

Repeating with necessary changes the arguments of Section 9.8, we get the following result.

Theorem 10.43. *As in Example 10.25, let $Q = A_l$, $\theta > 0$. Assume that $\mathbf{v}, r$ satisfy inequality (10.42). Then:*

(1)
$$\mathcal{M}(\mathbf{v}, \mathbf{w}) = T^* \mathcal{R}_\theta(\mathbf{v}, \mathbf{w}),$$

where $\mathcal{R}_\theta(\mathbf{v}, \mathbf{w}) = \mathcal{F}(\mathbf{v}; r)$ is the flag variety as in Example 10.25.

(2) *The action of $\operatorname{GL}(\mathbf{w})$ on $\mathcal{M}(\mathbf{v}, \mathbf{w})$ is Hamiltonian, and the moment map $\mu_{\mathbf{w}}$ is given by*

(10.43)
$$\mu_{\mathbf{w}} \colon \mathcal{M}(\mathbf{v}, \mathbf{w}) \to \mathfrak{gl}(\mathbf{w})$$
$$(F, y) \mapsto y$$

or, equivalently, in terms of the data (x, y, i, j),

$$\mu_{\mathbf{w}}(x, y, i, j) = ji.$$

(Note that it is the moment map for the action of $\operatorname{GL}(\mathbf{w})$, not $\operatorname{GL}(\mathbf{v})$.)

In fact, it turns out that the map $\mu_{\mathbf{w}}$ is closely related to the canonical projection $\pi\colon \mathcal{M}(\mathbf{v},\mathbf{w}) \to \mathcal{M}_0(\mathbf{v},\mathbf{w})$. Namely, let us extend the map $\mu_{\mathbf{w}}$ to a map
$$\mu_{\mathbf{w}}\colon \mu_{\mathbf{v},\mathbf{w}}^{-1}(0) \to \mathfrak{gl}(\mathbf{w})$$
given by $\mu_{\mathbf{w}}(x,y,i,j) = ji$. It is obvious that this map is constant on $\mathrm{GL}(\mathbf{v})$-orbits and thus defines a morphism
$$(10.44) \qquad \mathcal{M}_0(\mathbf{v},\mathbf{w}) = \mu_{\mathbf{v},\mathbf{w}}^{-1}(0)/\!\!/\mathrm{GL}(\mathbf{v}) \to \mathfrak{gl}(\mathbf{w}).$$

Theorem 10.44. *Under the assumptions of Theorem* 10.43, *the following results hold*:

(1) *The diagram*

$$\begin{array}{ccc} \mathcal{M}(\mathbf{v},\mathbf{w}) & \xrightarrow{\mu_{\mathbf{w}}} & \\ \pi\downarrow & & \mathfrak{gl}(\mathbf{w}) \\ \mathcal{M}_0(\mathbf{v},\mathbf{w}) & \xrightarrow{\mu_{\mathbf{w}}} & \end{array}$$

is commutative.

(2) *Morphism* (10.44) *is a closed embedding, so it identifies* $\mathcal{M}_0(\mathbf{v},\mathbf{w})$ *with an affine algebraic subvariety in* $\mathfrak{gl}(\mathbf{w})$. *Moreover, the image of $\mu_{\mathbf{w}}$ is the closure of the orbit of a certain nilpotent conjugacy class.*

(3) *Assume that* $(\mathbf{v},r)$ *satisfy the following inequality (which is stronger than* (10.42)*):*

$$(10.45) \qquad r - \mathbf{v}_l \geq \mathbf{v}_l - \mathbf{v}_{l-1} \geq \cdots \geq \mathbf{v}_2 - \mathbf{v}_1 \geq \mathbf{v}_1 > 0.$$

Then $\mathcal{M}(\mathbf{v},r) \to \mathcal{M}_0(\mathbf{v},r)$ *is a symplectic resolution of singularities.*

Proof. The first part is immediate from the definition.

To prove the second part, note that $y \in \mu_{\mathbf{w}}(\mathcal{M}_0)$ iff, for every k, we have
$$\dim \mathrm{Im}(y^k) \leq \mathbf{v}_{l-k+1}$$
or, equivalently, $\mathrm{rank}(y^k) \leq \mathbf{v}_{l-k+1}$, which is a closed condition. Thus, the image is closed.

Injectivity of $\mu_{\mathbf{w}}$ (and the fact that the image is a closure of an orbit) was proved in [**KP1979**] under a stronger assumption (10.45). The fact that it also holds under weaker assumption (10.42) was shown in a preprint of Shmelkin [**Shm2012**]. We skip the proof, referring the reader to the original papers.

For part (3), note that if this inequality holds, then for any partial flag F, one can choose y so that

(10.46) $$\operatorname{Im}(y^k) = V_{l-k+1}.$$

Namely, it is easy to see that for a nilpotent $r \times r$ matrix y, the dimensions $d_k = \dim \operatorname{Im}(y^k)$ are computed by the formula

$$d_0 = r, \qquad d_k = d_{k-1} - \# \text{ (blocks of size } \geq k).$$

Thus, if we choose a matrix such that

$$(\# \text{ blocks of size } \geq k) = \mathbf{v}_{l-k+2} - \mathbf{v}_{l-k+1},$$

then we will have $d_k = \dim \operatorname{Im}(y^k) = \mathbf{v}_{l-k+1}$.

This is possible if the sequence $\mathbf{v}_{l-k+2} - \mathbf{v}_{l-k+1}$ is decreasing, which is equivalent to condition (10.45).

On the other hand, it is obvious that if $\operatorname{Im}(y^k) = V_{l-k+1}$, then y is regular. Now the result follows from Theorem 10.40. □

The theorem implies that identifying $\mathcal{M}_0$ with $\mu_{\mathbf{w}}(\mathcal{M}_0) \subset \mathfrak{gl}(\mathbf{w})$, we see that the morphism π is identified with the restriction of the moment map $\mu_{\mathbf{w}}$. For example, in the Springer case ($\mathbf{v} = (1, 2, \ldots, l)$, $\mathbf{w} = r = l+1$) the morphism π is exactly the Springer resolution $\mu \colon \tilde{\mathcal{N}} \to \mathcal{N}$ (see Theorem 9.57).

Example 10.45. Consider a special case of the previous example, namely the quiver A_1. In this case the flag variety $\mathcal{F}(n,r)$ is the Grassmannian of n-dimensional subspaces in $\mathbb{C}^r$, so for $\theta > 0$ we get

$$\mathcal{M}(n,r) = T^*G(n,r),$$

and so $\dim \mathcal{M}(n,r) = 2n(r-n)$.

We can also describe $\mathcal{M}(n,r)$ as the set of pairs (y, V), where

$$V \subset \mathbb{C}^r, \qquad \dim V = n,$$
$$y \in \operatorname{End}(\mathbb{C}^r), \qquad \operatorname{Im}(y) \subset V, \qquad y(V) = 0.$$

In this case,

$$\mathcal{M}_0(n,r) = \{y \in \operatorname{End}(\mathbb{C}^r) \mid y^2 = 0, \operatorname{rank}(y) \leq n\}$$

and the morphism $\pi \colon \mathcal{M}(n,r) \to \mathcal{M}_0(n,r)$ is given by $(y, V) \mapsto y$. In particular,

$$\pi^{-1}(y) = \{V \subset \mathbb{C}^r \mid \operatorname{Im}(y) \subset V \subset \operatorname{Ker}(y)\} \simeq G(n-k, r-2k), \quad k = \operatorname{rank}(y).$$

If $r \geq 2k$, then $\mathcal{M}_0(n,r)$ is the closure of the orbit of $y_0 = J_{0,2}^{\oplus n}$, where $J_{0,2}$ is the 2×2 Jordan block with eigenvalue 0:

$$J_{0,2} = \begin{bmatrix} 0 & 1 \\ 0 & 0 \end{bmatrix}.$$

In the special case $n = 1$, $r = 2$, we get
$$\mathcal{M}(1, 2) = T^*\mathbb{P}^1$$
and $\mathcal{M}_0(1, 2) = \mathcal{N}$ is the set of nilpotent 2×2 matrices. In this case the symplectic resolution $\mathcal{M}(1, 2) \to \mathcal{M}_0(1, 2)$ is the resolution $T^*\mathbb{P}^1 \to Q$ constructed in Examples 9.33 and 9.56, which is exactly the Springer resolution for $\mathfrak{sl}(2, \mathbb{C})$.

10.8. Hyperkähler construction of quiver varieties

In this section, we will give a different construction of the quiver varieties $\mathcal{M}_\theta(\mathbf{v}, \mathbf{w})$, based on the hyperkähler quotient construction described in Section 9.9.

We begin by defining the hyperkähler structure on the vector space $R(Q^\sharp, \mathbf{v}, \mathbf{w})$. This is a complex vector space, so it already has a complex structure I. Let us now choose an orientation of Q, so that we have
$$R(Q^\sharp, V, W) = R(\vec{Q}, V) \oplus R(\vec{Q}^{opp}, V) \oplus L(V, W) \oplus L(W, V)$$
(compare with (10.20)).

Let us now choose a positive definite Hermitian form $(,)$ in each of the spaces V_i, W_i and define an $\mathbb{R}$-linear operator $J \colon R(Q^\sharp, V, W) \to R(Q^\sharp, V, W)$ by
$$(10.47) \qquad J(x, y, i, j) = (-y^*, x^*, j^*, -i^*),$$
where for a linear operator $a \colon V_1 \to V_2$, we denote by $a^* \colon V_2 \to V_1$ the adjoint operator: $(av_1, v_2) = (v_1, a^*v_2)$.

We also define the Hermitian form on $R(Q^\sharp, V, W)$ in the usual way, by
$$(10.48) \quad ((x, y, i, j), (x', y', i', j')) = \operatorname{tr}(xx'^*) + \operatorname{tr}(yy'^*) + \operatorname{tr}(ii'^*) + \operatorname{tr}(jj'^*).$$

Lemma 10.46. *Formulas (10.47), (10.48), together with the standard complex structure I, define on $R(Q^\sharp, V, W)$ the structure of a hyperkähler vector space. Corresponding symplectic forms ω_I, $\omega_\mathbb{C} = \omega_J + i\omega_K$ are given by*
$$\omega_I((x, y, i, j), (x', y', i', j')) = \operatorname{Im}\bigl(\operatorname{tr}(xx'^*) + \operatorname{tr}(yy'^*) + \operatorname{tr}(ii'^*) + \operatorname{tr}(jj'^*)\bigr),$$
$$\omega_\mathbb{C}((x, y, i, j), (x', y', i', j')) = \operatorname{tr}\biggl(\sum_{h \in \Omega}(y_{\bar{h}} x'_h - y'_{\bar{h}} x_h) + \sum_{k \in I} i_k j'_k - i'_k j_k\biggr).$$

Note that the form $\omega_\mathbb{C}$ coincides with the holomorphic symplectic form ω defined by (10.22).

We can now consider the hyperkähler quotient. Namely, consider the group $U(V) = \prod_k U(V_k)$ which acts on $R(Q^\sharp, V, W)$ in the obvious way. It is easy to see that this action commutes with I, J and thus is Hamiltonian.

10.8. Hyperkähler construction of quiver varieties

Lemma 10.47. *The moment maps μ_I, $\mu_\mathbb{C} = \mu_J + i\mu_K$ for the action of $U(V)$ are given by*

$$\mu_I(x, y, i, j) = \tfrac{i}{2}([x, x^*] + [y, y^*] + ii^* - j^*j)$$
$$= \tfrac{i}{2}\Big(\sum_{h \in H}(z_h z_h^* - z_{\bar{h}}^* z_{\bar{h}}) + ii^* - j^*j\Big),$$
$$\mu_\mathbb{C}(x, y, i, j) = [x, y] - ij.$$

The proof of this lemma is given by an explicit computation, similar to the one in Example 9.62.

Let us now consider hyperkähler quotients. For $\zeta \in \mathbb{R}^I$, we can consider $i\zeta$ as an element in $\mathfrak{u}(V)$ by $i\zeta = i \sum_{k \in I} \zeta_k \operatorname{id}_{V_k}$. Thus, any triple $\boldsymbol{\zeta} = (\zeta^1, \zeta^2, \zeta^3) \in \mathbb{R}^3 \otimes \mathbb{R}^I$ gives an element $i\boldsymbol{\zeta} \in \mathbb{R}^3 \otimes \mathfrak{u}(V)$ which is obviously $U(V)$-invariant, so we can consider hyperkähler quotients

$$\boldsymbol{\mu}^{-1}(i\boldsymbol{\zeta})/U(V)$$

as in Section 9.10.

Similar to Definition 10.33, let us call an element $\boldsymbol{\zeta} = (\zeta^1, \zeta^2, \zeta^3) \in \mathbb{R}^3 \otimes \mathbb{R}^I$ **v**-generic if for any $\alpha \in R'_+(\mathbf{v})$, we have

$$\boldsymbol{\zeta} \cdot \alpha = (\zeta^1 \cdot \alpha, \zeta^2 \cdot \alpha, \zeta^3 \cdot \alpha) \neq (0, 0, 0).$$

Theorem 10.48. *Let $\boldsymbol{\zeta} = (\zeta^1, \zeta^2, \zeta^3) \in \mathbb{R}^3 \otimes \mathbb{R}^I$ be **v**-generic. Then the action of $U(V)$ on $\boldsymbol{\mu}^{-1}(i\boldsymbol{\zeta})$ is free, so that the hyperkähler quotient*

$$\boldsymbol{\mu}^{-1}(i\boldsymbol{\zeta})/U(V)$$

is smooth.

Proof. This is the differential geometry analog of Theorem 10.35. Our proof follows [**Nak1994**, Theorem 2.8].

Assume that $m = (z, i, j) \in \boldsymbol{\mu}^{-1}(i\boldsymbol{\zeta})$ has a nontrivial stabilizer. Let $g \in U(V)$ be such that $g.m = m$, $g \neq 1$. Then action of g on the corresponding framed representation V^m commutes with z, i, j. Since $g \in U(V)$, it is diagonalizable: we can write $V = \bigoplus V(\lambda)$, $g|_{V(\lambda)} = \lambda$. Then $zV(\lambda) \subset V(\lambda)$, $\operatorname{Im}(i) \subset V(1)$, and $j(V(\lambda)) = 0$ for $\lambda \neq 1$.

Take $\lambda \neq 1$ such that $V(\lambda) \neq \{0\}$. Then $V(\lambda)$ is an (unframed) subrepresentation of V, of dimension $\mathbf{v}' = \dim V(\lambda) \leq \mathbf{v}$. Let $z' \in R(Q^\sharp, \mathbf{v}')$ be the corresponding element in the representation space. If the action of the group $U(\mathbf{v}')/U(1)$ on z' is not free, then we can choose a nontrivial element in $U(\mathbf{v}')/U(1)$ stabilizing z' and repeat the process; thus, without loss of generality, we may assume that the stabilizer of z' in $K = U(\mathbf{v}')/U(1)$ is trivial. Applying Theorem 9.70 to the set $R^{reg}(Q^\sharp, \mathbf{v}')$ of elements with

trivial stabilizer in K, we get the inequality
$$\dim_{\mathbb{R}} R(Q^\sharp, \mathbf{v}') - 4\dim_{\mathbb{R}} K \geq 0$$
or
$$0 \leq \dim_{\mathbb{C}} R(Q^\sharp, \mathbf{v}') - 2\dim_{\mathbb{R}} K = \sum_{i,j} 2n_{ij}\mathbf{v}'_i\mathbf{v}'_j - 2\sum \mathbf{v}'^2_i + 2 = 2 - 2\langle \mathbf{v}', \mathbf{v}'\rangle$$
so $\mathbf{v}' \in R'_+(\mathbf{v})$ (see (10.35)).

Let $a \in \mathfrak{gl}(\mathbf{v})$ be the orthogonal projection on $V(\lambda) \subset V$. Then $ia \in \mathfrak{u}(V)$ commutes with z, and $ja = 0$, $ai = 0$; by results of Example 9.45, we get
$$\langle \boldsymbol{\mu}(z,i,j), ia\rangle = 0 \in \mathbb{R}^3.$$
On the other hand, $\boldsymbol{\mu}(z,i,j) = i\boldsymbol{\zeta}$, so we get
$$\langle i\boldsymbol{\zeta}, ia\rangle = -\sum_{k\in I}((\zeta^1)_k, (\zeta^2)_k, (\zeta^3)_k)\operatorname{tr}(a_k) = -\boldsymbol{\zeta}\cdot \mathbf{v}' = 0.$$

This contradicts the assumption that $\boldsymbol{\zeta}$ is $\mathbf{v}$-generic.

Smoothness of the quotient now follows from Theorem 9.70. □

Moreover, we can relate these hyperkähler quotients with the quiver varieties constructed before by algebraic geometry methods. Let $\boldsymbol{\zeta} \in \mathbb{R}^3 \otimes \mathbb{R}^I$. Define the hyperkähler quiver variety
$$(10.49) \qquad \mathcal{M}_{\boldsymbol{\zeta}}(\mathbf{v}, \mathbf{w}) = \boldsymbol{\mu}^{-1}(i\boldsymbol{\zeta})/U(V).$$

We will also use the notation
$$(10.50)\quad \mathcal{M}_{\theta,\lambda}(\mathbf{v}, \mathbf{w}) = \mathcal{M}_{\boldsymbol{\zeta}}(\mathbf{v}, \mathbf{w}), \qquad \theta = \boldsymbol{\zeta}^1 \in \mathbb{R}^I, \lambda = \boldsymbol{\zeta}^2 + i\boldsymbol{\zeta}^3 \in \mathbb{C}^I.$$

Theorem 10.49.

(1) *If $\boldsymbol{\zeta}$ is $\mathbf{v}$-generic, then $\mathcal{M}_{\boldsymbol{\zeta}}(\mathbf{v}, \mathbf{w})$ is smooth and has a canonical structure of a hyperkähler manifold of dimension*
$$\dim_{\mathbb{R}} \mathcal{M}_{\boldsymbol{\zeta}}(\mathbf{v}, \mathbf{w}) = 4\mathbf{v}\cdot \mathbf{w} - 4\langle \mathbf{v}, \mathbf{v}\rangle.$$

(2) *If $\boldsymbol{\zeta}$ is $\mathbf{v}$-generic and $\theta = \boldsymbol{\zeta}^1 \in \mathbb{Z}^I$, then we have an isomorphism of complex manifolds*
$$\mathcal{M}_{\theta,\lambda}(\mathbf{v}, \mathbf{w}) \simeq \mu_{\mathbb{C}}^{-1}(i\lambda) /\!\!/_{\chi_\theta} \operatorname{GL}(\mathbf{v}),$$
where $\mathcal{M}_{\theta,\lambda}$ is considered as a complex manifold with complex structure I.

Proof. Part (1) follows from general construction of hyperkähler quotients (Theorem 9.70) and the fact that action of $U(V)$ is free (Theorem 10.48).

Part (2) follows from Theorem 9.71. □

Corollary 10.50. *Let $\theta \in \mathbb{Z}^I$ be $\mathbf{v}$-generic. Let $\zeta = (\theta, 0, 0) \in \mathbb{R}^3 \otimes \mathbb{R}^I$. Then one has an isomorphism of complex manifolds*
$$\mathcal{M}_\zeta(\mathbf{v}, \mathbf{w}) \simeq \mathcal{M}_\theta(\mathbf{v}, \mathbf{w}),$$
where $\mathcal{M}_\theta(\mathbf{v}, \mathbf{w})$ is the quiver variety defined in Definition 10.27, considered with analytic topology.

We also give without a proof the following result, which generalizes Corollary 10.39.

Theorem 10.51. *For any $\mathbf{v}$-generic ζ, ζ', varieties $\mathcal{M}_\zeta(\mathbf{v}, \mathbf{w})$, $\mathcal{M}_{\zeta'}(\mathbf{v}, \mathbf{w})$ are diffeomorphic.*

A proof of this theorem can be found in [**Nak1994**, Corollary 4.2].

Corollary 10.52. *For any $\mathbf{v}$-generic ζ, $\mathcal{M}_\zeta(\mathbf{v}, \mathbf{w})$ is diffeomorphic to a nonsingular affine algebraic variety.*

Indeed, if $\zeta_1 = 0$, then the corollary follows from Theorem 10.49. The general case now follows from Theorem 10.51.

10.9. $\mathbb{C}^\times$ action and exceptional fiber

Throughout this section, we assume that Q is a graph without edge loops and that Ω is an orientation of $\vec{Q}$ such that the quiver $\vec{Q} = (Q, \Omega)$ has no oriented cycles. We choose the skew-symmetric function ε used in the definition of the symplectic form on $R(Q^\sharp, \mathbf{v}, \mathbf{w})$ to be defined by orientation Ω so that
$$R(Q^\sharp, \mathbf{v}, \mathbf{w}) = T^* R(\vec{Q}, \mathbf{v}, \mathbf{w})$$
as symplectic manifolds.

We will write elements of $R(Q^\sharp)$ as quadruples $m = (x, y, i, j)$, $x \in R(\vec{Q}, \mathbf{v})$, $y \in R(\vec{Q}^{opp}, \mathbf{v})$.

Define the $\mathbb{C}^\times$ action on $R(Q^\sharp, \mathbf{v}, \mathbf{w})$ by
(10.51) $$\varphi_t(x, y, i, j) = (x, ty, ti, j), \quad t \in \mathbb{C}^\times.$$
The following properties of the action are immediate from the definition.

Lemma 10.53.
(1) *The action (10.51) commutes with the action of $\mathrm{GL}(\mathbf{v})$.*
(2) *Both the symplectic form and the moment map are homogeneous of degree one:*
$$\varphi_t^*(\omega) = t\omega,$$
$$\mu(\varphi_t(m)) = t \cdot \mu(m).$$

Corollary 10.54. *For any $\theta \in \mathbb{Z}^I$, the $\mathbb{C}^\times$ action descends to an algebraic action of $\mathbb{C}^\times$ on $\mathcal{M}_\theta(\mathbf{v},\mathbf{w})$, and the projective morphism $\pi\colon \mathcal{M}_\theta(\mathbf{v},\mathbf{w}) \to \mathcal{M}_0(\mathbf{v},\mathbf{w})$ is $\mathbb{C}^\times$-equivariant.*

Let us now study the fixed points of the $\mathbb{C}^\times$ action.

Theorem 10.55.
  (1) *For any $[m] \in \mathcal{M}_0(\mathbf{v},\mathbf{w})$, we have $\lim_{t\to 0}[m] = 0$.*
  (2) *The only fixed point of the $\mathbb{C}^\times$ action on $\mathcal{M}_0(\mathbf{v},\mathbf{w})$ is $[m] = 0$.*
  (3) *For $[m] \in \mathcal{M}_0(\mathbf{v},\mathbf{w})$, the limit $\lim_{t\to\infty}[m] = 0$ exists only if $[m] = 0$ is the fixed point.*

Proof. Let $m \in \mu^{-1}(0)$, and let $[m]$ be the corresponding point in $\mathcal{M}_0(\mathbf{v},\mathbf{w})$. Then $\lim_{t\to 0} t.[m] = \lim_{t\to 0}[t.m] = [(x,0,0,j)]$. In this case, it is easy to see that the closure of the orbit of $(x,0,0,j)$ contains the point $(x,0,0,0)$; since $\vec{Q}$ has no oriented cycles, the closure also contains the point $(0,0,0,0)$ (see Corollary 2.11). This proves part (1). The second part immediately follows.

To prove the last part, note that if $\lim_{t\to\infty}[m] = [n]$ exists, then it is a fixed point, so $[n] = [0]$. This shows that for any $GL(V)$-invariant polynomial f on $\mu_{\mathbf{v},\mathbf{w}}^{-1}(0)$, we have $\lim_{t\to\infty} f(t.m) = f(0)$. Since $f(t.m)$ is a polynomial in t, existence of limit implies that for all t, we have $f(t.m) = f(0)$ so, by definition, $[t.m] = [0]$. □

Let us now fix $\theta \in \mathbb{Z}^I$ which is $\mathbf{v}$-generic, so that the quiver variety $\mathcal{M}_\theta(\mathbf{v},\mathbf{w})$ is smooth. Let
$$F = \mathcal{M}_\theta(\mathbf{v},\mathbf{w})^{\mathbb{C}^\times}$$
be the set of fixed points of the $\mathbb{C}^\times$ action on $\mathcal{M}_\theta(\mathbf{v},\mathbf{w})$; it is immediate from Theorem 10.55 that $F \subset \pi^{-1}(0)$. Since $\mathcal{M}_\theta$ is smooth and $\mathbb{C}^\times$ is reductive, F is also smooth (see [**Ive1972**] or [**CG1997**, Lemma 5.11.1]). Let
$$F = \bigsqcup F_s$$
be the decomposition of F into connected components.

Theorem 10.56. *Assume that $\theta \in \mathbb{Z}^I$ is $\mathbf{v}$-generic. Define the exceptional fiber $\mathcal{L}_\theta(\mathbf{v},\mathbf{w})$ by*
$$\mathcal{L}_\theta(\mathbf{v},\mathbf{w}) = \pi^{-1}(0) \subset \mathcal{M}_\theta(\mathbf{v},\mathbf{w}).$$
Then:
  (1) $\mathcal{L}_\theta(\mathbf{v},\mathbf{w}) = \bigcup \mathcal{L}_s$, *where*
  $$\mathcal{L}_s = \{m \in \mathcal{M}_\theta(\mathbf{v},\mathbf{w}) \mid \lim_{t\to\infty} t.m \text{ exists and is in } F_s\}.$$
  (2) *Each $\mathcal{L}_s$ is a smooth connected locally closed Lagrangian subvariety in $\mathcal{M}_\theta(\mathbf{v},\mathbf{w})$.*

10.9. $\mathbb{C}^\times$ action and exceptional fiber

Proof. First, note that by Theorem 10.55 if $m \in \mathcal{M}_\theta(\mathbf{v}, \mathbf{w})$ is such that $\lim_{t \to \infty} t.m$ exists, then $\pi(m) = 0$; thus, $\mathcal{L}_s \subset \mathcal{L}_\theta(\mathbf{v}, \mathbf{w})$.

Conversely, since π is a projective morphism, $\mathcal{L}_\theta(\mathbf{v}, \mathbf{w}) = \pi^{-1}(0)$ is projective, so any morphism $\mathbb{C}^\times \to \mathcal{L}_\theta(\mathbf{v}, \mathbf{w})$ extends to a morphism $\mathbb{P}^1 \to \mathcal{L}_\theta(\mathbf{v}, \mathbf{w})$. Applying this to the map $t \mapsto t.m$, we see that for any $m \in \mathcal{L}_\theta(\mathbf{v}, \mathbf{w})$, the limit $\lim_{t \to \infty} t.m$ exists. Since the limit must be a fixed point, it must belong to one of F_s. This proves the first part of the theorem.

The fact that each $\mathcal{L}_s$ is a smooth connected locally closed subvariety follows from general result of Białynicki-Birula [**BB1973**, Theorem 4.3]. The varieties $\mathcal{L}_s$ are usually called Białynicki-Birula pieces.

To prove that each $\mathcal{L}_s$ is Lagrangian, let $m \in F_s$ be a fixed point of the $\mathbb{C}^\times$ action. Then the tangent space $T_m \mathcal{M}_\theta$ has a weight decomposition: $T_m \mathcal{M}_\theta = \bigoplus_{k \in \mathbb{Z}} T^k$, where $t \in \mathbb{C}^\times$ acts on T^k by multiplication by t^k. It is shown in [**BB1973**] that $T_m \mathcal{L}_s = \bigoplus_{l \leq 0} T^l$.

On the other hand, since the symplectic form ω has weight 1 (see Lemma 10.53), we see that for $x \in T^k, y \in T^l$ we have $\omega(x, y) = 0$ unless $k + l = 1$. Nondegeneracy of ω implies that it gives a nondegenerate pairing $T^k \otimes T^{1-k} \to \mathbb{C}$. Thus, $\bigoplus_{l \leq 0} T^l$ is a Lagrangian subspace in $T_m \mathcal{M}_\theta$. In particular, $\dim \mathcal{L}_s = \frac{1}{2} \dim \mathcal{M}_\theta$.

Finally, for any point $x \in \mathcal{L}_s$ (not necessarily a fixed point), and any tangent vector $v \in T_x L$, the limit $\lim_{t \to \infty} t.v$ exists. Therefore, the limit $\lim_{t \to \infty} \omega(t.v, t.w)$ also exists for any $v, w \in T_x L$. On the other hand, $\omega(t.v, t.w) = t\omega(v, w)$. Thus, $\omega(v, w) = 0$, so $T_x L$ is an isotropic subspace in $T_x \mathcal{M}_\theta$. Since $\dim \mathcal{L}_s = \frac{1}{2} \dim \mathcal{M}_\theta$, this proves that $\mathcal{L}_s$ is Lagrangian. $\square$

Note that this implies that the irreducible components of $\mathcal{L}$ are exactly the Zariski closures of $\mathcal{L}_s$.

Remark 10.57. The statement of the theorem can be false for quivers with oriented loops, as is easy to see in the example of the Jordan quiver (see Corollary 11.9 in the next chapter).

For $\theta > 0$, one can give an explicit description of $\mathcal{L}_\theta(\mathbf{v}, \mathbf{w})$.

Theorem 10.58. *Let $\theta > 0$ and let $m = (x, y, i, j) \in \mu_{\mathbf{v}, \mathbf{w}}^{-1}$ be θ-semistable. Then*

$$[m] \in \mathcal{L}_\theta(\mathbf{v}, \mathbf{w}) \iff (i = 0, z = (x, y) \text{ is nilpotent}).$$

Similarly, if $\theta < 0$, $m = (x, y, i, j) \in \mu_{\mathbf{v}, \mathbf{w}}^{-1}$ is θ-semistable, then

$$[m] \in \mathcal{L}_\theta(\mathbf{v}, \mathbf{w}) \iff (j = 0, z = (x, y) \text{ is nilpotent}).$$

Proof. If $m = (z, i, j)$, $\pi(m) = 0$, then, by Lemma 2.9, there exists a one-parameter subgroup $\lambda(t)$ such that $\lim_{t \to 0} \lambda(t).m = 0$. By Lemma 10.30, this means that V admits a decreasing filtration such that $zV^{\geq m} \subset V^{\geq m+1}$, and $V^{\geq 0} \subset \operatorname{Ker} j$, $\operatorname{Im} i \subset V^{\geq 1}$. This implies that z is nilpotent.

For $\theta < 0$, stability condition (10.36) implies that $V^{\geq 0} = 0$, so $\operatorname{Im} i = \{0\}$. Similarly, for $\theta > 0$ stability condition (10.37) implies that $V^{\geq 1} = V$, so $\operatorname{Ker} j = V$. $\square$

Comparing this description of $\mathcal{L}$ with the variety $\Lambda(\mathbf{v})$ of nilpotent representations of the double quiver, given in (5.9), we see that for $\theta > 0$, one has
$$\mathcal{L}_\theta(\mathbf{v}, \mathbf{w}) = (\Lambda(\mathbf{v}) \times \operatorname{Hom}(\mathbb{C}^\mathbf{v}, \mathbb{C}^\mathbf{w}))/\!\!/_\theta \operatorname{GL}(V).$$

The following result is a reformulation of [Sai2002, Lemma 4.6.2].

Corollary 10.59. *Denote by* $\operatorname{Irr}(\mathcal{L})$ *the set of irreducible components of* $\mathcal{L}$. *Assume that* $\theta > 0$. *Then we have a natural injective map*
$$\operatorname{Irr}(\mathcal{L}_\theta(\mathbf{v}, \mathbf{w})) \hookrightarrow \operatorname{Irr}(\Lambda(\mathbf{v})).$$

Proof. Denote $\tilde{\Lambda} = \Lambda(\mathbf{v}) \times \operatorname{Hom}(\mathbb{C}^\mathbf{v}, \mathbb{C}^\mathbf{w})$. Obviously, we have a natural bijection $\operatorname{Irr} \tilde{\Lambda} \simeq \operatorname{Irr}(\Lambda(\mathbf{v}))$. On the other hand, $\operatorname{Irr}(\mathcal{L}) = \operatorname{Irr}(\tilde{\Lambda}^s/\operatorname{GL}(\mathbf{v})) = \operatorname{Irr}(\tilde{\Lambda}^s)$. Since $\tilde{\Lambda}^s \subset \tilde{\Lambda}$ is open, we have an inclusion $\operatorname{Irr}(\tilde{\Lambda}^s) \subset \operatorname{Irr}(\tilde{\Lambda})$. $\square$

This corollary has a natural interpretation in terms of representation theory of Kac–Moody Lie algebras, which will be discussed in Chapter 13 (see discussion following the proof of Theorem 13.19).

Theorem 10.60. *Let θ be $\mathbf{v}$-generic. Then the inclusion* $\mathcal{L}_\theta(\mathbf{v}, \mathbf{w}) \hookrightarrow \mathcal{M}_\theta(\mathbf{v}, \mathbf{w})$ *is a homotopy equivalence.*

Proof. The proof is a modification of the proof in [Slo1980, Section 4.3]. Recall that by Theorem 10.49, one has a diffeomorphism $\mathcal{M}_0(\mathbf{v}, \mathbf{w}) \simeq \mu^{-1}(0) \cap \mu_\mathbb{C}^{-1}(0)/U(\mathbf{v})$. Thus, $\mathcal{M}_0(\mathbf{v}, \mathbf{w})$ has a norm $|\cdot|$ inherited from the norm on $R(Q^\sharp, \mathbf{v}, \mathbf{w})$:
$$|(x, y, i, j)|^2 = |x|^2 + |y|^2 + |i|^2 + |j|^2, \quad (x, y, i, j) \in \mu^{-1}(0) \cap \mu_\mathbb{C}^{-1}(0),$$
where $|x|^2 = \operatorname{tr}(xx^*)$.

The following properties of this norm are immediate:

(1) $|\cdot|$ is continuous.

(2) $|m| \geq 0$, with equality only for $m = 0$.

(3) Consider the action of $\mathbb{R}_+ \subset \mathbb{C}^\times$ on $\mathcal{M}_0$ defined by (10.51). Then for any $m \in \mathcal{M}_0$, $m \neq 0$, the function $|t.m|$ is a strictly increasing function of t.

The last property implies that for small enough $\varepsilon > 0$, each nonzero $\mathbb{R}_+$-orbit in $\mathcal{M}_\theta$ has a unique intersection with the set $\{m \in \mathcal{M}_\theta \mid |\pi(m)| = \varepsilon\}$, where π is the canonical morphism $\mathcal{M}_\theta \to \mathcal{M}_0$. It easily follows from this that the inclusion $B_\varepsilon(\mathcal{L}) \subset \mathcal{M}_\theta$, where $B_\varepsilon(\mathcal{L}) = \{m \in \mathcal{M}_\theta \mid |\pi(m)| \leq \varepsilon\}$, is a homotopy equivalence. On the other hand by standard results of algebraic topology, $\mathcal{L}$ is homotopy equivalent to its tubular neighborhood. This completes the proof. □

Chapter 11

Jordan Quiver and Hilbert Schemes

In this section, we study the quiver varieties for the Jordan quiver and discuss connections between these varieties, Hilbert schemes of points, and instanton moduli spaces.

As before, we only consider the case when the ground field $\mathbf{k} = \mathbb{C}$.

11.1. Hilbert schemes

Let X be an affine algebraic variety over $\mathbb{C}$. Recall that the symmetric powers of X are defined by

$$S^n X = X^n /\!/ S_n = \operatorname{Spec}(\mathbb{C}[X]^{\otimes n})^{S_n}.$$

For example, if $X = \mathbb{C}$ is the one-dimensional affine space, then

$$S^n X = \operatorname{Spec}(\mathbb{C}[x_1, \ldots, x_n]^{S_n}).$$

By the classical result about symmetric polynomials, $\mathbb{C}[x_1, \ldots, x_n]^{S_n} \simeq \mathbb{C}[s_1, \ldots, s_n]$, so in this case $S^n X \simeq \mathbb{C}^n$.

However, in general $S^n X$ is singular, even for smooth X. An example of this was given in Example 10.18, where it was shown that for $X = \mathbb{C}^2$, one has $S^2 X \simeq Q \times \mathbb{C}^2$, where Q is a quadric in $\mathbb{C}^3$.

A closely related, and in many ways more useful, object is the so-called Hilbert scheme of points in X.

Definition 11.1. Let X be an affine algebraic variety. Then the Hilbert scheme of n points in X is defined by

$$\operatorname{Hilb}^n X = \{J \subset A \mid J \text{ is an ideal in } A,\ \dim(A/J) = n\}, \qquad A = \mathbb{C}[X].$$

This defines $\operatorname{Hilb}^n X$ as a set; it can be shown that in fact $\operatorname{Hilb}^n X$ is a scheme. This was first shown by Grothendieck [**Gro1995**] in a much more general context. For Hilbert schemes of points (which is the only kind of Hilbert schemes considered in this book), a relatively simple proof can be found in [**IN1999**, Theorem 6.4]. We refer the reader to [**Leh2004**], [**Nak1999**] for a review of theory of Hilbert schemes.

Alternatively, one can define $\operatorname{Hilb}^n X$ as the set of isomorphism classes of pairs
(11.1)
$$(M, v), \qquad M \text{ a } \mathbb{C}[X]\text{-module of dimension } n, \quad v \text{ a cyclic vector in } M$$

(recall that for an A-module M, a vector $v \in M$ is called cyclic if $Av = M$; in such a case, M is called a cyclic module).

The correspondence between two descriptions is given by

$$J \mapsto (M = \mathbb{C}[X]/J, v = 1 \mod J).$$

We list some of the most important properties of Hilbert schemes; proofs can be found in [**Leh2004**].

Theorem 11.2.

(1) *One has a canonical projective morphism (Hilbert–Chow morphism)*

$$\pi \colon \operatorname{Hilb}^n X \to S^n X$$
$$J \mapsto \operatorname{supp}(\mathbb{C}[X]/J),$$

where for a $\mathbb{C}[X]$-module M, we denote by $\operatorname{supp}(X)$ the support of M, considered as a set of points in X with multiplicities.

(2) *Let*
$$S_0^n X = \{(t_1, \ldots, t_n) \in X^n \mid t_i \neq t_j\}/S_n,$$
and let $\operatorname{Hilb}_0^n X = \pi^{-1}(S_0^n X)$. Then $\operatorname{Hilb}_0^n X$ is open in $\operatorname{Hilb}^n X$, and the restriction of the Hilbert–Chow morphism

$$\pi \colon \operatorname{Hilb}_0^n X \to S_0^n X$$

is an isomorphism: for every unordered n-tuple $\mathbf{t} = (t_1, \ldots, t_n) \in S_0^n X$, the corresponding ideal $J_\mathbf{t} = \pi^{-1}(\mathbf{t})$ is given by

$$J_\mathbf{t} = \{f \in \mathbb{C}[X] \mid f(t_1) = f(t_2) = \cdots = f(t_n) = 0\}.$$

Moreover, if $\dim X = 2$, then $\operatorname{Hilb}_0^n X$ is dense in $\operatorname{Hilb}^n X$.

In general, a Hilbert scheme of points is not smooth, even for smooth X. However, the following important result was proved by Fogarty [**Fog1968**].

Theorem 11.3. *Let X be a nonsingular variety of dimension 2. Then the Hilbert scheme $\operatorname{Hilb}^n X$ is smooth, and $\pi\colon \operatorname{Hilb}^n X \to S^n X$ is a resolution of singularities.*

Example 11.4. Let $X = \mathbb{C}^2$. Consider the Hilbert scheme $\operatorname{Hilb}^2 X$ and the Hilbert–Chow morphism $\pi\colon \operatorname{Hilb}^2 X \to S^2 X$.

If $\mathbf{t} = (t_1, t_2) \in S_0^2 X$, i.e. $t_1 \neq t_2$, then $\pi^{-1}(\mathbf{t}) = J_{\mathbf{t}}$ is a single point. To study the variety $\pi^{-1}(\mathbf{t})$ for a double point $\mathbf{t} = (t, t)$, consider the case $\mathbf{t} = (0, 0)$. Instead of describing the ideal $J \in \pi^{-1}(\mathbf{t})$, let us describe the quotient $M = \mathbb{C}[X]/J$. This is a 2-dimensional module over $\mathbb{C}[X] = \mathbb{C}[z_1, z_2]$, generated by a single vector and on which operators z_1, z_2 act nilpotently. It is easy to check that any such M must be of the form

$$M = \mathbb{C}[z_1, z_2]/(z_1^2, z_2^2, z_1 z_2, \alpha z_1 + \beta z_2)$$

for some $(\alpha, \beta) \neq (0, 0)$. Thus, we see that $\pi^{-1}(0, 0) \simeq \mathbb{P}^1$. Similarly, for any other double point $\mathbf{t} = (t, t)$, we also have $\pi^{-1}(\mathbf{t}) = \mathbb{P}^1$.

Intuitively, one should think of an ideal $J \in \pi^{-1}(0, 0)$ as a cluster of two points, infinitely close to each other; the projective space $\mathbb{P}^1$ parametrizes direction of the vector connecting these two points, up to rescaling.

Recalling that in this case $S^2 X = Q \times \mathbb{C}^2$, where Q is a quadric (see Example 10.18), we see that one can write $\operatorname{Hilb}^2 X = \widehat{Q} \times \mathbb{C}^2$, where $\widehat{Q}$ is a resolution of singularities of Q, with $\pi^{-1}(0) = \mathbb{P}^1$. In fact, one can show that $\widehat{Q} = T^* \mathbb{P}^1$ (compare with the Springer resolution of the nilpotent cone for Lie algebra $\mathfrak{sl}(2)$ described in Example 9.58).

11.2. Quiver varieties for the Jordan quiver

Let us now consider the quiver varieties for the Jordan quiver:

$$Q = \;\raisebox{-2pt}{\begin{tikzpicture}\node[circle,fill,inner sep=1.5pt] (a){}; \draw (a) to[out=60,in=120,loop,looseness=20] (a);\end{tikzpicture}}$$

The corresponding (unframed) varieties $\mathcal{R}_0(n)$ and $\mathcal{M}_0(n)$ were discussed in Example 10.18, where it was shown that

$$\mathcal{R}_0(n) = \mathbb{C}^n / S_n \simeq \mathbb{C}^n,$$
$$\mathcal{M}_0(n) = \mathbb{C}^{2n} / S_n.$$

Let us now study the varieties associated with framed representations, as shown below:

$$W \underset{j}{\overset{i}{\rightleftarrows}} V x \circlearrowright y$$

Theorem 11.5. *Let $\vec{Q}$ be the Jordan quiver, and let $\theta < 0$, $\mathbf{v} = n$, $\mathbf{w} = 1$. Then one has isomorphisms*

$$\mathcal{M}_0(n,1) \simeq S^n \mathbb{C}^2,$$
$$\mathcal{M}_\theta(n,1) \simeq \mathrm{Hilb}^n \mathbb{C}^2.$$

Under this isomorphism, the canonical morphism $\pi\colon \mathcal{M}_\theta(n,1) \to \mathcal{M}_0(n,1)$ is identified with the Hilbert–Chow morphism $\mathrm{Hilb}^n \mathbb{C}^2 \to S^n \mathbb{C}^2$.

Proof. Using the formula for the moment map given in Lemma 10.26, we see that

$$\mu_{n,1}^{-1}(0) = \{x, y\colon \mathbb{C}^n \to \mathbb{C}^n, i\colon \mathbb{C} \to \mathbb{C}^n, j\colon \mathbb{C}^n \to \mathbb{C} \mid [x,y] - ij = 0\}.$$

Thus, $\mu(x,y,i,j) = [x,y] - ij = 0$ implies that $[x,y] = ij$. Therefore, $[x,y]$ has rank ≤ 1. We can now use the following remarkable lemma.

Lemma 11.6. *Let $x, y\colon V \to V$ be linear operators in a finite-dimensional space over an algebraically closed field such that $\mathrm{rank}([x,y]) \leq 1$. Then there exists a basis in which both x, y are upper-triangular.*

This lemma was first proved in [**Gur1979**]; a shorter proof can be found in [**EG2002**, Lemma 12.7].

This implies that $V = \mathbb{C}^n$ has a decreasing filtration

$$\cdots \supset V^{\geq k} \supset V^{\geq k+1} \supset \cdots$$

which is stable under x, y. Shifting the index if necessary, we can assume that $\mathrm{Im}(i) \subset V^{\geq 0}$, $\mathrm{Im}(i) \not\subset V^{\geq 1}$.

On the other hand, since $ij = [x,y]$, we see that $ij(V^{\geq k}) \subset V^{\geq k+1}$. Since W is one-dimensional, this implies that $ij(V^{\geq 0}) = 0$.

The same reasoning as in the proof of Theorem 2.10 then shows that the orbit of (x,y,i,j) is closed if and only if x, y are diagonalizable and $i = 0$, $j = 0$. Therefore, $\mathcal{M}_0(n,1) = \mathcal{M}_0(n,0) = S^n \mathbb{C}^2$ (see Example 10.18).

To study the moduli space $\mathcal{M}_\theta(n,1)$, note that by Example 10.36, the set of θ-semistable elements in $\mu^{-1}(0)$ is given by

$$(\mu^{-1}(0))^s = \{(x,y,i,j) \mid [x,y] - ij = 0,\ \mathrm{Im}(i) \text{ generates } V$$
$$\text{under action of } x, y\}.$$

Therefore, i is injective and $V = V^{\geq 0}$, so $j = 0$, which implies $[x, y] = 0$. Thus, in this case V is a module of dimension n over the algebra $\mathbb{C}[x,y]$, generated by the one-dimensional subspace $\operatorname{Im} i$. Comparing it with the description of the Hilbert scheme given in (11.1), we see that
$$\mathcal{M}_\theta(n,1) = (\mu^{-1}(0))^s/\operatorname{GL}(n,\mathbb{C}) = \operatorname{Hilb}^n \mathbb{C}^2.$$
The projective morphism $\pi\colon \mathcal{M}_\theta \to \mathcal{M}_0$ is given by $(x,y,i,0) \mapsto (x^s, y^s, 0, 0)$, where x^s, y^s are the semisimple parts of x, y. Identifying $\mathcal{M}_0(n,1) \simeq \mathbb{C}^{2n}/S_n$, we see that π is given by $(x,y,i,0) \mapsto (\lambda, \mu)$, where $\lambda = (\lambda_1, \ldots, \lambda_n)$, $\mu = (\mu_1, \ldots, \mu_n)$ are eigenvalues of x, y respectively. On the other hand, this is exactly the support of V considered as a $\mathbb{C}[x,y]$-module. $\square$

Note that by Lemma 10.29, for $\theta > 0$ we also have $\mathcal{M}_\theta(n,1) \simeq \operatorname{Hilb}^n \mathbb{C}^2$.

Note that by Example 10.36, $\mathcal{M}_\theta(n,1)$ is smooth for $\theta \neq 0$. Thus, this gives a new proof that the Hilbert scheme $\operatorname{Hilb}^n \mathbb{C}^2$ is smooth. Moreover, by Theorem 10.35, the quiver variety $\mathcal{M}_\theta$ is symplectic, and the morphism $\pi\colon \mathcal{M}_\theta \to \mathcal{M}_0$ is Poisson. Since we also know that $\operatorname{Hilb}^n \mathbb{C}^2 \to S^n\mathbb{C}^2$ is a resolution of singularities, we get the following result.

Theorem 11.7. *The Hilbert scheme $\operatorname{Hilb}^n \mathbb{C}^2$ is a symplectic manifold, and the Hilbert–Chow morphism $\operatorname{Hilb}^n \mathbb{C}^2 \to S^n\mathbb{C}^2$ is a symplectic resolution of singularities.*

Remark 11.8. Note that we cannot use Theorem 10.40 (which requires that the set of regular points in $\mathcal{M}_0$ be nonzero) to prove this theorem since, in this case, there are no regular points: for every closed orbit, the stabilizer contains the subgroup of scalars $\mathbb{C}^\times \subset \operatorname{GL}(\mathbf{v})$ and thus is never trivial. The set of closed orbits for which the stabilizer of the orbit is exactly $\mathbb{C}^\times$ can be identified with the set $S_0^n(\mathbb{C}^2) \subset \operatorname{Hilb}^n \mathbb{C}^2$ defined in Theorem 11.2.

Corollary 11.9. *The zero fiber $\pi^{-1}(n[0]) \subset \operatorname{Hilb}^n(\mathbb{C}^2)$ is an isotropic subvariety.*

Note that this result shows that $\dim \pi^{-1}(n[0]) \leq n$. In fact, more careful analysis shows that $\dim \pi^{-1}(n[0]) = n - 1$ (see [**Nak1999**, Theorem 1.13]).

This theorem can be generalized: for any smooth 2-dimensional X over $\mathbb{C}$ with a holomorphic symplectic form, the Hilbert scheme $\operatorname{Hilb}^n X$ has a symplectic structure. A proof can be found in [**Nak1999**, Theorem 1.17].

For the Hilbert scheme of points in $\mathbb{C}^2$, one has a stronger statement.

Corollary 11.10. *The Hilbert scheme $\operatorname{Hilb}^n \mathbb{C}^2$ is a hyperkähler manifold.*

Indeed, by Theorem 10.49 any quiver variety $\mathcal{M}_\theta(\mathbf{v}, \mathbf{w})$, with $\mathbf{v}$-generic θ, has a hyperkähler structure.

11.3. Moduli space of torsion free sheaves

We now consider the more general quiver variety $\mathcal{M}_\theta(n,r)$ for the Jordan quiver, with $r \geq 1$. The goal of this section is to show that this variety can also be interpreted as a certain moduli space, namely the moduli space of torsion free sheaves.

Recall that a quasicoherent sheaf $\mathcal{F}$ on an algebraic variety X is called *torsion free* if for every affine open subset $U \subset X$, the space of local sections $\Gamma(U, \mathcal{F})$ is torsion free as a module over $\mathcal{O}(U)$: for any nonzero section $s \in \Gamma(U, \mathcal{F})$ and nonzero $f \in \mathcal{O}(U)$, we have $fs \neq 0$.

A typical example of a torsion free sheaf is a subsheaf in the locally free sheaf: if $\mathcal{V}$ is a locally free sheaf (i.e. the sheaf of sections of a vector bundle V), then any subsheaf $\mathcal{F} \subset \mathcal{V}$ is torsion free.

From now on, we will only be interested in coherent torsion free sheaves on smooth varieties. It turns out that in this case, if dimension is 2, then any torsion free sheaf is a subsheaf in a locally free sheaf.

Recall that for any quasicoherent sheaf $\mathcal{F}$, we denote by $\mathcal{F}^\vee$ the dual sheaf:
$$\mathcal{F}^\vee = \mathcal{H}om(\mathcal{F}, \mathcal{O}),$$
where $\mathcal{H}om$ is the sheaf of homomorphisms of $\mathcal{O}$-modules. It is easy to see that we have a canonical morphism $\mathcal{F} \to \mathcal{F}^{\vee\vee}$. If $\mathcal{F}$ is locally free of finite rank, then the morphism $\mathcal{F} \to \mathcal{F}^{\vee\vee}$ is an isomorphism; in general, however, it is not so.

The following result is well known (see, e.g., [**Bar2000**, Section 2]).

Theorem 11.11. *Let X be a nonsingular algebraic variety and let $\mathcal{F}$ be a coherent torsion free sheaf on X.*

(1) *There exists a Zariski open subset $U \subset X$ of codimension ≥ 2 such that the restriction $\mathcal{F}|_U$ is locally free.*

(2) *If $\dim X = 2$, then the sheaf $\mathcal{F}^{\vee\vee}$ is locally free of finite rank, and the morphism $\mathcal{F} \to \mathcal{F}^{\vee\vee}$ is injective. Restriction of this morphism to U is an isomorphism $\mathcal{F}|_U \simeq (\mathcal{F}^{\vee\vee})|_U$.*

Thus, we see that if X is a smooth variety of dimension 2, then any coherent torsion free sheaf over X is isomorphic over an open dense subset $U \subset X$ to the sheaf of sections of a vector bundle. We denote the rank of this vector bundle by $r = \mathrm{rank}(\mathcal{F})$. We also note that since every coherent sheaf admits a resolution by vector bundles, one can define Chern classes $c_i(\mathcal{F}) \in H^{2i}(X)$.

Example 11.12. Let X be a smooth affine variety, $\dim X > 1$, and let $\mathbf{t} = (t_1, \ldots, t_n)$, $t_i \in X$, $t_i \neq t_j$. Consider the subsheaf $\mathcal{F}_\mathbf{t} \subset \mathcal{O}_X$ whose

11.3. Moduli space of torsion free sheaves

sections are given by the equations $f(t_i) = 0$. This subsheaf is coherent and torsion free but not free. In this case $\mathcal{F}_\mathbf{t}^{\vee\vee} \simeq \mathcal{O}$, and the quotient $\mathcal{F}_\mathbf{t}^{\vee\vee}/\mathcal{F}_\mathbf{t}$ is the direct sum of skyscraper sheaves at t_i.

More generally, for every point in the Hilbert scheme $J \in \text{Hilb}^n X$, let $\mathcal{F}_J$ be the corresponding subsheaf of $\mathcal{O}$ so that $\Gamma(X, \mathcal{F}_J) = J$. Then $\mathcal{F}_J$ is a torsion free sheaf, and $\mathcal{O}/\mathcal{F}_J$ is a torsion sheaf:

$$\Gamma(X, \mathcal{O}/\mathcal{F}_J) = \mathbb{C}[X]/J.$$

Example 11.13. Consider the morphism of sheaves on $\mathbb{C}^2$

$$f: \mathcal{O} \to \mathcal{O} \oplus \mathcal{O}$$
$$s \mapsto (z_1 s, z_2 s),$$

where z_1, z_2 are coordinates on $\mathbb{C}^2$. Then it is obvious that f is injective (as a sheaf morphism) and that

$$\text{Im}(f) = \{(s_1, s_2) \in \mathcal{O} \oplus \mathcal{O} \mid z_2 s_1 = z_1 s_2\}.$$

In this case, the quotient $\mathcal{F} = \mathcal{O} \oplus \mathcal{O}/\text{Im}(f)$ is a torsion free sheaf. Indeed, consider the morphism

$$g: \mathcal{O} \oplus \mathcal{O} \to \mathcal{O}$$
$$(s_1, s_2) \mapsto z_2 s_1 - z_1 s_2.$$

It is immediate that $\text{Ker}(g) = \text{Im}(f)$, so

$$\mathcal{O} \oplus \mathcal{O}/\text{Im}(f) \simeq \text{Im}(g) = \{s \in \mathcal{O} \mid s(0,0) = 0\}$$

which is clearly a torsion free sheaf.

The result of the last example can be generalized. We will need the following lemma later.

Lemma 11.14. *Let V be a finite-dimensional complex vector space and let $A_1, A_2: V \to V$ be linear operators. Let $\mathcal{V} = V \otimes \mathcal{O}$ be the sheaf of V-valued functions on $\mathbb{C}^2$.*

Consider the morphism of sheaves

$$A: \mathcal{V} \to \mathcal{V} \oplus \mathcal{V}$$
$$v \mapsto ((A_1 - z_1)v, (A_2 - z_2)v),$$

where z_1, z_2 are coordinates on $\mathbb{C}^2$. Then A is injective and the quotient sheaf $\mathcal{V} \oplus \mathcal{V}/\text{Im}(A)$ is torsion free.

Proof. Let

$$D_1 = \{(z_1, z_2) \in \mathbb{C}^2 \mid (A_1 - z_1) \text{ is not invertible}\},$$
$$D_2 = \{(z_1, z_2) \in \mathbb{C}^2 \mid (A_2 - z_2) \text{ is not invertible}\}.$$

Then D_1, D_2 are codimension one subvarieties in $\mathbb{C}^2$ and $D = D_1 \cap D_2$ is a codimension 2 subvariety.

If $v \in \mathcal{V}$ is a nonvanishing local section, then $Av \neq 0$ on $\mathbb{C}^2 \setminus D$. This implies that A is injective.

To prove that $\mathcal{V} \oplus \mathcal{V}/\operatorname{Im}(A)$ is torsion free, assume that $s = (v_1, v_2)$ is a local section of $\mathcal{V} \oplus \mathcal{V}$ such that $fs \in \operatorname{Im}(A)$ for some $f \in \mathcal{O}$, i.e. $fv_i = (A_i - z_i)v$ for some $v \in \mathcal{V}$. Then $v/f = (A_1 - z_1)^{-1}v_1$ is regular on $\mathbb{C}^2 \setminus D_1$; similarly, v/f is regular on $\mathbb{C}^2 \setminus D_2$. Therefore, v/f is regular outside of the codimension 2 subvariety D, so it is regular everywhere. □

Let us now consider torsion free sheaves on $X = \mathbb{P}^2$. Let $l_\infty = \{(0 : z_1 : z_2)\} \subset \mathbb{P}^2$ be the line at infinity, so that $\mathbb{P}^2 \setminus l_\infty \simeq \mathbb{C}^2$.

Definition 11.15. Let $\mathcal{F}$ be a torsion free sheaf of rank r on $\mathbb{P}^2$. A *framing* of $\mathcal{F}$ is an isomorphism $\Phi \colon \mathcal{F}|_{l_\infty} \simeq \mathcal{O}_{l_\infty}^{\oplus r}$.

The moduli space of framed torsion free sheaves on $\mathbb{P}^2$ is defined by

$$\mathcal{M}^{fr}(n, r) = \{\text{isomorphism classes of pairs } (\mathcal{F}, \Phi)\},$$

$\mathcal{F}$: torsion free sheaf of rank r on $\mathbb{P}^2$, with $c_2(\mathcal{F}) = n$,

Φ : a framing of $\mathcal{F}$.

This defines $\mathcal{M}^{fr}(n, r)$ as a set; however, it can be shown that this set has a natural structure of a scheme, which is a fine moduli space of framed torsion free sheaves.

Note that it follows from existence of framing that $c_1(\mathcal{F}) = 0$.

Remark 11.16. In [**Nak1999**], this moduli space is denoted by $\mathcal{M}(r, n)$. We switched the order of arguments to match the notation used for quiver varieties (see Theorem 11.18 below).

Example 11.17. Let $r = 1$. Then condition $c_1(\mathcal{F}) = 0 = c_1(\mathcal{F}^{\vee\vee}) = 0$ implies $\mathcal{F}^{\vee\vee} \simeq \mathcal{O}$. Since $\mathcal{F} \hookrightarrow \mathcal{F}^{\vee\vee} = \mathcal{O}$ is an embedding, the quotient sheaf $\mathcal{O}/\mathcal{F}$ is a coherent sheaf on $\mathbb{P}^2$ which is zero in a neighborhood of line l_∞ and which has $\dim \Gamma(\mathbb{C}^2, \mathcal{O}/\mathcal{F}) = n$. Thus, $M = \Gamma(\mathbb{C}^2, \mathcal{O}/\mathcal{F})$ is an n-dimensional module over the algebra $\mathbb{C}[\mathbb{C}^2] = \mathbb{C}[z_1, z_2]$, with a cyclic vector given by constant section $1 \in \mathcal{O}$. Therefore, we see that in this case we have an isomorphism

$$\mathcal{M}^{fr}(n, 1) \simeq \operatorname{Hilb}^n \mathbb{C}^2.$$

We can now formulate the main result of this section.

Theorem 11.18. *For any $n, r \geq 1$, we have an isomorphism $\mathcal{M}^{fr}(n, r) \simeq \mathcal{M}_\theta(n, r)$, where $\mathcal{M}_\theta(n, r)$ is the quiver variety for the Jordan quiver, with $\theta < 0$.*

11.3. Moduli space of torsion free sheaves

Note that in the case $r = 1$, by results of Example 11.17, this theorem becomes Theorem 11.5:
$$\mathcal{M}^{fr}(n, 1) \simeq \mathrm{Hilb}^n \mathbb{C}^2 \simeq \mathcal{M}_\theta(n, 1).$$

The full proof of this theorem can be found in [**Nak1999**, Theorem 2.1]. We only give part of the proof here, namely construction of the map
$$\mathcal{M}_\theta(n, r) \to \mathcal{M}^{fr}(n, r).$$

Let $V = \mathbb{C}^n$, $W = \mathbb{C}^r$, and let $m = [(x, y, i, j)] \in \mathcal{M}_\theta(n, r)$. Let $\mathcal{V} = V \otimes \mathcal{O}_{\mathbb{P}^2}$, $\mathcal{W} = W \otimes \mathcal{O}_{\mathbb{P}^2}$ be the corresponding locally free sheaves on $\mathbb{P}^2$.

Consider the following complex of locally free sheaves on $\mathbb{P}^2$:

(11.2)
$$\mathcal{V}(-1) \xrightarrow{a} \mathcal{V} \oplus \mathcal{V} \oplus \mathcal{W} \xrightarrow{b} \mathcal{V}(1),$$
$$a(v) = ((z_0 x - z_1)v, (z_0 y - z_2)v, z_0 j v),$$
$$b(v_1, v_2, w) = (-(z_0 y - z_2)v_1 + (z_0 x - z_1)v_2 - z_0 i w).$$

Here z_0, z_1, z_2 are coordinates on $\mathbb{P}^2$. Then we have the following facts:

- (11.2) is a complex: $ba = [z_0 x - z_1, z_0 y - z_2] - z_0^2 ij = z_0^2([x, y] - ij) = 0$.
- Restriction of this complex to the line at infinity is a direct sum of an exact complex $\mathcal{V}(-1) \xrightarrow{a} \mathcal{V} \oplus \mathcal{V} \xrightarrow{b} \mathcal{V}(1)$ and $\mathcal{W}$ (in degree zero).
- b is surjective. Indeed, surjectivity of b on the line l_∞ has already been checked. Thus it suffices to check that for every $t = (z_1, z_2) \in \mathbb{C}^2$, the map of the fibers at t
$$V \oplus V \oplus W \xrightarrow{b_t} V$$
$$(v_1, v_2, w) \mapsto -(y - z_2)v_1 + (x - z_1)v_2 - iw$$
is surjective.
  Let $V' = \mathrm{Im}(b_t) \subset V$. Then this subspace is closed under the action of x, y: indeed, for $v \in V'$, we have
$$xv = (x - z_1)v + z_1 v = b_t(0, v, 0) + z_1 v \in V'$$
and similarly for y. In addition, this subspace contains $\mathrm{Im}(i)$. Thus, the stability condition (10.37) implies that $V' = V$, so $\mathrm{Im}(b_t) = V$.
- a is injective (as a morphism of sheaves), and the sheaf $\mathcal{F}$ defined by
$$\mathcal{F} = \mathrm{Ker}(b)/\mathrm{Im}(a)$$
is a torsion free sheaf of rank $r = \dim W$. Indeed, restriction of $\mathcal{F}$ to the line at infinity is identified with $\mathcal{W}$, and restriction to $\mathbb{P}^2 \setminus l_\infty$ is torsion free — by the same reasoning as in Lemma 11.14.

- $c_2(\mathcal{F}) = \dim V = n$. Indeed, let $c(\mathcal{F}) = 1 + c_1(\mathcal{F}) + c_2(\mathcal{F}) \in H^*(\mathbb{P}^2)$. Then $c(\mathcal{V}) = c(\mathcal{W}) = 1$, $c(\mathcal{V}(1)) = (1+t)^n$, $c(\mathcal{V}(-1)) = (1-t)^n$, and $c(\mathcal{F}) = c(\mathcal{V}(1))c(\mathcal{V}(-1)) = ((1+t)(1-t))^n = 1 - nt^2$, where $t \in H^2(\mathbb{P}^1)$ is the generator of cohomology.

This gives us a map $\mathcal{M}_\theta(n,r) \to \mathcal{M}^{fr}(n,r)$; it can be shown that in fact this is a morphism of schemes. To prove that it is an isomorphism, we need to construct an inverse map, i.e. to show that every framed torsion free sheaf of rank r and Chern class $c_2(\mathcal{F}) = n$ is quasi-isomorphic to a complex of the form (11.2). We will not give a proof of this fact here, referring the reader to [**Nak1999**, Theorem 2.1]; a crucial step in the proof is Beĭlinson's resolution of the structure sheaf of the diagonal $\Delta \subset \mathbb{P}^2 \times \mathbb{P}^2$ by locally free sheaves on $\mathbb{P}^2 \times \mathbb{P}^2$.

As a corollary of this theorem, we also get the following result.

Theorem 11.19. *Under the isomorphism of Theorem* 11.18, *the subset of regular elements $\mathcal{M}_\theta^{reg}(n,r)$ is identified with the moduli space of framed vector bundles of rank r with $c_2(\mathcal{F}) = n$.*

Proof. Indeed, the sheaf $\mathcal{F} = \mathrm{Ker}(b)/\mathrm{Im}(a)$ defined by (11.2) is locally free if and only if for any $t = (z_1, z_2) \in \mathbb{C}^2$, the map $a_t \colon V \to V \oplus V \oplus W$ of fibers at t is injective. This in turn is equivalent to the condition

(11.3) $$\mathrm{Ker}(x - z_1) \cap \mathrm{Ker}(y - z_2) \cap \mathrm{Ker}\, j = \{0\}.$$

It is an easy exercise to show that if $[x,y] - ij = 0$, then (11.3) is equivalent to the condition that any subspace $V' \subset \mathrm{Ker}\, j$ which is invariant under x, y is zero. By results of Theorem 10.41, this (together with the condition that (x,y,i,j) is θ-stable) is equivalent to the condition that (x,y,i,j) is regular. $\square$

Remark 11.20. Note that the theorem also holds for $r = 1$, but in this case it is empty: there are no nontrivial framed line bundles on $\mathbb{P}^2$, and the set of regular points $\mathcal{M}_\theta^{reg}(n,1)$ is empty (see Remark 11.8).

Using this theorem, we can also give an explicit description of the projective morphism $\mathcal{M}_\theta(n,r) \to \mathcal{M}_0(n,r)$. Namely, recall that by Theorem 10.42, we have

$$\mathcal{M}_0(\mathbf{v}, \mathbf{w}) = \bigsqcup_{\mathbf{v}' \leq \mathbf{v}} \mathcal{M}_0^{reg}(\mathbf{v}', \mathbf{w}) \times \mathcal{M}_0(\mathbf{v} - \mathbf{v}', 0).$$

For the Jordan quiver, $\mathcal{M}_0^{reg}(n,r) = \mathcal{M}_\theta^{reg}(n,r)$ is the moduli space of framed vector bundles, and $\mathcal{M}_0(k,0) = S^k\mathbb{C}^2$.

Theorem 11.21. *Let $\theta < 0$. Then the canonical morphism*

$$\pi \colon \mathcal{M}_\theta(n, r) \to \mathcal{M}_0(n, r) = \bigsqcup_{k \geq 0} \mathcal{M}_0^{reg}(n - k, r) \times S^k \mathbb{C}^2$$

is given by

(11.4) $$\mathcal{F} \mapsto \big(\mathcal{F}^{\vee\vee}, \operatorname{supp}(\mathcal{F}^{\vee\vee}/\mathcal{F})\big),$$

where $\mathcal{F}$ is a framed torsion free sheaf on $\mathbb{P}^2$, considered as a point of $\mathcal{M}_\theta(n, r)$ via Theorem 11.18, and $\mathcal{F}^{\vee\vee}$ is the corresponding framed vector bundle on $\mathbb{P}^2$, considered as a point in $\mathcal{M}_0^{reg}(n - k, r)$ via Theorem 11.19.

A proof of this theorem (for the more general case of torsion free sheaves on $\mathbb{C}^2/G$, which we will consider in the next chapter) can be found in [**VV1999**, Theorem 1].

11.4. Anti-self-dual connections

In the remaining sections, we give yet one more interpretation of the (open part of the) quiver variety $\mathcal{M}_0(n, r)$ for the Jordan quiver, in terms of moduli spaces of instantons, or anti-self-dual connections on $\mathbb{R}^4$. This description was first introduced in the paper [**AHDM1978**] by Atiyah, Drinfel'd, Hitchin, and Manin and is usually called *ADHM construction*. Our exposition follows the books [**Nak1999**, Section 3.2] and [**DK1990**].

We begin with a short review of anti-self-dual connections. We only give the main statements here, referring the reader to [**DK1990**] for details.

Let X be a Riemannian manifold and let E be a complex vector bundle of rank r over X with a Hermitian metric. Let $\mathcal{A}$ be the space of metric connections on E; for each such connection A, we denote by F_A its curvature. It is a 2-form on X with values in the bundle $\mathfrak{u}(E)$, where $\mathfrak{u} = \mathfrak{u}(r) = \operatorname{Lie}(U(r))$ is the Lie algebra of skew-Hermitian matrices, and $\mathfrak{u}(E)$ is the associated bundle.

Assume that X is oriented and $\dim X = 4$. Then one can compute Chern classes $c_i(E) \in H^{2i}(X)$ in terms of F_A. In particular, if X is compact and connected, so that we can identify $H^4(X, \mathbb{R}) \simeq \mathbb{R}$, we have

(11.5) $$c_2(E) - \tfrac{1}{2}c_1^2(E) = \frac{1}{8\pi^2} \int_X \operatorname{tr}(F_A \wedge F_A).$$

This combination of Chern classes is also known as the degree 2 part of the Chern character of E (up to a minus sign):

$$c_2(E) - \tfrac{1}{2}c_1^2(E) = -ch_2(E).$$

In the case when the structure group of the sheaf can be reduced from U(n) to SU(n) so that the curvature takes values in $\Omega^2(X, \mathfrak{su}(E))$, we have $c_1(E) = 0$ and thus

$$\tag{11.6} \int_X \operatorname{tr}(F_A \wedge F_A) = 8\pi^2 c_2(E).$$

The Riemannian metric on X gives rise to the Hodge operator $*\colon \Omega^2(X) \to \Omega^2(X)$ such that $*^2 = \mathrm{id}$ and thus to a decomposition

$$\tag{11.7} \Omega^2(X) = \Omega^+ \oplus \Omega^-, \qquad \Omega^\pm = \{\omega \in \Omega^2(X) \mid *\omega = \pm\omega\}.$$

For a 2-form $\omega \in \Omega^2(X)$, we will denote by $\omega^\pm$ its projection to $\Omega^\pm$:

$$\omega = \omega^+ + \omega^-, \quad \omega^\pm \in \Omega^\pm.$$

Forms $\omega \in \Omega^+$ will be called self-dual, and $\omega \in \Omega^-$ will be called anti-self-dual.

This can be extended in an obvious way to a 2-form with values in any vector bundle, in particular to $\Omega^2(X, \mathfrak{u}(E))$.

Definition 11.22. Let X be an oriented Riemannian manifold of dimension 4 and let E be a Hermitian vector bundle on X. A metric connection A on E is called anti-self-dual (ASD) if $F_A^+ = 0$, i.e. $*F_A = -F_A$.

It is easy to show that the notion of ASD connection depends only on the conformal class of the metric on X.

Example 11.23. Let $X = \mathbb{R}^4$. Then a connection A is ASD if and only if the curvature F_A satisfies the following equations:

$$F_{12} + F_{34} = 0,$$
$$F_{14} + F_{23} = 0,$$
$$F_{13} + F_{42} = 0.$$

A motivation for the definition of ASD connections comes from physics. Let us introduce the L^2 norm on the space $\Omega^2(X, \mathfrak{u}(E))$:

$$\tag{11.8} \|F\|^2 = \int_X -\operatorname{tr}(F \wedge *F)$$

(recall that the bilinear form $(a, b) = \operatorname{tr}(ab)$ on the Lie algebra $\mathfrak{u}(n)$ is negative definite). In physics, the quantity $\|F_A\|^2$ is interpreted as the energy of connection A.

It is easy to see that $\|F\|^2 = \|F^+\|^2 + \|F^-\|^2$. On the other hand, (11.6) implies that $\|F^-\|^2 - \|F^+\|^2 = 8\pi^2(c_2(E) - \frac{1}{2}c_1(E))$. Thus, for a

11.4. Anti-self-dual connections

metric connection A in a Hermitian vector bundle E on a compact oriented connected 4-manifold X, with $c_1(E) = 0$, we have

$$\tag{11.9} \|F_A\|^2 = 8\pi^2 c_2(E) + 2\|F_A^+\|^2 \geq 8\pi^2 c_2(E)$$

and the equality is achieved iff A is an ASD connection. In particular, this shows that ASD connections are solutions of Yang–Mills equations of motion in classical gauge field theory.

Note that it is immediate from (11.9) that for $c_2(E) < 0$, there are no ASD connections, and for $c_2(E) = 0$, all ASD connections are flat. However, for $c_2(E) > 0$, there exist nonflat ASD connections.

Definition 11.24. Let X be an oriented Riemannian 4-manifold. Then for any Hermitian vector bundle E on X, we define the moduli space of ASD connections $\mathcal{M}_{ASD}(E)$ to be the set of gauge equivalence classes of metric ASD connections on E.

The definition above defines $\mathcal{M}_{ASD}(E)$ as a set. However, it can be shown that for compact X, this set has a structure of a finite-dimensional real analytic set; in particular, it contains an open set $\mathcal{M}_{ASD}^{reg}(E)$ of so-called regular connections, which is a smooth finite-dimensional manifold (see [**DK1990**, Section 4.3.3]).

Let us now consider the situation when X is Kähler and thus, in addition to a Riemannian metric g, also has a complex structure I and a Kähler form ω (see Definition 9.59). In this case, we have the Hodge decomposition of the space of forms:

$$\Omega^n(X, \mathbb{C}) = \bigoplus_{p+q=n} \Omega^{p,q}(X).$$

Then we have the following fundamental result, the proof of which can be found in [**DK1990**, Proposition 2.1.56].

Theorem 11.25. *Let X be a Kähler manifold, with $\dim_{\mathbb{R}}(X) = 4$, and let E be a holomorphic vector bundle on X with a Hermitian metric. Then a metric connection A on E is compatible with the holomorphic structure on E if and only if we have*

$$F_A \in \Omega^{1,1}(X, \mathfrak{u}(E)).$$

Moreover, such a connection exists and is unique.

Conversely, given a Hermitian vector bundle E on X and a metric connection A on E such that $F_A \in \Omega^{1,1}(X, \mathfrak{u}(E))$, there exists a unique holomorphic structure on E compatible with A.

Thus, we see that a holomorphic structure uniquely determines a connection, and vice versa.

Let us now describe the relation between connections coming from holomorphic structure and ASD connections. An easy explicit calculation given in [**DK1990**, Proposition 2.1.56] shows that $\Omega^-(X,\mathbb{C}) \subset \Omega^{1,1}(X)$; moreover,

$$\Omega^{1,1}(X) = \Omega^-(X,\mathbb{C}) \oplus \Omega^0(X,\mathbb{C})\omega, \tag{11.10}$$

where ω is the Kähler form on X.

In particular, it shows that for any ASD connection A, the curvature $F_A \in \Omega^{1,1}(X, \mathfrak{u}(E))$ and thus defines a holomorphic structure on E; conversely, given a holomorphic structure on E, there is a unique metric connection A on E compatible with the holomorphic structure, and this connection is ASD iff the projection of F_A on $\Omega^0(X,\mathbb{C})\omega$ in decomposition (11.10) is zero.

Finally, let us consider the case when X is hyperkähler, with complex structures I, J, K and corresponding 2-forms $\omega_I, \omega_J, \omega_K$ (see Definition 9.67). In this case, the situation is actually easier than in the Kähler case. Namely, an explicit computation similar to the one in the Kähler case shows that

$$\Omega^-(X,\mathbb{C}) = \Omega_I^{1,1}(X) \cap \Omega_J^{1,1}(X) \cap \Omega_K^{1,1}(X), \tag{11.11}$$

where $\Omega_I^{1,1}(X)$ is the space of $(1,1)$ forms with respect to the complex structure I, and similarly for J, K. Combining this with Theorem 11.25, we get the following result.

Theorem 11.26. *Let X be a hyperkähler manifold of real dimension 4, and let E be a Hermitian vector bundle on X with a metric connection A. Assume that for each of the complex structures I, J, K on X, bundle E admits a holomorphic structure compatible with the connection A. Then A is ASD.*

11.5. Instantons on $\mathbb{R}^4$ and ADHM construction

Let us now study the moduli spaces of ASD connections on $\mathbb{R}^4$. Since $\mathbb{R}^4$ is not compact, considering all ASD connections would give infinite-dimensional moduli space. Following [**DK1990**, Section 3.3], we will instead consider finite-energy ASD connections, i.e. ASD connections satisfying the additional condition

$$\|F_A\|^2 = \int_{\mathbb{R}^4} \operatorname{tr}(F_A \wedge F_A) < \infty. \tag{11.12}$$

We will also call finite energy ASD connections *instantons*.

For such connections, we can apply Uhlenbeck's removable singularity theorem (see [**DK1990**, Theorem 4.4.12]), which says that such a connection can be extended to an ASD connection in a bundle E on $S^4 = \mathbb{R}^4 \cup \infty$.

11.5. Instantons on $\mathbb{R}^4$ and ADHM construction

In particular, it shows that every finite energy connection on $\mathbb{R}^4$ is asymptotically flat: there exists a trivialization of E near infinity (i.e. outside of a ball of radius R for large enough R) such that in this trivialization the connection matrix $A \in \Omega^1(\mathfrak{u}(r))$ satisfies

$$(11.13) \qquad |\nabla^{(l)} A(x)| = O(|x|^{-3-l}), \qquad l \geq 0.$$

Note that this implies $|F_A(x)| = O(|x|^{-4})$.

It should be noted that even though any vector bundle on $\mathbb{R}^4$ is trivial, the extension of a vector bundle with a finite energy connection to S^4 can be topologically nontrivial. Indeed, if we choose a trivialization of E on $\mathbb{R}^4$ and another trivialization in a neighborhood of infinity satisfying (11.13), then the comparison of these two trivializations gives a map $f \colon B_R(\infty) \to \mathrm{U}(r)$, where $B_R(\infty) = \{x \in \mathbb{R}^4 \mid |x| \geq R\}$. Since $B_R(\infty)$ is homotopic to the 3-sphere, one sees that any finite energy ASD connection gives an element in the homotopy group $\pi_3(\mathrm{U}(r))$. It is well known that for $r \geq 2$ we have

$$\pi_3(\mathrm{U}(r)) = \pi_3(\mathrm{SU}(r)) = \mathbb{Z}.$$

For example, for $r = 2$, this is obvious since $\mathrm{SU}(2) \simeq S^3$.

The same argument shows that in fact any Hermitian vector bundle E of rank $r \geq 2$ on S^4 defines an integer invariant $k(E) \in \pi_3(\mathrm{SU}(r)) = \mathbb{Z}$.

Lemma 11.27. *Let E be a Hermitian bundle on S^4 of rank $r \geq 2$. Then the invariant $k(E) \in \pi_3(\mathrm{SU}(r)) = \mathbb{Z}$ defined above coincides with the Chern class $c_2(E) \in H^4(S^4) = \mathbb{Z}$. Moreover, this invariant completely determines the topological type of E.*

In particular, for a finite energy ASD connection on $\mathbb{R}^4$ this invariant is given by

$$c_2(E) = \frac{1}{8\pi^2} \int_{S^4} F_A \wedge F_A = \frac{1}{8\pi^2} \int_{\mathbb{R}^4} F_A \wedge F_A.$$

Our goal in this section will be to describe the moduli space of n-instantons, i.e. finite energy ASD connections on $\mathbb{R}^4$ with $c_2(E) = n$. It will be more convenient to work with *framed instantons*, which are defined as follows.

Definition 11.28. The moduli space of rank r framed n-instantons on $\mathbb{R}^4$ is

$$\mathcal{M}_{ASD}^{fr}(n, r) = \{\text{isomorphism classes of pairs } (A, \Phi)\},$$

where A is an ASD connection in a rank r Hermitian vector bundle E on S^4 with $c_2(E) = n$, and $\Phi \colon E_\infty \to \mathbb{C}^r$ is a trivialization of the fiber at ∞ (framing).

Theorem 11.29. *Let $\mathcal{M}_\theta(n,r)$ be the quiver variety for the Jordan quiver, with $\theta < 0$. Denote by $\mathcal{M}^{reg}(n,r) \subset \mathcal{M}_\theta(n,r)$ the regular part as defined in Definition 9.26. Then one has an isomorphism*
$$\mathcal{M}^{fr}_{ASD}(n,r) \simeq \mathcal{M}^{reg}(n,r).$$

A proof of this result is quite long and will not be reproduced here. The main step is showing that any ASD connection on $\mathbb{R}^4$ (up to gauge equivalence) can be described by a certain finite collection of algebraic data, so-called ADHM data. A precise statement can be found in the original paper [**AHDM1978**]; a detailed exposition is given in [**DK1990**, Theorem 3.3.8] (where it is formulated for unframed instantons). This construction uses a choice of spinor structure on $\mathbb{R}^4$. The proof involves some rather hard analytical arguments.

After this, the second (and much easier) step is establishing an isomorphism between the moduli space of framed ADHM data and the quiver variety $\mathcal{M}^{reg}(n,r)$. This can be found in [**Nak1999**, Theorem 3.48]; to construct this isomorphism, we need to choose an identification $\mathbb{R}^4 \simeq \mathbb{C}^2$. Again, we will not give the proof here; instead, we will give a description of the isomorphism $\mathcal{M}^{reg} \to \mathcal{M}^{fr}_{ASD}(n,r)$. Namely, choose an identification $\mathbb{R}^4 \simeq \mathbb{C}^2$. For any point $m \in \mathcal{M}^{reg}(n,r)$, let $\mathcal{F}$ be the corresponding vector bundle on $\mathbb{C}^2$ constructed in Theorem 11.19; choosing an inner product in $\mathcal{V}, \mathcal{W}$ gives an inner product in $\mathcal{F}$. Thus, $\mathcal{F}$ is a Hermitian holomorphic vector bundle on $\mathbb{C}^2 \simeq \mathbb{R}^4$. By Theorem 11.25, it has a unique metric connection A compatible with the holomorphic structure. It can be shown that this connection is ASD and has finite energy; thus, it can be extended to a connection in a vector bundle $\tilde{\mathcal{F}}$ on S^4. Moreover, using framing on $\mathcal{F}$ (considered as a vector bundle on $\mathbb{P}^2$), we can construct a framing of $\tilde{\mathcal{F}}$ at $\infty \in S^4$. Thus, we get a framed instanton on $\mathbb{R}^4$. We refer the reader to [**Nak1999**, Theorem 3.48] for details.

Remark 11.30. This shows that the quiver variety $\mathcal{M}_\theta(n,r)$ is a partial compactification of the instanton moduli space $\mathcal{M}^{fr}_{ASD}(n,r)$. It is also worth noting that the instanton moduli space uses compactification of $\mathbb{R}^4$ to S^4, whereas the moduli space of framed torsion free bundles described in Section 11.3 used the compactification of $\mathbb{C}^2$ to $\mathbb{P}^2$.

Chapter 12

Kleinian Singularities and Geometric McKay Correspondence

In this chapter we study a generalization of the results of the previous chapter, giving a geometric construction of the quiver varieties for Euclidean quivers. This construction is based on the results of Chapter 8, establishing a bijection between Euclidean (or Dynkin) graphs and finite subgroups in SU(2).

Throughout this chapter, G is a nontrivial finite subgroup in SU(2), and ρ_i, $i \in I$, are the irreducible representations of G. As in Chapter 8, we use index 0 for the trivial representation: $\rho_0 = \mathbb{C}$.

Inclusion $G \subset \mathrm{SU}(2)$ gives an action of G on $\mathbb{C}^2$ and thus on the algebra $A = \mathbb{C}[\mathbb{C}^2]$ of polynomials on $\mathbb{C}^2$. Choosing coordinates z_1, z_2 on $\mathbb{C}^2$, we can write
$$A = \mathbb{C}[z_1, z_2].$$
We will denote by ρ the two-dimensional representation of G:
$$\rho = \mathbb{C}^2 = \langle z_1, z_2 \rangle.$$
This representation is self-dual: $\rho^* \simeq \rho$.

12.1. Kleinian singularities

Consider the quotient space

(12.1) $$\mathbb{C}^2/G.$$

This is a singular algebraic variety. For example, for $G = \mathbb{Z}_2$, this variety is isomorphic to a quadric in $\mathbb{C}^3$ (see Example 10.18). Singularities of the form (12.1) are called *Kleinian singularities*; they are also called rational double points or simple singularities. We briefly review some facts about these singularities, referring the reader to review articles by Slodowy [**Slo1983**] and Durfee [**Dur1979**] for details.

The following description of these singularities goes back to Felix Klein ([**Kle1884**]).

Theorem 12.1. *Each singularity of the form (12.1) is isomorphic to a hypersurface in $\mathbb{C}^3$, given by a single equation $f(X, Y, Z) = 0$. This hypersurface has only one singular point, $X = Y = Z = 0$.*

Table 12.1 (taken from [**Slo1983**]) gives the equation of each of the Kleinian singularities. It also lists, for every G, the corresponding Dynkin graph Γ as described in Section 8.2; recall that vertices of Γ are in bijection with nonidentity conjugacy classes in G, and branches of Γ correspond to G-orbits in $\mathbb{P}^1$ with a nontrivial stabilizer.

Table 12.1. Defining equations for Kleinian singularities.

G	Γ	$|G|$	equation
$\mathbb{Z}_n$, $n \geq 2$	A_{n-1}	n	$X^n - YZ = 0$
binary dihedral group BD_{4n}, $n \geq 2$	D_{n+2}	$4n$	$X(Y^2 - X^n) + Z^2 = 0$
binary tetrahedral	E_6	24	$X^4 + Y^3 + Z^2 = 0$
binary octahedral	E_7	48	$X^3 + XY^3 + Z^2 = 0$
binary icosahedral	E_8	120	$X^5 + Y^3 + Z^2 = 0$

Example 12.2. Consider the case $G = \mathbb{Z}_n$; the corresponding Dynkin graph is $\Gamma = A_{n-1}$. The generator of this group acts on $\mathbb{C}^2$ by $(z_1, z_2) \mapsto (\zeta z_1, \zeta^{-1} z_2)$, where $\zeta = e^{2\pi i/n}$. In this case, the algebra $\mathbb{C}[z_1, z_2]^G$ is generated by $X = z_1 z_2$, $Y = z_1^n$, $Z = z_2^n$, subject to the relation $X^n - YZ = 0$.

As always, given a singular variety, one would like to study its resolution. In the case of Kleinian singularities, these resolutions were constructed by Du Val in 1934 ([**DV1934**]) and have been extensively studied. The following theorem, taken from [**Slo1983**, Section 6], summarizes the properties of these resolutions.

Theorem 12.3. *Let $Y = \mathbb{C}^2/G$ be a Kleinian singularity. Then*

(1) *$\mathbb{C}^2/G$ has a resolution of singularities $\pi\colon X \to \mathbb{C}^2/G$ which is minimal: every other resolution of singularities factors through π. Such a resolution is unique.*

(2) The exceptional fiber $\mathcal{L} = \pi^{-1}(0)$ of the minimal resolution is a union of copies of $\mathbb{P}^1$:
$$\pi^{-1}(0) = \bigcup_{i \in \mathrm{Irr}(\mathcal{L})} C_i, \qquad C_i \simeq \mathbb{P}^1.$$
The set of irreducible components $\mathrm{Irr}(\mathcal{L})$ is in bijection with the vertices of the Dynkin graph Γ corresponding to G as in Theorem 8.9.

(3) The irreducible components C_i of $\pi^{-1}(0)$ intersect transversally, and the intersection pairing is given by
$$C_i \cdot C_j = -(\alpha_i, \alpha_j),$$
where $(\,,\,)$ is the symmetrized Euler form of Γ. In particular, $C_i^2 = -2$, so the normal bundle to C_i in X is isomorphic to the cotangent bundle $T^*\mathbb{P}^1$.

Thus, we see that the irreducible components C_i of the exceptional fiber $\pi^{-1}(0)$ are indexed by vertices of the Dynkin graph Γ, and two components either do not intersect at all (if the corresponding vertices are not connected by an edge in Γ) or intersect transversally at a single point (if the corresponding vertices are connected). The picture below shows the configuration of irreducible components for the minimal resolution of singularity of type A:

We do not give Du Val's construction of the resolution here; the interested reader can find a good exposition in [**Lam1986**, IV§6–§9]. Instead, we will construct a resolution by other means in the next section.

We only mention that (at least in the case when $G \supset \{\pm I\}$) Du Val's resolution can be obtained in two steps: first, resolving $T^*\mathbb{P}^1 \to \mathbb{C}^2/\mathbb{Z}_2$ (see Example 9.33) and then resolving $T^*\mathbb{P}^1/\overline{G}$, where $\overline{G} = G/\{\pm I\}$. Since singular points of $T^*\mathbb{P}^1/\overline{G}$ correspond to points in $\mathbb{P}^1$ with a nontrivial stabilizer in $\overline{G}$, this explains the relation with the graph Γ: the branches of Γ are also indexed by orbits with nontrivial stabilizer.

12.2. Resolution of Kleinian singularities via Hilbert schemes

In this section, we construct a resolution of the Kleinian singularity $\mathbb{C}^2/G$ using Hilbert schemes. Recall the Hilbert scheme $\mathrm{Hilb}^n X$ defined in Section 11.1. The action of G on $\mathbb{C}^2$ gives a natural action of G on $\mathrm{Hilb}^n \mathbb{C}^2$; in particular, we can consider the fixed point set

(12.2) $$\mathrm{Hilb}^G \mathbb{C}^2 = (\mathrm{Hilb}^n \mathbb{C}^2)^G, \qquad n = |G|.$$

Note that for every ideal $J \in \mathrm{Hilb}^G \mathbb{C}^2$, the quotient A/J is a representation of G and thus can be decomposed into irreducible representations (here $A = \mathbb{C}[\mathbb{C}^2] = \mathbb{C}[z_1, z_2]$).

Lemma 12.4.

(1) *The fixed point set* $\mathrm{Hilb}^G \mathbb{C}^2$ *is a nonsingular algebraic variety.*

(2) *One has a decomposition*
$$\mathrm{Hilb}^G \mathbb{C}^2 = \bigsqcup_{\mathbf{v}} X^{\mathbf{v}},$$
where the sum is over all $\mathbf{v} \in \mathbb{Z}_+^I$ *with* $\sum \mathbf{v}_i \dim \rho_i = |G|$ *and*
$$X^{\mathbf{v}} = \{ J \in \mathrm{Hilb}^G \mathbb{C}^2 \mid A/J \simeq \bigoplus \mathbf{v}_i \rho_i \text{ as a representation of } G \}.$$
Each $X^{\mathbf{v}}$ *is a nonsingular subvariety of* $\mathrm{Hilb}^G \mathbb{C}^2$.

Proof. The first part follows from a general result: if a reductive algebraic group acts on a nonsingular algebraic variety, then the fixed point set is nonsingular (see [**Ive1972**]). The second part is obvious. □

Consider now a special case when we take $\mathbf{v} = \delta$, i.e. $\mathbf{v}_i = \delta_i = \dim \rho_i$, so that $\bigoplus \mathbf{v}_i \rho_i$ is the regular representation of G. For example, let $\mathbf{t} \in \mathbb{C}^2 \setminus (0,0)$. Consider the ideal $J_{G\mathbf{t}}$ consisting of all functions on $\mathbb{C}^2$ which vanish on the orbit $G\mathbf{t}$. Since for any nonzero $\mathbf{t} \in \mathbb{C}^2$, its stabilizer in $\mathrm{SL}(2, \mathbb{C})$ (and thus in G) is trivial, the quotient $A/J_{G\mathbf{t}}$ is isomorphic to the regular representation of G and thus $J_{G\mathbf{t}} \in X^\delta$.

We will need the following lemma, the proof of which will be given later (see Corollary 12.10).

Lemma 12.5. X^δ *is connected.*

Consider the Hilbert–Chow morphism $\mathrm{Hilb}^n \mathbb{C}^2 \to S^n \mathbb{C}^2$. Restricting it to the subset of G-fixed points, we get a morphism $\mathrm{Hilb}^G \mathbb{C}^2 \to (S^n \mathbb{C}^2)^G$, where $n = |G|$. We can now use the following lemma.

Lemma 12.6. *One has an isomorphism* $(S^n \mathbb{C}^2)^G = \mathbb{C}^2/G$, $n = |G|$.

Proof. Our proof follows [**IN1999**, Lemma 9.2]. Since the stabilizer in G of every $t \in \mathbb{C}^2 - \{0\}$ is trivial, every G-orbit in $\mathbb{C}^2$ is either a single point 0, or consists of n distinct points. This shows that we have an isomorphism $(S^n(\mathbb{C}^2 - \{0\}))^G \to (\mathbb{C}^2 - \{0\})/G$, which extends to a bijective morphism $(S^n \mathbb{C}^2)^G \to \mathbb{C}^2/G$. Since $\mathbb{C}^2/G$ is normal, it is an isomorphism. □

Therefore, we see that the Hilbert–Chow morphism gives a morphism $X^\delta \to \mathbb{C}^2/G$.

From now on, we use the notation

(12.3) $$\widehat{\mathbb{C}^2/G} = X^\delta.$$

Theorem 12.7.
  (1) *The Hilbert–Chow morphism $\pi\colon \widehat{\mathbb{C}^2/G} \to \mathbb{C}^2/G$ is a resolution of singularities.*
  (2) *One has $\pi^*(K_{\mathbb{C}^2/G}) = K_{\widehat{\mathbb{C}^2/G}}$, where K_X is the canonical class of X.*

A proof of this theorem can be found in Ito–Nakamura [**IN1999**].

Proof. Let $Y_0 = (\mathbb{C}^2 - \{0\})/G \subset \mathbb{C}^2/G$. By Lemma 12.6, the restriction of the Hilbert–Chow morphism to $\pi^{-1}(Y_0) \to Y_0$ is an isomorphism. Since Y_0 is open and dense in $\mathbb{C}^2/G$, and $\widehat{\mathbb{C}^2/G}$ is connected and nonsingular, this shows that $\pi^{-1}(Y_0)$ is dense in $\widehat{\mathbb{C}^2/G}$; thus, $\pi\colon \widehat{\mathbb{C}^2/G} \to \mathbb{C}^2/\Gamma$ is a resolution of singularities.

To prove part (2), note that by Corollary 11.9 we have a holomorphic symplectic structure on $\mathrm{Hilb}^n \mathbb{C}^2$ which restricts to a holomorphic symplectic structure on $\widehat{\mathbb{C}^2/G} = X^\delta$. In particular, the canonical bundle is trivial: $K_X = \mathcal{O}_X = \pi^*(\mathcal{O}_{\mathbb{C}^2/G})$. $\square$

Resolutions $\pi\colon X \to Y$ satisfying condition $\pi^* K_Y = K_X$ are called *crepant* (as there is no discrepancy between K_X and K_Y). It can be shown that in dimension 2, if a crepant resolution exists, it is minimal: any other resolution factors through it. However, in general existence of crepant resolutions is not guaranteed.

Corollary 12.8. *The resolution $\pi\colon \widehat{\mathbb{C}^2/G} \to \mathbb{C}^2/G$ coincides with the minimal resolution of singularities described in Theorem 12.3. In particular, irreducible components of the exceptional fiber $\pi^{-1}(0)$ are indexed by vertices of Dynkin graph Γ.*

12.3. Quiver varieties as resolutions of Kleinian singularities

Let us now relate the resolutions of Kleinian singularities constructed in the previous section with quiver varieties. As before, let G be a nontrivial finite subgroup in $SU(2)$. Let Q be the Euclidean graph corresponding to G under McKay correspondence (Theorem 8.15); recall that the set I of vertices of Q is in bijection with the set of isomorphism classes of irreducible finite-dimensional representations of G, and the number of edges A_{ij} between i and j in Q is given by the equation

$$\rho \otimes \rho_i = \bigoplus A_{ij} \rho_j,$$

where $\rho = \mathbb{C}^2$ is the two-dimensional representation given by embedding $G \subset \mathrm{SU}(2)$ (note that ρ is self-dual: $\rho^* \simeq \rho$, but not canonically). We will denote by C the corresponding Cartan matrix:

(12.4) $$c_{ij} = (\alpha_i, \alpha_j) = 2\delta_{ij} - A_{ij}, \qquad i, j \in I.$$

Note that C has a one-dimensional kernel spanned by the vector $\delta_i = \dim \rho_i$ (see Lemma 8.12).

Throughout this section, we assume that we have chosen an orientation Ω of Q and denote by $\vec{Q}$ the correspondent quiver. As before, we will use notation $\mathcal{M}_\theta(\mathbf{v}, \mathbf{w})$ for the quiver varieties for $\vec{Q}$.

The following theorem is a generalization of Theorem 11.5, establishing a relation between quiver varieties and Hilbert schemes.

Theorem 12.9. *Let* $\mathbf{v}, \mathbf{w} \in \mathbb{Z}^I$ *be defined by* $\mathbf{v} = \delta$ *(i.e.* $\mathbf{v}_i = \dim \rho_i$*) and* $\mathbf{w}_0 = 1$, $\mathbf{w}_i = 0$ *for* $i \neq i_0$ *(where* i_0 *is the extending vertex, corresponding to the trivial representation of* G*). Then one has isomorphisms*

$$\mathcal{M}_0(\mathbf{v}, \mathbf{w}) \simeq (S^n \mathbb{C}^2)^G = \mathbb{C}^2/G, \qquad n = |G|,$$
$$\mathcal{M}_\theta(\mathbf{v}, \mathbf{w}) \simeq \widehat{\mathbb{C}^2/G}, \qquad \theta < 0.$$

Under this isomorphism, the canonical projective morphism $\mathcal{M}_\theta(\mathbf{v}, \mathbf{w}) \to \mathcal{M}_0(\mathbf{v}, \mathbf{w})$ *is identified with the Hilbert–Chow morphism* $\widehat{\mathbb{C}^2/G} \to \mathbb{C}^2/G$.

Proof. Recall that for any two vertices $i, j \in I$, one can construct a basis φ_h in the space $\mathrm{Hom}_G(\rho_i, \rho_j \otimes \mathbb{C}^2)$ indexed by edges $h \colon i \to j$ in $Q^\sharp$ (see (8.13)). Therefore, every $z \in R(Q^\sharp, \mathbf{v})$ defines a morphism of representations of G

$$f_z = \sum_h z_h \otimes \varphi_h \colon M \to M \otimes \mathbb{C}^2,$$

where $M = \bigoplus V_i \otimes \rho_i$, considered with a natural structure of a representation of G; note that $\dim M = n = |G|$. Moreover, this gives an isomorphism

$$R(Q^\sharp, \mathbf{v}) \simeq \mathrm{Hom}_G(M, M \otimes \mathbb{C}^2).$$

The same argument as in the proof of Theorem 11.5 shows that if $\mu_{\mathbf{v},\mathbf{w}}(x, y, i, j) = 0$, then the orbit of (x, y, i, j) is closed if and only if x, y are diagonalizable and $i = 0, j = 0$. Therefore, we have a bijection

$$\{(x, y, i, j) | \mu_{\mathbf{v},\mathbf{w}}(x, y, i, j) = 0, \text{ orbit is closed}\}$$
$$\simeq \{f \in \mathrm{Hom}_G(M, M \otimes \mathbb{C}^2) \mid [f_1, f_2] = 0, \ f_i \text{ are diagonalizable}\},$$

where $f_1, f_2 \colon M \to M$ are components of f.

This easily implies that we have an isomorphism

$$\mathcal{M}_0(\mathbf{v}, \mathbf{w}) \simeq (S^n \mathbb{C}^2)^G$$

12.3. Quiver varieties as resolutions of Kleinian singularities

given by
$$(x, y) \mapsto (\lambda, \mu),$$
where $\lambda = (\lambda_1, \ldots, \lambda_n)$ are eigenvalues of f_1 and $\mu = (\mu_1, \ldots, \mu_n)$ are eigenvalues of f_2 (compare with Example 10.18).

For $\mathcal{M}_\theta(\mathbf{v}, \mathbf{w})$, the same argument as in the proof of Theorem 11.5 shows that a point $m = (x, y, i, j)$ is θ-stable iff $j = 0$, $[x, y] = 0$, and $i(W) = V_0$ generates $M = \bigoplus V_i \otimes \rho_i$ under the action of x, y. Thus, $\mathcal{M}_\theta$ is the moduli space of pairs (M, v), where M is a representation of $\mathbb{C}[G] \ltimes \mathbb{C}[\mathbb{C}^2]$, which is isomorphic to the regular representation of G, and v is a G-invariant cyclic vector. This coincides with the definition of $\widehat{\mathbb{C}^2/G} \subset \mathrm{Hilb}^G \mathbb{C}^2$.

Since the Hilbert–Chow map is given by $\pi(M) = \mathrm{supp}(M)$, and for a finite-dimensional $\mathbb{C}[x, y]$-module, $\mathrm{supp}(M) = (\lambda_1, \ldots, \lambda_n; \mu_1, \ldots, \mu_n) \in S^n \mathbb{C}^2$, where λ_i, μ_i are eigenvalues of x, y respectively, we see that the diagram

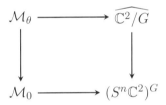

is commutative. □

As a corollary, we get the proof of connectivity of $\widehat{\mathbb{C}^2/G}$, thus proving Lemma 12.5.

Corollary 12.10. $\widehat{\mathbb{C}^2/G}$ *is connected.*

This follows from Theorem 12.9 and the connectivity of $\mathcal{M}_\theta(\mathbf{v}, \mathbf{w})$ (Theorem 10.37).

This theorem can be generalized.

Theorem 12.11. *Let* $\mathbf{v}, \mathbf{w} \in \mathbb{Z}^I$ *be as in Theorem* 12.9, *and let* $\theta \in \mathbb{Z}^I$ *be* $\mathbf{v}$-*generic. Then we have an isomorphism of algebraic varieties*

(12.5) $$\mathcal{M}_\theta(\mathbf{v}, \mathbf{w}) \simeq \widehat{\mathbb{C}^2/G}.$$

Proof. Indeed, in this case $\mathcal{M}_\theta(\mathbf{v}, \mathbf{w})$ is a resolution of singularities of $\mathcal{M}_0(\mathbf{v}, \mathbf{w}) = \mathbb{C}^2/G$. Therefore, it must factor through the minimal resolution, so we have a canonical morphism of algebraic varieties $\mathcal{M}_\theta(\mathbf{v}, \mathbf{w}) \to \widehat{\mathbb{C}^2/G}$. On the other hand, by Theorem 10.49, $\mathcal{M}_\theta(\mathbf{v}, \mathbf{w})$ is diffeomorphic to $\mathcal{M}_{\theta'}(\mathbf{v}, \mathbf{w})$ for any generic θ, θ'; thus, the morphism $\mathcal{M}_\theta(\mathbf{v}, \mathbf{w}) \to \widehat{\mathbb{C}^2/G}$ is an isomorphism. □

Corollary 12.12. *The space $\widehat{\mathbb{C}^2/G}$ has a hyperkähler structure. In particular, it has a holomorphic symplectic structure, and $\widehat{\mathbb{C}^2/G} \to \mathbb{C}^2/G$ is a symplectic resolution.*

Indeed, by Theorem 10.49 any quiver variety $\mathcal{M}_\theta(\mathbf{v}, \mathbf{w})$, with $\mathbf{v}$-generic θ, has a canonical hyperkähler structure.

Theorem 12.9 can be generalized. Namely, for $\mathbf{v} \in \mathbb{Z}_+^I$, define
(12.6)
$$X^\mathbf{v} = \{J \in (\operatorname{Hilb}^n \mathbb{C}^2)^G \mid A/J \simeq \bigoplus \mathbf{v}_i \rho_i \text{ as a representation of } G\},$$
$$n = \sum \mathbf{v}_i \dim \rho_i$$

(this is the same definition as in Lemma 12.4, except that we no longer require $n = |G|$). The same argument as in the proof of Lemma 12.4 shows that $X^\mathbf{v}$ is a nonsingular algebraic variety.

Then we have the following result.

Theorem 12.13. *Let $\mathbf{v} \in \mathbb{Z}_+^I$. Then we have an isomorphism of algebraic varieties*
$$X^\mathbf{v} \simeq \mathcal{M}_\theta(\mathbf{v}, \mathbf{w}),$$
where $\theta < 0$ and $\mathbf{w}$ is as in Theorem 12.9.

The proof of this theorem repeats with minor changes the proof of Theorem 12.9. We leave the details to the reader.

We also mention that it was shown in [**Kuz2007**] that for any integer $k > 0$ it is possible to choose the parameter θ so that
$$\mathcal{M}_\theta(k\delta, \mathbf{w}) \simeq \operatorname{Hilb}^k(\widehat{\mathbb{C}^2/G}),$$
where $\mathbf{w}$ is as in Theorem 12.9. The proof of this result will not be given here.

12.4. Exceptional fiber and geometric McKay correspondence

As was discussed is Chapter 8, there are two ways to relate a finite subgroup $G \subset \operatorname{SU}(2)$ (and thus a Kleinian singularity $\mathbb{C}^2/G$) with a Euclidean or Dynkin graph. The first correspondence, described in Theorem 8.9, produces, for each G, a Dynkin graph Γ whose vertices are in bijection with nontrivial conjugacy classes in G and also with the irreducible components of the zero fiber $\pi^{-1}(0)$ where $\pi \colon \widehat{\mathbb{C}^2/G} \to \mathbb{C}^2/G$ is the minimal resolution of singularities. On the other hand, we have the McKay correspondence, which identifies simple representations of G with vertices of a Euclidean graph Q.

12.4. Exceptional fiber and geometric McKay correspondence

As mentioned in Section 8.3, in all cases the Dynkin graph Γ is exactly the graph obtained from the Euclidean graph Q by removing the extending vertex; however, at that time we did not provide any explanation of this fact. In this section, we fill this gap, by establishing an explicit bijection between the set of irreducible components of $\mathcal{L} = \pi^{-1}(0)$ and the set $I - \{0\}$, where I is the set of vertices of Q (and also the set of irreducible representations of G) and 0 is the extending vertex, corresponding to the trivial representation. This construction was found in [**Nak1996**] and independently rediscovered in [**IN1999**]. We give an outline of Nakajima's construction, which does not require a case-by-case analysis. Our exposition follows [**Nak2001b**, Section 6.1].

Recall that we have defined a subvariety $X^{\mathbf{v}} \subset (\mathrm{Hilb}^n \mathbb{C}^2)^G$ for any $\mathbf{v} \in \mathbb{Z}_+^I$, where $n = \sum \mathbf{v}_i \dim \rho_i$ (see (12.6)).

For any $i \in I$, define

$$(12.7) \quad B_i(\mathbf{v}) = \{J_1 \in X^{\mathbf{v}-\alpha_i}, J_2 \in X^{\mathbf{v}} \mid J_1 \supset J_2\} \subset X^{\mathbf{v}-\alpha_i} \times X^{\mathbf{v}}.$$

This is a special case of a *Hecke correspondence*, which will be discussed in a more general situation in Section 13.3. It follows from general results about Hecke correspondences (see Theorem 13.14) that that $B_i(\mathbf{v})$ is a smooth Lagrangian subvariety in $X^{\mathbf{v}-\alpha_i} \times X^{\mathbf{v}}$. In our case, it is also possible to prove it in a more elementary way; see [**Nak1996**, §5].

Consider now a special case when $\mathbf{v} = \delta$, i.e. $\mathbf{v}_i = \dim \rho_i$. In this case, by Theorem 12.9, we have $X^{\mathbf{v}} = \widehat{\mathbb{C}^2/G}$. It is easy to show that for $i = 0$, $X^{\delta-\alpha_i} = \varnothing$. It can also be shown (see [**Nak2001b**, Theorem 4.6]; we will discuss it later, in Example 13.20) that for $i \neq 0$, $X^{\delta-\alpha_i}$ is a point, so in this case $B_i(\delta)$ is a Lagrangian subvariety in $X^{\delta} = \widehat{\mathbb{C}^2/G}$. The following theorem summarizes results of [**Nak2001b**, Example 6.3].

Theorem 12.14. *Let $i \in I - \{0\}$, and let $B_i(\delta) \subset \widehat{\mathbb{C}^2/G}$ be defined by (12.7). Then*

(1) $B_i(\delta) \simeq \mathbb{P}^1$.

(2) $B_i(\delta) \subset \mathcal{L} = \pi^{-1}(0)$. *Moreover,* $\mathcal{L} = \bigcup_{i \in I - \{0\}} B_i(\delta)$.

(3) *For any $i, j \in I - \{0\}$, the intersection pairing in homology is given by*

$$[B_i(\delta)] \cdot [B_j(\delta)] = -c_{ij},$$

where c_{ij} is the Cartan matrix (12.4) of Q. For $i \neq j$, $B_i(\delta), B_j(\delta)$ intersect transversally.

Thus, $B_i(\delta)$ are exactly the irreducible components of the exceptional fiber $\mathcal{L}$ (compare with Theorem 12.3). This theorem gives a bijection $I - \{0\} \to \mathrm{Irr}(\mathcal{L})$.

Example 12.15. Let $G = \mathbb{Z}_{n+1}$ so that the graph $Q = \widehat{A_n}$ is the cyclic graph with $n+1$ vertices.

In this case, it is easy to see that for any $i = 1, \ldots, n$, the set $B_i(\delta)$ consists of isomorphism classes of representations of $Q^\sharp$ shown in Figure 12.1. Up to an isomorphism, such a representation only depends on $[\lambda : \mu] \in \mathbb{P}^1$.

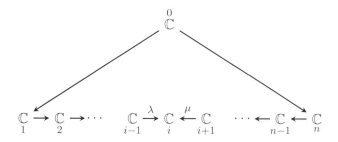

Figure 12.1. Construction of $B_i(\delta)$ for type A. All arrows which are not labeled are identity maps; all arrows not shown are zero maps. At least one of λ, μ must be nonzero.

Thus, each $B_i(\delta) \simeq \mathbb{P}^1$. We leave it to the reader to check that $[B_i(\delta)] \cdot [B_j(\delta)] = 1$ if $i = j \pm 1$ and is zero otherwise.

The bijection $I - \{0\} \simeq \mathrm{Irr}(\mathcal{L})$ can also be described in terms of the K-group of $\widehat{\mathbb{C}^2/G}$. This was found by Gonzalez-Sprinberg and Verdier [**GSV1983**] and further refined by Ito and Nakajima [**IN2000**] and Kapranov and Vasserot [**KV2000**].

To explain this construction, recall that for any variety Y with an action of a linear algebraic group G, one can consider the category $\mathrm{Coh}_G(Y)$ of equivariant coherent sheaves on Y. We let $K_G(Y)$ be the Grothendieck group of this category; this group is usually called the equivariant K-group of Y. We list here some properties of this group, referring the reader to [**CG1997**, Chapter 5] for detailed discussion and proofs of these results.

(1) The equivariant K-group $K_G(Y)$ is a module over the ring $R(G) = K(\mathrm{Rep}\, G)$. If Y is a point, then $K_G(Y) = R(G)$.

(2) ([**CG1997**, Theorem 5.4.17]) If V is a vector space with a linear action of G, then we have an isomorphism

$$R(G) \simeq K_G(V)$$
$$X \mapsto X \otimes \mathcal{O}_V.$$

(This is a special case of the Thom isomorphism theorem, relating the K-groups of a manifold X and of the total space of a vector bundle over X.)

12.4. Exceptional fiber and geometric McKay correspondence

(3) ([**CG1997**, Theorem 5.1.28]) If Y is smooth, then any G-equivariant coherent sheaf has a finite resolution by locally free equivariant sheaves; thus, in this case $K_G(Y)$ is the K-group of G-equivariant algebraic vector bundles on Y.

(4) For every G-equivariant algebraic vector bundle $\mathcal{E}$ on Y, one has a well-defined operation of tensoring by $[\mathcal{E}]$ on $K_G(Y)$. In particular, if Y is smooth, then $K_G(Y)$ is a ring.

We will be interested in the case $Y = \mathbb{C}^2$, $G \subset \mathrm{SU}(2)$. By (2), we see that we have an isomorphism $K_G(\mathbb{C}^2) \simeq R(G)$.

Let us now consider the (usual, not equivariant) K-group $K(X)$, where $X = \widehat{\mathbb{C}^2/G}$ is the minimal resolution of $\mathbb{C}^2/G$ constructed in Section 12.2. Our goal is to establish a relation between equivariant sheaves on $\mathbb{C}^2$ and sheaves on $\widehat{\mathbb{C}^2/G}$.

For a point $x \in X$, let $Z_x \subset \mathbb{C}^2$ be the corresponding subscheme in $\mathbb{C}^2$, so that $\mathbb{C}[Z_x] = \mathbb{C}[z_1, z_2]/J_x$. Note that this subscheme is generally not reduced and thus cannot be described without using the scheme theoretic language. Let

$$Z \subset X \times \mathbb{C}^2$$

be the tautological family: the fiber of Z over $x \in X$ is Z_x. We then have the following obvious projections:

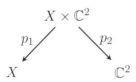

We will denote by q_1, q_2 the restrictions of these projections to Z.

Following [**KV2000**], for an equivariant sheaf $\mathcal{F} \in \mathrm{Coh}_G(\mathbb{C}^2)$, consider the complex of sheaves on X defined by

$$\tilde{\mathcal{F}} = Rq_{1*}Lq_2^*\mathcal{F} = Rp_{1*}(\mathcal{O}_Z \otimes p_2^*\mathcal{F}).$$

This is a G-equivariant complex of sheaves on X; since the action of G on X is trivial, we can consider the subcomplex of G-invariants $\tilde{\mathcal{F}}^G \subset \tilde{\mathcal{F}}$. Let $D_G^b(\mathbb{C}^2)$ be the derived category of G-equivariant coherent sheaves on $\mathbb{C}^2$ and define the functor $\Phi \colon D_G^b(\mathbb{C}^2) \to D^b(X)$ by

(12.8) $$\Phi(\mathcal{F}) = (\tilde{\mathcal{F}})^G.$$

Theorem 12.16 ([**KV2000**]). *The functor Φ defined by (12.8) is an equivalence of triangulated categories*

$$\Phi \colon D^b_G(\mathbb{C}^2) \simeq D^b(\widehat{\mathbb{C}^2/G}).$$

We do not give a proof of this theorem here, referring the reader to the original paper.

Example 12.17. Let $\mathcal{F} = \mathcal{O}_{\mathbb{C}^2}$. Then $\tilde{\mathcal{F}} = (p_1)_* \mathcal{O}_Z$. By definition, the fiber of this sheaf at a point $x \in X$ is given by $\tilde{\mathcal{F}}_x = \mathbb{C}[Z_x] = \mathbb{C}[z_1, z_2]/J_x$. This is a locally free sheaf on X, which is usually called the tautological sheaf and will be denoted by $\mathcal{V}$:

$$(12.9) \qquad \mathcal{V} = (p_1)_* \mathcal{O}_Z.$$

By definition of X, the fiber of the tautological sheaf at point $x \in X$ is isomorphic to the regular representation of G. Thus, we can write

$$(12.10) \qquad \mathcal{V} = \bigoplus_{k \in I} \rho_k \otimes \mathcal{V}_k,$$

where each $\mathcal{V}_k$ is a vector bundle on X of rank $\delta_k = \dim \rho_k$. Thus, for any representation V of G, we have

$$\Phi(V \otimes \mathcal{O}_{\mathbb{C}^2}) = \bigoplus (V \otimes \rho_k)^G \otimes \mathcal{V}_k.$$

In particular, taking $V = \rho_k$, we get $\Phi(\rho_k \otimes \mathcal{O}_{\mathbb{C}^2}) = \mathcal{V}_{k^\vee}$, where the involution $\vee \colon I \to I$ is defined by $\rho_{i^\vee} \simeq \rho_i^*$.

Corollary 12.18. *One has the isomorphisms*

$$R(G) \simeq K_G(\mathbb{C}^2) \simeq K(X)$$
$$\rho_k \mapsto [\rho_k \otimes \mathcal{O}_{\mathbb{C}^2}] \mapsto [\mathcal{V}_{k^\vee}].$$

In particular, this shows that classes $[\mathcal{V}_k]$, $k \in I$, form a basis of $K(X)$. This was first shown in [**GSV1983**], using case-by-case analysis; the above proof can be found in [**KV2000**].

Let us now consider the exceptional fiber:

$$\mathcal{L} = \pi^{-1}(0) = \bigcup C_i$$

(see Theorem 12.3). Recall that by Theorem 10.60, the embedding $\mathcal{L} \hookrightarrow \widehat{\mathbb{C}^2/G}$ is a homotopy equivalence; thus, we see that $H_2(\widehat{\mathbb{C}^2/G}) = H_2(\mathcal{L}) = \bigoplus \mathbb{Z} \cdot [C_i]$.

Let $\mathcal{F} \in \mathrm{Coh}(\widehat{\mathbb{C}^2/G})$ be a coherent sheaf on $\widehat{\mathbb{C}^2/G}$ and let $c_1(\mathcal{F}) \in H^2(\widehat{\mathbb{C}^2/G})$ be the first Chern class. Then for any irreducible component C of $\mathcal{L}$ we can define the degree $\deg(\mathcal{F}|_C) = \langle c_1(\mathcal{F}), [C] \rangle$.

Theorem 12.19. *For any $i \in I - \{0\}$, let $B_i = B_i(\delta) \subset \mathcal{L}$ be the corresponding irreducible component of $\mathcal{L}$ as constructed in Theorem 12.14. Then for a nontrivial irreducible representation $\rho = \rho_k$ of G, we have*
$$\deg(\mathcal{V}_k|_{B_i}) = \delta_{ik}.$$
Moreover,

(1) *if $i = k$, then $\mathcal{V}_i|_{B_i} = \mathcal{O}(1) \oplus \mathcal{O}^{\dim \rho_i - 1}$;*
(2) *if $i \neq k$, then $\mathcal{V}_k|_{B_i} = \mathcal{O}^{\dim \rho_k}$.*

The proof of the first part can be found in [**Nak2001b**, Example 6.3]; the second part follows from it, as was shown in [**KV2000**, Lemma 2.1].

Remark 12.20. The geometric McKay correspondence of this section, which uses G-equivariant sheaves on $\mathbb{C}^2$, is related to constructions of Section 8.4, which used G-equivariant sheaves on $\mathbb{P}^1$. This is discussed in [**Bra2009**].

For a generalization of results of this section to other singularities of the form M/G, see [**BKR2001**].

12.5. Instantons on ALE spaces

As before, let $X = \widehat{\mathbb{C}^2/G}$ be the minimal resolution of $\mathbb{C}^2/G$ and let Q be the Euclidean graph corresponding to G. In this section, we give an overview of results of [**KN1990**], giving a description of the moduli space of instantons (anti-self-dual finite energy connections) on X; this generalizes the results of Section 11.5. For the most part, we will be omitting the proofs, referring the interested readers to the original paper.

We begin by giving the construction of a family of hyperkähler structures on $\widehat{\mathbb{C}^2/G}$, due to Kronheimer [**Kro1989**]. Recall that for any $\boldsymbol{\zeta} \in \mathbb{R}^3 \otimes \mathbb{R}^I$, we can define the quiver variety $\mathcal{M}_{\boldsymbol{\zeta}}(\mathbf{v}, \mathbf{w})$ as the hyperkähler quotient $\boldsymbol{\mu}_{\mathbf{v},\mathbf{w}}^{-1}(i\boldsymbol{\zeta})/\mathrm{U}(\mathbf{v})$ (see Section 10.8). In particular, this variety can be defined in the unframed case $\mathbf{w} = 0$. Note that in this case, $\mathrm{U}(1) \subset \mathrm{U}(\mathbf{v})$ acts trivially on $R(Q^\sharp, \mathbf{v})$, which implies that for any $z \in R(Q^\sharp, \mathbf{v})$,
$$\boldsymbol{\mu}_{\mathbf{v}}(z) \in i\mathbb{R}^3 \otimes D_{\mathbf{v}}, \qquad D_{\mathbf{v}} = \{\theta \in \mathbb{R}^I \mid \theta \cdot \mathbf{v} = 0\}.$$
In particular, $\mathcal{M}_{\boldsymbol{\zeta}}(\mathbf{v}, 0) = \varnothing$ unless $\boldsymbol{\zeta} \in \mathbb{R}^3 \otimes D_{\mathbf{v}}$.

Let
$$(12.11) \quad (\mathbb{R}^3 \otimes D_{\mathbf{v}})^\circ = \left\{ \boldsymbol{\zeta} \in \mathbb{R}^3 \otimes D_{\mathbf{v}} \,\middle|\, \begin{array}{l} \text{for any } \alpha \in \mathbb{Z}_+^I, \alpha \leq \mathbf{v}, \alpha \neq \mathbf{v} \\ \text{we have } \boldsymbol{\zeta} \cdot \alpha \neq (0,0,0) \end{array} \right\}.$$

Lemma 12.21. *Let $\mathbf{v} \in \mathbb{Z}_+^I$. Assume that $\boldsymbol{\zeta} = (\zeta^1, \zeta^2, \zeta^3) \in (\mathbb{R}^3 \otimes D_{\mathbf{v}})^\circ$. Then the action of the group $\mathrm{U}'(\mathbf{v}) = \mathrm{U}(\mathbf{v})/\mathrm{U}(1)$ on $\boldsymbol{\mu}_{\mathbf{v}}^{-1}(i\boldsymbol{\zeta})$ is free, and $\mathcal{M}_{\boldsymbol{\zeta}}(\mathbf{v}, 0)$ is a smooth hyperkähler manifold of real dimension $4 - 4\langle \mathbf{v}, \mathbf{v} \rangle$.*

The proof is parallel to the proof of Theorem 10.48 and is left to the reader.

Let us now consider the special case when $\mathbf{v} = \delta$ is the same as in Theorem 12.9, i.e. $\mathbf{v}_i = \delta_i = \dim \rho_i$. For any $\zeta^\circ \in (\mathbb{R}^3 \otimes D_\delta)^\circ$, define

(12.12) $$X_{\zeta^\circ} = \mathcal{M}_{\zeta^\circ}(\delta, 0).$$

Theorem 12.22 ([**Kro1989**]). *For any $\zeta^\circ \in (\mathbb{R}^3 \otimes D_\delta)^\circ$, the variety X_{ζ° is diffeomorphic to the resolution of singularities $\widehat{\mathbb{C}^2/G}$. Thus, we have a canonical morphism*
$$\pi \colon X_{\zeta^\circ} \to \mathbb{C}^2/G.$$

We do not give a proof of this theorem here, referring the readers to the original paper. Instead, we explain the relation of this result with the construction of $\widehat{\mathbb{C}^2/G}$ given in Theorem 12.9.

Lemma 12.23. *Let $\mathbf{v}, \mathbf{w}$ be as in Theorem 12.9 and let $\theta^\circ \in D_\mathbf{v}$. Choose any $k \in \mathbb{R}$ and consider $\theta = \theta^\circ + k\alpha_0$. Then we have a natural map*
$$\mu_{\mathbf{v},\mathbf{w}}^{-1}(i\theta, 0, 0) \to \mu_{\mathbf{v},0}^{-1}(i\theta^\circ, 0, 0)$$
$$(z, i, j) \mapsto z.$$

If $k \neq 0$, then this map is a $\mathrm{U}(1)$ bundle, where we consider $\mathrm{U}(1)$ as the subgroup of scalar matrices in $\mathrm{U}(\mathbf{v})$.

The proof of this lemma is given by an easy explicit computation, which we leave to the reader (see also [**Kuz2007**, Theorem 19]).

Note that this lemma immediately implies that if $\zeta^\circ = (\theta^\circ, 0, 0) \in (\mathbb{R}^3 \otimes D_\delta)^\circ$, then we have a diffeomorphism

(12.13) $$\mathcal{M}_\theta(\mathbf{v}, \mathbf{w}) \simeq \mathcal{M}_{\zeta^\circ}(\mathbf{v}, 0), \qquad \theta = \theta^\circ + k\alpha_0, \qquad k \neq 0.$$

In particular, choosing θ°, k so that $\theta = \theta^\circ + k\alpha_0 \in \mathbb{Z}^I$, $\theta < 0$, we see that in this case, by Theorem 12.9, we have
$$X_{\zeta^0} \simeq \mathcal{M}_\theta(\mathbf{v}, \mathbf{w}) \simeq \widehat{\mathbb{C}^2/G}, \qquad \zeta^\circ = (\theta^\circ, 0, 0).$$

Let us now study the structure of X_{ζ° at infinity. To do so, we choose $R > 0$, and define
$$X_{\zeta^\circ}(R) = \{x \in X_{\zeta^\circ} \mid |\pi(x)| > R\},$$
where $|\cdot|$ is the usual metric on $\mathbb{C}^2$ and $\pi \colon X_{\zeta^\circ} \to \mathbb{C}^2/G$ is the canonical morphism defined in Theorem 12.22. Following [**KN1990**], we will refer to the subsets of the form $X_{\zeta^\circ}(R)$ as "ends" of X_{ζ°. Obviously, π gives an isomorphism

(12.14) $$B(R)/G \simeq X_{\zeta^\circ}(R),$$

12.5. Instantons on ALE spaces

where $B(R) = \{x \in \mathbb{C}^2 \mid |x| > R\}$ is a neighborhood of infinity in $\mathbb{C}^2$. In particular, this implies $\pi_1(X_{\zeta^\circ}(R)) \simeq G$.

The morphism $\pi\colon X_{\zeta^\circ} \to \mathbb{C}^2/G$ does not preserve the metric. However, we have the following result, due to Kronheimer ([**Kro1989**, Proposition 3.14]).

Theorem 12.24. *As $r = |\pi(x)| \to \infty$, the metric $g(x)$ on X_{ζ° approaches the flat Euclidean metric:*
$$|g(x) - g_0(x)| = O(r^{-4}),$$
where g_0 is the pullback of the standard metric on $\mathbb{C}^2$.

For this reason, these spaces are called asymptotically locally Euclidean (ALE).

Let us now consider connections on X_{ζ°. As in Section 11.5, let E be a Hermitian vector bundle on X_{ζ° and let A be a metric connection on E. We denote by F_A the curvature of this connection. We will call the pair (E, A) an instanton if A is anti-self-dual ($*F_A = -F_A$) and has finite energy:

$$(12.15) \qquad \|F_A\|^2 = \int_{X_{\zeta^\circ}} -\operatorname{tr}(F_A \wedge *F_A) < \infty.$$

Using (12.14), we can lift any connection on X_{ζ° to a G-equivariant connection on $B(R)$. If A has finite energy, then so is the lifting, and by Uhlenbeck's removable singularity theorem, this connection can be extended to a connection on $B(R) \cup \infty$; we will denote by E_∞ the fiber of this extension at ∞. G-equivariance gives an action of G on E_∞. This implies that A is asymptotically flat: there exists a flat connection A_0 on $X_{\zeta^\circ}(R)$ such that
$$\nabla_{A_0}^{(l)}(A - A_0) = O(r^{-3-l}).$$
In particular, this implies that $F_A = O(r^{-4})$ (compare with (11.13)). The action of G on E_∞ can then be described as the monodromy of the connection A_0.

For an instanton (E, A), we can define a number of topological invariants as follows. First, decomposing E_∞ into irreducible representations of G
$$E_\infty \simeq \bigoplus m_i \rho_i$$
gives multiplicities $m_i \in \mathbb{Z}_+$, $i \in I$ (recall that I is the set of irreducible representations of G). In particular, the dimension of E can be expressed in terms of m_i: $\dim E = \sum m_i \dim \rho_i$.

We can also consider the first Chern class $c_1(E) \in H^2(X_{\zeta^\circ})$, which gives numerical invariants
$$u_i = \langle c_1(E), [C_i] \rangle, \qquad i \in \operatorname{Irr}(\mathcal{L}),$$

indexed by irreducible components of the exceptional fiber $\mathcal{L} = \pi^{-1}(0) \subset X_{\zeta^\circ}$.

Finally, there is an analog of the second Chern class $c_2(E)$ or the Chern character $ch_2(E) = \frac{1}{2}c_1^2(E) - c_2(E)$. Recall that if X is a compact, oriented 4-dimensional manifold, then for a vector bundle E with a connection, the second Chern character $ch_2(E) = \frac{1}{2}c_1^2(E) - c_2(E) \in H^2(X)$ can be computed by

$$\langle ch_2(E), [X] \rangle = -\frac{1}{8\pi^2} \int_X F_A \wedge F_A.$$

In our case, X_{ζ° is not compact, but the finite energy condition shows that this integral converges, so we define the topological charge of E by

$$k(E) = \frac{1}{8\pi^2} \int_{X_{\zeta^\circ}} F_A \wedge F_A$$

(this differs by sign from the conventions of [**KN1990**]). Note that because the one-point compactification of X_{ζ° is not a manifold but an orbifold, the invariant $k(E)$ is not necessarily an integer: it can be shown that $k(E) \in \frac{1}{|G|}\mathbb{Z}$.

Example 12.25. Let $\mathcal{V}_k$ be the vector bundle defined by (12.10), and let A be the metric connection defined by the holomorphic structure of $\mathcal{V}_k$ (in [**KN1990**], notation $\mathcal{R}_i = \mathcal{V}_i^\vee$ is used). Then it is shown in [**KN1990**, Proposition 2.2] that $E_\infty \simeq \rho_{k^\vee}$, so the invariants m_i are defined by $m_i(\mathcal{V}_k) = \delta_{i,k^\vee}$, and $u_i = \langle c_1(\mathcal{V}_k), [C_i] \rangle = \delta_{i,k}$ (where we used the bijection $I - \{0\} \simeq \mathrm{Irr}(\mathcal{L})$ defined in Theorem 12.14). The topological charges $k(\mathcal{V}_k)$ can be computed from the following formula, which can be found in [**KN1990**, Proof of Lemma A.4]:

$$\sum_{j \in I} c_{ij} \int_{X_{\zeta^\circ}} ch_2(\mathcal{V}_j) = \delta_{i0} - \frac{\dim \rho_i}{|G|},$$

where $c_{ij} = (\alpha_i, \alpha_j)$, $i, j \in I$, is the Cartan matrix of Q (see (12.4)).

Note that the Cartan matrix has a one-dimensional kernel, so this equation alone is not sufficient to determine the numbers $k(\mathcal{V}_i) = -\int_{X_{\zeta^\circ}} ch_2(\mathcal{V}_i)$. To determine the solution uniquely, we also need to use $k(\mathcal{V}_0) = 0$, which easily follows from the fact that $\mathcal{V}_0$ is the trivial line bundle.

We can now try to describe the moduli space of instantons on X_{ζ°. As before, it is convenient to consider framed instantons, i.e. an instanton (E, A) together with a unitary isomorphism of vector spaces

(12.16) $$\Phi \colon E_\infty \to \mathbb{C}^{\dim E}.$$

We denote by $\mathcal{M}^{fr}_{ASD}(X_{\zeta^\circ})$ the set of isomorphism classes of framed instantons on X_{ζ°.

12.5. Instantons on ALE spaces

Theorem 12.26. *Let* $\mathbf{v}, \mathbf{w} \in \mathbb{Z}_+^I$ *and let* $\zeta^\circ \in (\mathbb{R}^3 \times D_\delta)^\circ$. *Then:*

(1) *There exists a map*
$$\mathcal{M}^{reg}_{-\zeta^\circ}(\mathbf{v}, \mathbf{w}) \to \mathcal{M}^{fr}_{ASD}(X_{\zeta^\circ})$$
such that for a point $m \in \mathcal{M}^{reg}_{\zeta^\circ}(\mathbf{v}, \mathbf{w})$, *the topological invariants of the corresponding instanton are given by*

(12.17)
$$m_i(E) = \mathbf{w}_i, \qquad i \in I,$$
$$u_i(E) = \mathbf{w}_i - \sum_{j \in I} c_{ij} \mathbf{v}_j, \qquad i \in I - \{0\},$$
$$k(E) = \sum_{i \in I} u_i k(\mathcal{V}_i) - \frac{\sum \mathbf{v}_i \dim \rho_i}{|G|},$$

where $c_{ij} = (\alpha_i, \alpha_j)$ *is the Cartan matrix of* Q, *and in the formula for* k, *we also use* $u_0 = \mathbf{w}_0 - \sum_{j \in I} c_{0j} \mathbf{v}_j$.

(2) *The map of part (1) is a bijection between* $\mathcal{M}^{reg}_{-\zeta^\circ}(\mathbf{v}, \mathbf{w})$ *and the set of isomorphism classes of framed instantons on* X_{ζ° *with topological invariants* m_i, u_i, k *given by* (12.17).

The proof of this theorem is given in [**KN1990**, Theorem 9.1].

Remark 12.27. The meaning of the formula for $u_i(E)$ becomes clearer if we rewrite it in the form
$$u_i(E) = \langle \Lambda_\mathbf{w} - \alpha_\mathbf{v}, \alpha_i^\vee \rangle,$$
where $\alpha_\mathbf{v} = \sum \mathbf{v}_i \alpha_i$ and $\Lambda_\mathbf{w} = \sum \mathbf{w}_i \Lambda_i$; here Λ_i are the fundamental weights. We will return to this in Section 13.4.

Example 12.28. Let $\mathbf{v} = 0$, $\mathbf{w} = \alpha_i$. Then the $\mathcal{M}^{reg}_{\zeta^\circ}(\mathbf{v}, \mathbf{w})$ is a point; the corresponding instanton is $\mathcal{V}_{i^\vee}$.

Remark 12.29. As in the case of Jordan quivers, it is also possible to give a description of the moduli space of framed torsion free sheaves on X_{ζ° in terms of quiver varieties $\mathcal{M}_\zeta(\mathbf{v}, \mathbf{w})$. However, in this case the situation is more complicated; the parameter ζ is not equal to ζ° (since ζ° lies on a wall D_δ) but must be taken from an adjacent chamber, and the choice of ζ depends on $\mathbf{v}, \mathbf{w}$. Precise statements can be found in [**Nak2007**].

Chapter 13

Geometric Realization of Kac–Moody Lie Algebras

In this chapter, we present the geometric construction of Kac–Moody Lie algebras using an appropriate homology theory of quiver varieties. This result is due to Nakajima. Our exposition follows [**Nak1998**], [**Gin2012**].

Throughout this chapter, Q is a connected graph without edge loops, and the ground field is the field $\mathbb{C}$ of complex numbers.

13.1. Borel–Moore homology

We begin by describing the appropriate homology theory, namely the Borel–Moore homology. We give the statements of main results and constructions, but omit the proofs. We refer the reader to [**CG1997**, Section 2.6] and references therein for details.

Throughout this section, the word "space" will mean a topological space which has the following properties:

- X is locally compact.
- X has the homotopy type of a finite CW complex.
- X admits a closed embedding into a C^∞ manifold M (countable at infinity), and there exists an open neighborhood $U \supset X$ such that X is a homotopy retract of U.

It is known that any real or complex algebraic variety (with analytic topology) has these properties.

For such a space X, we denote by $C_*^{\mathrm{BM}}(X)$ the space of infinite singular chains $c = \sum_{i=1}^{\infty} a_i \sigma_i$, where $a_i \in \mathbb{C}$ and σ_i are singular simplexes, such that c is locally finite: for every compact $K \subset X$, only finitely many simplexes σ_i intersect K. We define the boundary operator $\partial \colon C_*^{\mathrm{BM}}(X) \to C_{*-1}^{\mathrm{BM}}(X)$ in the usual way.

Definition 13.1. For a space X, its Borel–Moore homology (with complex coefficients) is defined by
$$H_i^{\mathrm{BM}}(X) = H_i(C_*^{\mathrm{BM}}(X), \partial).$$

We will also denote by $H_i(X)$ the ordinary singular homology of X with complex coefficients.

For a complex algebraic variety M, we will use the notation
$$(13.1) \quad H_{top}^{\mathrm{BM}}(M) = H_{2n}^{\mathrm{BM}}(M), \quad H_{mid}^{\mathrm{BM}}(M) = H_n^{\mathrm{BM}}(M), \qquad n = \dim_{\mathbb{C}} M,$$
and similarly for usual (co)homology.

The following theorem lists the basic properties of Borel–Moore homology. The proofs can be found in [**CG1997**, Section 2.6] and references therein.

Theorem 13.2.

(1) We have a natural map $H_*(X) \to H_*^{\mathrm{BM}}(X)$. If X is compact, this map is an isomorphism.

(2) If X is a smooth, oriented (not necessarily compact) manifold of real dimension n, then we have an intersection pairing
$$\cap \colon H_i^{\mathrm{BM}}(X) \times H_{n-i}(X) \to \mathbb{C}$$
which is nondegenerate and thus defines an isomorphism
$$H_i^{\mathrm{BM}}(X) \simeq H^{n-i}(X).$$

(3) If X is a smooth, oriented (not necessarily compact) manifold of real dimension n, and $Z', Z'' \subset X$ are closed subspaces, then we have an intersection pairing
$$\cap \colon H_i^{\mathrm{BM}}(Z') \times H_j^{\mathrm{BM}}(Z'') \to H_{i+j-n}^{\mathrm{BM}}(Z' \cap Z'').$$

(4) Fundamental class: if X is a smooth, oriented (not necessarily compact) manifold of real dimension n, then we have a well-defined fundamental class $[X] \in H_n^{\mathrm{BM}}(X)$. If X is connected, then $H_n^{\mathrm{BM}}(X) = \mathbb{C} \cdot [X]$.

(5) Kunneth formula: for any spaces M_1, M_2, we have a natural isomorphism
$$\boxtimes \colon H_*^{\mathrm{BM}}(M_1) \otimes H_*^{\mathrm{BM}}(M_2) \to H_*^{\mathrm{BM}}(M_1 \times M_2).$$

Borel–Moore homology theory also has pullback and pushforward operations.

Theorem 13.3.

(1) If $f\colon X \to Y$ is a proper map, then one has a pushforward map
$$f_*\colon H_i^{\mathrm{BM}}(X) \to H_i^{\mathrm{BM}}(Y).$$

(2) If $U \subset X$ is an open subset, then we have a natural restriction map $H_i^{\mathrm{BM}}(X) \to H_i^{\mathrm{BM}}(U)$.

(3) If $p\colon \tilde{X} \to X$ is a locally trivial fibration with a smooth oriented fiber F and $i\colon X \to \tilde{X}$ is a section of $\tilde{X}$, then one can define natural pullback maps
$$p^*\colon H_*^{\mathrm{BM}}(X) \to H_{*+d}^{\mathrm{BM}}(\tilde{X}), \qquad d = \dim_{\mathbb{R}} F,$$
$$i^*\colon H_*^{\mathrm{BM}}(\tilde{X}) \to H_{*-d}^{\mathrm{BM}}(X)$$

such that $i^* p^* = \mathrm{id}$.

Using the restriction map, we can generalize the notion of fundamental class to nonsmooth algebraic varieties.

Theorem 13.4. Let X be a (possibly singular) complex algebraic variety, $\dim_{\mathbb{C}}(X) = n$.

(1) Let $X^{ns} \subset X$ be the set of nonsingular points in X. Then the restriction map
$$H_{top}^{\mathrm{BM}}(X) \to H_{top}^{\mathrm{BM}}(X^{ns})$$
is an isomorphism. In particular, we have a well-defined fundamental class
$$[X] \in H_{top}^{\mathrm{BM}}(X)$$
such that its restriction to X^{ns} coincides with $[X^{ns}]$.

(2) If $X_1, \ldots, X_k$ are the n-dimensional irreducible components of X, then $[X] = \sum [X_i]$ and
$$H_{top}^{\mathrm{BM}}(X) = \bigoplus \mathbb{C} \cdot [X_i].$$

13.2. Convolution algebras

Let M_1, M_2, M_3 be oriented C^∞ manifolds and let
$$Z_{12} \subset M_1 \times M_2, \qquad Z_{23} \subset M_2 \times M_3$$

be closed subspaces. Define the set-theoretic "composition" $Z_{12} \circ Z_{23} \subset M_1 \times M_3$ by
(13.2)
$$Z_{12} \circ Z_{23} = \{(m_1, m_3) \mid \exists m_2 \in M_2, (m_1, m_2) \in Z_{12}, (m_2, m_3) \in Z_{23}\}$$
$$= p_{13}(p_{12}^{-1}(Z_{12}) \cap p_{23}^{-1}(Z_{23})),$$

where p_{ij} are natural projections
$$p_{ij} \colon M_1 \times M_2 \times M_3 \to M_i \times M_j.$$

Example 13.5. If Z_{12}, Z_{23} are graphs of functions $f \colon M_1 \to M_2$, $g \colon M_2 \to M_3$, then $Z_{12} \circ Z_{23}$ is the graph of composition $g \circ f$.

Definition 13.6. Let M_1, M_2, M_3 be oriented C^∞ manifolds. Closed subspaces
$$Z_{12} \subset M_1 \times M_2, \qquad Z_{23} \subset M_2 \times M_3$$
are called *composable* if the map
(13.3)
$$p_{13} \colon p_{12}^{-1}(Z_{12}) \cap p_{23}^{-1}(Z_{23}) \to M_1 \times M_3$$
is proper.

Note that it is easy to show that if projection $Z_{12} \to M_2$ is proper, then Z_{12} is composable with any closed Z_{23}.

The following lemma immediately follows from the definition.

Lemma 13.7. *If* $Z_{12} \subset M_1 \times M_2$, $Z_{23} \subset M_2 \times M_3$ *are composable, then* $Z_{12} \circ Z_{23}$ *is closed.*

We can now define the convolution in Borel–Moore homology, first introduced by Ginzburg in [**Gin1991**].

Definition 13.8. In the assumptions of Lemma 13.7, let $d = \dim_{\mathbb{R}} M_2$. Then we define the convolution morphism
(13.4)
$$* \colon H_i^{\mathrm{BM}}(Z_{12}) \otimes H_j^{\mathrm{BM}}(Z_{23}) \to H_{i+j-d}^{\mathrm{BM}}(Z_{12} \circ Z_{23})$$
by
$$c_{12} * c_{23} = (p_{13})_* \big((c_{12} \boxtimes [M_3]) \cap ([M_1] \boxtimes c_{23})\big),$$
where $[M_i]$ is the fundamental class and $\cap$ is the intersection pairing from Theorem 13.2(3).

Note that this definition uses the pushforward map $(p_{13})_*$ for Borel–Moore homology, which is well defined since by assumption the restriction of p_{13} to $p_{12}^{-1}(Z_{12}) \cap p_{23}^{-1}(Z_{23})$ is proper.

The following result is proved in [**CG1997**, Section 2.7.18].

13.2. Convolution algebras

Theorem 13.9. *In the assumptions of Definition 13.8, assume additionally that we are given a smooth oriented manifold M_4 and a closed subset $Z_{34} \subset M_3 \times M_4$ such that $(Z_{12} \circ Z_{23})$ and Z_{34} are composable. Then convolution in Borel–Moore homology is associative:*
$$(c_{12} * c_{23}) * c_{34} = c_{12} * (c_{23} * c_{34}),$$
where $c_{ij} \in H_*^{\mathrm{BM}}(Z_{ij})$.

Example 13.10. Let M be compact. Then convolution defines a structure of an associative algebra on the space $H = H_*^{\mathrm{BM}}(M \times M)$. The unit for this algebra is the fundamental class $[\Delta]$ of the diagonal $\Delta = \{(m,m) \mid m \in M\} \subset M \times M$.

We will need a generalization of this example. Namely, assume that we have a collection M_α of smooth oriented (but not necessarily compact) C^∞ manifolds, indexed by $\alpha \in A$. We assume that for each α, all connected components of M_α have the same dimension $d_\alpha = \dim_\mathbb{R} M_\alpha$.

Assume additionally that for any $\alpha, \beta \in A$ we have a space (not necessarily smooth) $Y_{\alpha,\beta}$ and proper maps $M_\alpha \to Y_{\alpha,\beta}$, $M_\beta \to Y_{\alpha,\beta}$. Let

(13.5) $$Z_{\alpha\beta} = M_\alpha \times_{Y_{\alpha,\beta}} M_\beta \subset M_\alpha \times M_\beta.$$

Theorem 13.11.

(1) *Under the assumptions above, convolution defines an associative algebra structure on*
$$H_* = \bigoplus_{\alpha,\beta} H_*^{\mathrm{BM}}(Z_{\alpha\beta}).$$

If the set of indices α is finite, then this algebra is unital, with the unit given by
$$\bigoplus [\Delta_\alpha],$$
where $\Delta_\alpha \subset M_\alpha \times M_\alpha$ is the diagonal. This algebra is graded with respect to the grading given by

$\deg c = \frac{d_\alpha + d_\beta}{2} - i, \qquad c \in H_i^{\mathrm{BM}}(Z_{\alpha\beta}), \qquad d_\alpha = \dim_\mathbb{R} M_\alpha$

(note that this grading in general is half-integer).

(2) *Choose some index $\beta \in A$ and a collection of elements $y = (y_\alpha)_{\alpha \in A}$, $y_\alpha \in Y_{\alpha,\beta}$. Let*
$$N_\alpha(y) = p_\alpha^{-1}(y_\alpha) \subset M_\alpha,$$
where p_α is the map $M_\alpha \to Y_{\alpha\beta}$.
Then the space $M = \bigoplus_\alpha H_*^{\mathrm{BM}}(N_\alpha(y))$ is a module over the algebra H_*. This module is graded, with the grading given by

$\deg c = \frac{d_\alpha}{2} - i, \qquad c \in H_i^{\mathrm{BM}}(N_\alpha(y)).$

Proof. The first part immediately follows from Theorem 13.9 if we notice that, in this case, $Z_{\alpha\beta} \circ Z_{\beta\gamma} \subset Z_{\alpha\gamma}$.

The second part also follows if we notice that one can write
$$N_\alpha(y) = M_\alpha \times_{Y_{\alpha\beta}} \{pt\}$$
with the map $pt \to Y_{\alpha\beta}$ given by $* \mapsto y_\alpha$. $\square$

13.3. Steinberg varieties

Throughout this section, let $\vec{Q}$ be a quiver without edge loops. We fix a choice of a stability parameter $\theta \in \mathbb{Z}^I$, $\theta > 0$, and denote by $\mathcal{M}(\mathbf{v}, \mathbf{w})$ the corresponding quiver variety; by Example 10.36, $\mathcal{M}(\mathbf{v}, \mathbf{w})$ is smooth, and we have a canonical projective morphism $\pi\colon \mathcal{M}(\mathbf{v}, \mathbf{w}) \to \mathcal{M}_0(\mathbf{v}, \mathbf{w})$.

Recall also that for any $\mathbf{v}', \mathbf{v}'' \in \mathbb{Z}^I_+$, we have a canonical $\mathrm{GL}(\mathbf{v}')$-equivariant morphism

$$(13.6) \qquad \begin{aligned} \mathcal{M}_0(\mathbf{v}', \mathbf{w}) &\to \mathcal{M}_0(\mathbf{v}' + \mathbf{v}'', \mathbf{w}) \\ (x, y, i, j) &\mapsto (x \oplus 0, y \oplus 0, i \oplus 0, j \oplus 0) \end{aligned}$$

(compare with (10.41)). It is easy to see that this morphism is a closed embedding. Combining it with the canonical morphism $\pi\colon \mathcal{M}(\mathbf{v}', \mathbf{w}) \to \mathcal{M}_0(\mathbf{v}', \mathbf{w})$, we get a proper morphism

$$(13.7) \qquad \mathcal{M}(\mathbf{v}', \mathbf{w}) \to \mathcal{M}_0(\mathbf{v}' + \mathbf{v}'', \mathbf{w}).$$

Let us now define, for any $\mathbf{v}', \mathbf{v}'' \in \mathbb{Z}^I_+$, a subvariety

$$(13.8) \qquad Z(\mathbf{v}', \mathbf{v}'', \mathbf{w}) \subset \mathcal{M}(\mathbf{v}', \mathbf{w}) \times \mathcal{M}(\mathbf{v}'', \mathbf{w})$$

by
$$Z(\mathbf{v}', \mathbf{v}'', \mathbf{w}) = \mathcal{M}(\mathbf{v}', \mathbf{w}) \times_{\mathcal{M}_0(\mathbf{v}'+\mathbf{v}'',\mathbf{w})} \mathcal{M}(\mathbf{v}'', \mathbf{w}),$$

where the fiber product is defined using morphisms (13.7).

Following [**Gin2012**], we will call $Z(\mathbf{v}', \mathbf{v}'', \mathbf{w})$ a *Steinberg variety*. For example, if $\mathbf{v}' = \mathbf{v}'' = \mathbf{v}$, then one of the irreducible components of $Z(\mathbf{v}, \mathbf{v}, \mathbf{w})$ is the diagonal $\Delta_{\mathbf{v},\mathbf{w}} \subset \mathcal{M}(\mathbf{v}, \mathbf{w}) \times \mathcal{M}(\mathbf{v}, \mathbf{w})$.

Steinberg varieties are usually singular and reducible. The following result is proved in [**Nak1998**, Theorem 7.2].

Theorem 13.12.

(1) *For any irreducible component X of $Z(\mathbf{v}', \mathbf{v}'', \mathbf{w})$, we have*

$$\dim X \le \tfrac{1}{2}\Big(\dim \mathcal{M}(\mathbf{v}', \mathbf{w}) + \dim \mathcal{M}(\mathbf{v}'', \mathbf{w})\Big).$$

13.3. Steinberg varieties

(2) Define the set of regular elements in $Z(\mathbf{v}', \mathbf{v}'', \mathbf{w})$ by

$$Z^{reg} = Int\Big(\{x \in Z(\mathbf{v}', \mathbf{v}'', \mathbf{w}) \mid \pi(x) \in \mathcal{M}_0^{reg}(\mathbf{v}, \mathbf{w}) \text{ for some } \mathbf{v} \leq \mathbf{v}' + \mathbf{v}''\}\Big),$$

where Int stands for interior in Zariski topology on Z: $Int(X) = Z - \overline{(Z - X)}$, π is the canonical morphism $\pi \colon Z(\mathbf{v}', \mathbf{v}'', \mathbf{w}) \to \mathcal{M}_0(\mathbf{v}' + \mathbf{v}'', \mathbf{w})$, and we consider $\mathcal{M}_0(\mathbf{v}, \mathbf{w})$ as a subset in $\mathcal{M}_0(\mathbf{v}' + \mathbf{v}'', \mathbf{w})$ via embedding (13.6).

Then Z^{reg} is a locally closed Lagrangian subvariety in $\mathcal{M}(\mathbf{v}', \mathbf{w}) \times \mathcal{M}(\mathbf{v}'', \mathbf{w})$, where the symplectic structure on the product is defined by $\omega = \omega_1 \oplus (-\omega_2)$. In particular, each irreducible component of Z^{reg} has dimension

(13.9) $$\dim Z^{reg} = \tfrac{1}{2}\Big(\dim \mathcal{M}(\mathbf{v}', \mathbf{w}) + \dim \mathcal{M}(\mathbf{v}'', \mathbf{w})\Big).$$

Note that it follows from Theorem 10.42 that if Q is Dynkin, then $Z = Z^{reg}$. If Q is not Dynkin, it is no longer true; even for Euclidean quivers, it is possible that $Z^{reg} = \varnothing$ (compare with Remark 11.8). However, it can be shown that for Euclidean quivers, the dimension of each irreducible component of Z is given by (13.9) (see [**Nak2009**, Remark 2.24]).

Example 13.13. Let us consider the quiver of type A_1, consisting of a single vertex and no edges. By results of Example 10.45, in this case we have $\mathcal{M}(n, r) = T^*G(n, r) = \{(V, y)\}$, where $V \subset \mathbb{C}^r$ is a subspace of dimension n and $y \colon \mathbb{C}^r \to \mathbb{C}^r$ satisifes $\operatorname{Im}(y) \subset V \subset \operatorname{Ker}(y)$. Thus, in this case we have

$$Z(n_1, n_2, r) = \{(V_1, V_2, y)\},$$
$$V_1, V_2 \subset \mathbb{C}^r, \quad \dim V_i = n_i,$$
$$y \colon \mathbb{C}^r \to \mathbb{C}^r,$$
$$\operatorname{Im}(y) \subset V_i \subset \operatorname{Ker}(y).$$

Assuming that $n_2 \geq n_1$ and $r \geq 2n_2$, we see that in this case the projection $\pi_1 \colon Z(n_1, n_2, r) \to \mathcal{M}(n_1, r)$ is surjective and the fiber over (V_1, y) is isomorphic to $G(n_2 - k, r - 2k)$, where $k = \operatorname{rank}(y)$. In particular, the fiber over a generic point $x = (V_1, y) \in \mathcal{M}^{reg}(n_1, r)$ is isomorphic to $G(n_2 - n_1, r - 2n_1)$. In this case, it is easy to check that

$$\dim Z(n_1, n_2, r) = n_1(r - n_1) + n_2(r - n_2) = \tfrac{1}{2}(\dim \mathcal{M}(n_1, r) + \dim \mathcal{M}(n_2, r)).$$

Let us now consider a special case of Steinberg varieties. Namely, let $\mathbf{v} \in \mathbb{Z}_+^I$. Choose a vertex $k \in I$ and let $\alpha_k \in \mathbb{Z}_+^I$ be the corresponding simple root; denote $\mathbf{v}'' = \mathbf{v} + \alpha_k$. Define the variety

(13.10) $$B_k(\mathbf{v}, \mathbf{w}) = \{m', m'', \varphi\},$$

where $m' \in \mathcal{M}(\mathbf{v}, \mathbf{w})$, $m'' \in \mathcal{M}(\mathbf{v} + \alpha_k, \mathbf{w})$, and $\varphi \colon V^{m'} \to V^{m''}$ is a morphism of corresponding framed representations. It easily follows from stability of m', m'' that if such a φ exists, it is unique and is injective; therefore, $B_k(\mathbf{v}, \mathbf{w})$ is a subvariety in $\mathcal{M}(\mathbf{v}, \mathbf{w}) \times \mathcal{M}(\mathbf{v} + \alpha_k, \mathbf{w})$.

We can also describe $B_k(\mathbf{v}, \mathbf{w})$ as the set of isomorphism classes of pairs (V', V''), where V'' is a stable $\mathbf{w}$-framed representation of dimension $\dim V'' = \mathbf{v} + \alpha_k$, and $V' \subset V''$ is a subspace with $\dim V' = \mathbf{v}$ which is z-stable and satisfies $\operatorname{Im}(i_k) \subset V'$. In this case, the quotient $V''/V' \simeq S(k)$ is a one-dimensional representation of $Q^\sharp$.

Following Nakajima, we will call varieties $B_k(\mathbf{v}, \mathbf{w})$ *Hecke correspondences*. The following theorem, the proof of which can be found in [**Nak1998**, Section 5.i], summarizes some properties of these varieties.

Theorem 13.14. $B_k(\mathbf{v}, \mathbf{w})$ *is a smooth Lagrangian subvariety in* $\mathcal{M}(\mathbf{v}, \mathbf{w}) \times \mathcal{M}(\mathbf{v} + \alpha_k, \mathbf{w})$. *In particular,*
$$\dim B_k(\mathbf{v}, \mathbf{w}) = \tfrac{1}{2}\Big(\dim \mathcal{M}(\mathbf{v}, \mathbf{w}) + \dim \mathcal{M}(\mathbf{v} + \alpha_k, \mathbf{w})\Big).$$

As a corollary, we immediately see that $B_k(\mathbf{v}, \mathbf{w})$ is an irreducible component of $Z(\mathbf{v}, \mathbf{v} + \alpha_k, \mathbf{w})$.

13.4. Geometric realization of Kac–Moody Lie algebras

Let us now define the corresponding convolution algebras. As before, we choose $\mathbf{w} \in \mathbb{Z}_+^I$ and consider the space
$$H_*^{\mathrm{BM}}(\mathbf{w}) = \bigoplus_{\mathbf{v}', \mathbf{v}'' \in \mathbb{Z}_+^I} H_*^{\mathrm{BM}}(Z(\mathbf{v}', \mathbf{v}'', \mathbf{w})).$$

The following result immediately follows from the results of Section 13.2.

Theorem 13.15.

(1) *The convolution product* (13.4) *defines on* $H_*^{\mathrm{BM}}(\mathbf{w})$ *a structure of an associative algebra.*

(2) *The subspace*
$$H_{top}^{\mathrm{BM}}(\mathbf{w}) = \bigoplus_{\mathbf{v}', \mathbf{v}'' \in \mathbb{Z}_+^I} H_{d(\mathbf{v}', \mathbf{v}'')}^{\mathrm{BM}}(Z(\mathbf{v}', \mathbf{v}'', \mathbf{w})),$$
$$d(\mathbf{v}', \mathbf{v}'') = 2 \dim_{\mathbb{C}} Z(\mathbf{v}', \mathbf{v}'', \mathbf{w}) = \dim_{\mathbb{C}} \mathcal{M}(\mathbf{v}', \mathbf{w}) + \dim_{\mathbb{C}} \mathcal{M}(\mathbf{v}', \mathbf{w}),$$
is a subalgebra in $H_*^{\mathrm{BM}}(\mathbf{w})$ *with respect to the convolution product.*

Note that by Theorem 13.4, $H_{top}^{\mathrm{BM}}(\mathbf{w})$ has a natural basis given by the fundamental classes of irreducible components of $Z(\mathbf{v}', \mathbf{v}'', \mathbf{w})$. In particular,

13.4. Geometric realization of Kac–Moody Lie algebras

we have the following elements in $H^{BM}_{top}(\mathbf{w})$:

(13.11) $\qquad [\Delta_{\mathbf{v}}] \in H^{BM}_{top}(\mathbf{v}, \mathbf{v}, \mathbf{w}),$

(13.12) $\qquad [B_k(\mathbf{v}, \mathbf{w})] \in H^{BM}_{top}(\mathbf{v}, \mathbf{v} + \alpha_k, \mathbf{w}),$

(13.13) $\qquad [B_k^t(\mathbf{v}, \mathbf{w})] \in H^{BM}_{top}(\mathbf{v} + \alpha_k, \mathbf{v}, \mathbf{w}),$

where $\Delta_{\mathbf{v}} \subset \mathcal{M}(\mathbf{v}, \mathbf{w}) \times \mathcal{M}(\mathbf{v}, \mathbf{w})$ is the diagonal, $B_k(\mathbf{v}, \mathbf{w})$ is the Hecke correspondence (13.10), and $t\colon \mathcal{M}(\mathbf{v}', \mathbf{w}) \times \mathcal{M}(\mathbf{v}'', \mathbf{w}) \to \mathcal{M}(\mathbf{v}'', \mathbf{w}) \times \mathcal{M}(\mathbf{v}', \mathbf{w})$ is the permutation of factors.

Now let $\mathfrak{g} = \mathfrak{g}(Q)$ be the Kac–Moody algebra corresponding to Q, as described in Appendix A. We denote by P the weight lattice of $\mathfrak{g}$ and by $\Lambda_i \in P$ the fundamental weights (see Definition A.4). For $\mathbf{w} \in \mathbb{Z}^I$, we define

(13.14) $\qquad \Lambda_{\mathbf{w}} = \sum \mathbf{w}_i \Lambda_i \in P,$

and for $\mathbf{v} \in \mathbb{Z}^I$, we define

(13.15) $\qquad \alpha_{\mathbf{v}} = \sum \mathbf{v}_i \alpha_i \in L \subset P.$

Theorem 13.16. *Let Q be a quiver without edge loops, and let $\tilde{U} = \tilde{U}(\mathfrak{g}(Q))$ be the modified universal enveloping algebra of $\mathfrak{g}(Q)$ as defined in (A.20). Let $\mathbf{w} \in \mathbb{Z}^I_+$. Then there exists a unique algebra homomorphism*

$$\Phi\colon \tilde{U} \to H^{BM}_{top}(\mathbf{w})$$

such that

- *if $\lambda = \Lambda_{\mathbf{w}} - \alpha_{\mathbf{v}}$ for some $\mathbf{v} \in \mathbb{Z}^I_+$, then*

$$\Phi(a_\lambda) = [\Delta_{\mathbf{v}}],$$
$$\Phi(E_k a_\lambda) = [B_k(\mathbf{v} - \alpha_k, \mathbf{w})],$$
$$\Phi(F_k a_\lambda) = (-1)^{\langle \alpha_i^\vee, \lambda \rangle + 1}[B_k^t(\mathbf{v}, \mathbf{w})];$$

- *if λ cannot be written in the form $\lambda = \Lambda_{\mathbf{w}} - \alpha_{\mathbf{v}}$ for some $\mathbf{v} \in \mathbb{Z}^I_+$, then $\Phi(a_\lambda) = 0$.*

A proof of this theorem is given in [**Nak1998**, Section 9]. It is based on an explicit computation of multiplication in the convolution algebra H^{BM}_{top}.

In a similar way one can construct modules over $\tilde{U}$. Namely, let $\mathbf{v}^0 \in \mathbb{Z}^I_+$ and let $x \in \mathcal{M}_0^{reg}(\mathbf{v}^0, \mathbf{w})$. Then for every $\mathbf{v} \geq \mathbf{v}^0$, we can define

(13.16) $\qquad \mathcal{M}(\mathbf{v}, \mathbf{w})_x = \pi^{-1}(x) \subset \mathcal{M}(\mathbf{v}, \mathbf{w}),$

where $\pi\colon \mathcal{M}(\mathbf{v}, \mathbf{w}) \to \mathcal{M}_0(\mathbf{v}, \mathbf{w})$ is the canonical projective morphism, and we consider x as an element in $\mathcal{M}_0(\mathbf{v}, \mathbf{w})$ via embedding (13.6).

Theorem 13.17. *Let $\mathbf{v}^0 \in \mathbb{Z}_+^I$ be such that $\mathcal{M}_0^{reg}(\mathbf{v}^0, \mathbf{w})$ is nonempty. Then:*

(1) *Weight $\lambda \in P$ defined by*

(13.17) $$\lambda = \Lambda_\mathbf{w} - \alpha_{\mathbf{v}^0}$$

is dominant: $\lambda \in P_+$ (see Theorem A.28).

(2) *For any $x \in \mathcal{M}_0^{reg}(\mathbf{v}^0, \mathbf{w})$, the vector space*

$$\bigoplus_\mathbf{v} H_{top}^{BM}(\mathcal{M}(\mathbf{v}, \mathbf{w})_x)$$

is a module over the algebra $\tilde{U}$. As a $\tilde{U}$-module, it is isomorphic to the irreducible integrable highest weight module $L(\lambda)$, where λ is given by (13.17). Under this isomorphism, $H_{top}^{BM}(\mathcal{M}(\mathbf{v}, \mathbf{w})_x)$ is identified with the weight space $L(\lambda)_\mu$, $\mu = \Lambda_\mathbf{w} - \alpha_\mathbf{v}$.

The first part of this theorem is proved in [**Nak1998**, Lemma 4.7].

The fact that $\bigoplus_\mathbf{v} H_{top}^{BM}(\mathcal{M}(\mathbf{v}, \mathbf{w})_x)$ has a structure of a module over $\bigoplus H_{top}^{BM}(Z(\mathbf{v}', \mathbf{v}'', \mathbf{w}))$ (and thus over $\tilde{U}$) follows from Theorem 13.11. The proof that this module is isomorphic to $L(\lambda)$ is given in [**Nak1998**, Theorem 10.2].

In particular, we can take in the above theorem $\mathbf{v}^0 = 0$, in which case

$$\mathcal{M}(\mathbf{v}, \mathbf{w})_x = \pi^{-1}(0) = \mathcal{L}(\mathbf{v}, \mathbf{w})$$

is the exceptional fiber introduced in Section 10.9. This gives the following result.

Corollary 13.18. *As a $\tilde{U}$-module, the space*

$$\bigoplus_\mathbf{v} H_{top}^{BM}(\mathcal{L}(\mathbf{v}, \mathbf{w}))$$

is isomorphic to the irreducible integrable highest weight module with highest weight $\Lambda_\mathbf{w}$. Under this isomorphism, $H_{top}^{BM}(\mathcal{L}(\mathbf{v}, \mathbf{w}))$ is identified with the weight space $L(\Lambda_\mathbf{w})_\mu$, $\mu = \Lambda_\mathbf{w} - \alpha_\mathbf{v}$.

Note that since $\mathcal{L}(\mathbf{v}, \mathbf{w})$ is compact, in this case Borel–Moore homology coincides with the usual homology. By Theorem 10.60, we have

$$H_{top}^{BM}(\mathcal{L}(\mathbf{v}, \mathbf{w})) = H_{top}(\mathcal{L}(\mathbf{v}, \mathbf{w}), \mathbb{C}) = H_{mid}(\mathcal{M}(\mathbf{v}, \mathbf{w}), \mathbb{C}).$$

Thus, Corollary 13.18 implies that

(13.18) $$\dim H_{top}(\mathcal{L}(\mathbf{v}, \mathbf{w}), \mathbb{C}) = \dim H_{mid}(\mathcal{M}(\mathbf{v}, \mathbf{w}), \mathbb{C}) = \dim L(\Lambda_\mathbf{w})_{\Lambda_\mathbf{w} - \alpha_\mathbf{v}}.$$

These dimensions can be computed using the Kac–Peterson character formula (see [**Kac1990**, Chapter 10]).

13.4. Geometric realization of Kac–Moody Lie algebras

This allows us to give a criterion of nonemptiness for $\mathcal{M}(\mathbf{v}, \mathbf{w})$.

Theorem 13.19. *Let* $\mathbf{v}, \mathbf{w} \in \mathbb{Z}_+^I$, *and let* $\theta > 0$. *Then*:

(1) *The following three conditions are equivalent*:
   - $\mathcal{M}(\mathbf{v}, \mathbf{w})$ *is nonempty*;
   - $\mathcal{L}_\theta(\mathbf{v}, \mathbf{w})$ *is nonempty*;
   - $\lambda = \Lambda_\mathbf{w} - \alpha_\mathbf{v}$ *is a weight of* $L(\Lambda_\mathbf{w})$.

(2) *If* $\mathcal{M}_0^{reg}(\mathbf{v}, \mathbf{w})$ *is nonempty, then* $\lambda = \Lambda_\mathbf{w} - \alpha_\mathbf{v}$ *is in* P_+ *and is a weight of* $L(\mathbf{w})$.

(3) *Assume additionally that* $\mathbf{w}$ *satisfies the following condition*:

$$\text{for any } \alpha \in K, \, (\alpha, \Lambda_\mathbf{w}) \geq 2,$$

where K *is the set of positive imaginary roots defined in* (A.15). *Then the converse of the previous part also holds: if* $\lambda = \Lambda_\mathbf{w} - \alpha_\mathbf{v}$ *is in* P_+ *and is a weight of* $L(\Lambda_\mathbf{w})$, *then* $\mathcal{M}_0^{reg}(\mathbf{v}, \mathbf{w})$ *is nonempty*.

Proof. If $\mathcal{M}(\mathbf{v}, \mathbf{w})$ is nonempty, then, by Theorem 10.60, $\mathcal{L}(\mathbf{v}, \mathbf{w})$ is also nonempty and $H_{top}(\mathcal{L}(\mathbf{v}, \mathbf{w}), \mathbb{C}) = \bigoplus \mathbb{C}[X_i] \neq 0$, where the X_i's are irreducible components of $\mathcal{L}(\mathbf{v}, \mathbf{w})$. Thus, by (13.18), $\Lambda_\mathbf{w} - \alpha_\mathbf{v}$ is a weight of $L(\Lambda_\mathbf{w})$. Conversely, if $\Lambda_\mathbf{w} - \alpha_\mathbf{v}$ is a weight of $L(\Lambda_\mathbf{w})$, then (13.18) implies that $\mathcal{M}(\mathbf{v}, \mathbf{w})$ is nonempty. This proves part (1).

For part (2), notice that if $\mathcal{M}_0^{reg}(\mathbf{v}, \mathbf{w})$ is nonempty, then $\mathcal{M}(\mathbf{v}, \mathbf{w})$ is also nonempty (see Theorem 10.40), and thus by part (1), $\lambda = \Lambda_\mathbf{w} - \alpha_\mathbf{v}$ is a weight of $L(\Lambda_\mathbf{w})$; by Theorem 13.17, λ is dominant.

The proof of the last statement is more complicated, and will not be given here; interested readers can find it in [**Nak1998**, Proposition 10.5]. □

This geometric construction also gives a distinguished basis in $L(\Lambda_\mathbf{w})$. Namely, for every $\mathbf{v}$, we have a natural basis in $H_{top}(\mathcal{L}(\mathbf{v}, \mathbf{w}))$, given by classes $[X_i]$, where X_i are irreducible components of $\mathcal{L}$; recall that by Theorem 10.56, they are the closures of Białynicki-Birula pieces. This basis is called the semicanonical basis. It is shown in [**Sai2002**] that this basis has a special combinatorial structure; such bases are called *crystal bases*. Saito has also shown that there is a bijection between the semicanonical basis in $L(\Lambda_\mathbf{w})$ constructed above and a certain subset of the semicanonical basis for the positive (or negative) part of the universal enveloping algebra (see Theorem 5.18). This bijection coincides with the one described in Corollary 10.59. This shows that the semicanonical basis in $U\mathfrak{n}_-$ is compatible with all irreducible highest weight modules: if we write the irreducible highest weight module as $L(\Lambda) = M(\Lambda)/J$, where $M(\Lambda)$ is the Verma module and $J \subset M(\Lambda)$ is the maximal proper submodule, then there is a subset of the semicanonical basis in $M(\Lambda)$ which forms a basis in J, and the remaining basis elements form a basis in $L(\Lambda)$. The same property also holds for the canonical basis; however, as was mentioned earlier, these two bases do not coincide.

We refer the reader to [**Sai2002**], [**Lus2000**] for further details.

Example 13.20. Let $G \subset \mathrm{SU}(2)$ be a nontrivial finite subgroup, and let Q be the Euclidean graph which corresponds to G under McKay correspondence. As in Chapter 7 and Chapter 12, we denote by I the set of vertices of Q (which coincides with the set of irreducible representations of G) and by $0 \in I$ the extending vertex, corresponding to the trivial representation. The corresponding Kac–Moody algebra $\mathfrak{g}(Q)$ can be described as the affinization of the semisimple finite-dimensional Lie algebra $\mathfrak{g}(Q_f)$, where Q_f is obtained from Q by removing the extending vertex.

Now let $\mathbf{w} \in \mathbb{Z}_+^I$ be defined by $\mathbf{w}_i = \delta_{i0}$. Then the corresponding weight $\Lambda_{\mathbf{w}} = \Lambda_0$ is the fundamental weight. The irreducible integrable representation $L(\Lambda_0)$ is called the *basic representation* and its structure is well studied. In particular:

(1) For $\mathbf{v} = \delta$, $\dim L(\Lambda_0)_{\Lambda_0 - \delta} = |I_f| = |I| - 1$.

(2) For $\mathbf{v} = \delta + \alpha$, where $\alpha \in R_f$ is a root (not necessarily positive) of the finite root system $R_f = R(Q_f)$, we have $\dim L(\Lambda_0)_{\Lambda_0 - \delta - \alpha} = 1$.

For this choice of $\mathbf{w}$, the quiver varieties $\mathcal{M}(\mathbf{v}, \mathbf{w})$ can be described as irreducible components of the G-equivariant Hilbert scheme:

$$\mathcal{M}(\mathbf{v}, \mathbf{w}) \simeq X^{\mathbf{v}} \subset (\mathrm{Hilb}^n \mathbb{C}^2)^G, \quad n = \sum \mathbf{v}_i \dim \rho_i$$

(see Theorem 12.13). Note that in Theorem 12.13 we used $\theta < 0$; however, since $\mathcal{M}_\theta(\mathbf{v}, \mathbf{w}) \simeq \mathcal{M}_{-\theta}(\mathbf{v}, \mathbf{w})$ (Lemma 10.29,), the above isomorphism also holds for $\theta > 0$. Thus, we see that

(13.19) $$H_{mid}(X^{\mathbf{v}}, \mathbb{C}) \simeq L(\Lambda_0)_{\Lambda_0 - \alpha_{\mathbf{v}}}.$$

In particular, this implies

(13.20) $$H_{mid}(X^\delta, \mathbb{C}) = H_2(\widehat{\mathbb{C}^2/G}, \mathbb{C}) \simeq L(\Lambda_0)_{\Lambda_0 - \delta}$$

and $H_0(X^{\mathbf{v}}, \mathbb{C}) \simeq \mathbb{C}$, for $\mathbf{v} = \delta + \alpha$, $\alpha \in R_f$, so in this case $X^{\mathbf{v}}$ is a point.

Finally, we mention that the geometric construction of Kac–Moody Lie algebras and their representations, given above, is closely related to several other constructions. We list here some of them and provide references for the reader's convenience.

Heisenberg algebra. It was shown by Nakajima and Grojnowski that there is an analog of the constructions of this section, in which quiver varieties are replaced by Hilbert schemes $\mathrm{Hilb}^n(X)$, where X is a smooth quasiprojective surface. Namely, one can define homology classes

$$P_\alpha^i \in H_{mid}^{\mathrm{BM}}(\mathrm{Hilb}^n(X) \times \mathrm{Hilb}^{n+i} X)$$

indexed by $\alpha \in H_*(X)$, such that these classes (with respect to convolution) satisfy the relations of the (super) Heisenberg algebra $V \otimes \mathbb{C}[t, t^{-1}] \oplus \mathbb{C}K$, generated by the vector space $V = H_*(X)$. Moreover, $\bigoplus_i H_*^{\mathrm{BM}}(\mathrm{Hilb}^i(X))$ is a Fock module over this Heisenberg algebra.

In particular, taking $X = \mathbb{C}^2$ (in which case the Hilbert scheme is exactly the quiver variety for the Jordan quiver), and α being the generator of $H_0(\mathbb{C}^2)$, we get a geometric realization of the simplest Heisenberg algebra $\mathbb{C}[t, t^{-1}] \oplus \mathbb{C}K$. For more information, see [**Gro1996**] and [**Nak1999**, Chapter 8].

Quantum affine algebras. Note that the construction of the universal enveloping algebra of a Kac–Moody algebra $\mathfrak{g}(Q)$ given in Section 13.4 does not generalize easily to quantum univeral enveloping algebras (unlike the Hall algebra constructions discussed in Chapter 4).

Instead, it was shown by Nakajima that if we replace cohomology by equivariant K-theory, then we get the quantized universal enveloping algebra $U_q\widehat{\mathfrak{g}}$ of the *affinization* $\widehat{\mathfrak{g}}$ of $\mathfrak{g}(Q)$. Details of this construction can be found in [**Nak2001a**].

Appendix A

Kac–Moody Algebras and Weyl Groups

In this section we give a brief summary of the main definitions and results about Kac–Moody algebras and Weyl groups. We give no proofs, referring the reader to monographs such as [**Kac1990**] and [**Hum1990**] for proofs and details.

A.1. Cartan matrices and root lattices

Definition A.1. Given a finite set I, an integer matrix $C = (c_{ij})_{i,j \in I}$ is called a *generalized Cartan matrix* if the following conditions hold:

(1) $c_{ii} = 2$.
(2) $c_{ij} \leq 0$ for $i \neq j$.
(3) $c_{ij} = 0$ iff $c_{ji} = 0$.

A generalized Cartan matrix is called decomposable if there exists a decomposition $I = I_1 \sqcup I_2$ such that $C = C_1 \oplus C_2$, with $C_1 = (c_{ij})_{i,j \in I_1}$, and similarly for C_2. Otherwise, C is called *indecomposable*.

Example A.2. Let Q be a finite graph with the set of vertices I. Assume that Q has no edge loops. Let the matrix A be defined by

$$A_{ij} = \text{number of edges between } i, j.$$

Then the matrix $C_Q = 2 - A$ is a generalized Cartan matrix; moreover, it is symmetric. It is easy to see that in fact any symmetric generalized Cartan matrix can be obtained in this way.

The matrix C_Q is indecomposable if and only if Q is connected.

Given a generalized Cartan matrix C, we define the corresponding root lattice as the free abelian group with set of generators indexed by I:

(A.1) $$L = \bigoplus_{i \in I} \mathbb{Z}\alpha_i.$$

Elements $\alpha_i, i \in I$, will be called *simple roots*.

We denote by L_+ the positive cone:

$$L_+ = \Big\{\sum n_i \alpha_i \in L \mid n_i \geq 0\Big\}.$$

We write

(A.2) $$\alpha \geqslant \beta \text{ if } \alpha - \beta \in L_+.$$

We will also write $L_- = -L_+$.

Finally, let us consider the real vector space spanned by simple roots:

$$L_\mathbb{R} = L \otimes_\mathbb{Z} \mathbb{R}.$$

Definition A.3. A *realization* of a Cartan matrix C is a real vector space $\mathfrak{h}_\mathbb{R}$ and collections of vectors $\alpha_i^\vee \in \mathfrak{h}_\mathbb{R}$, $\alpha_i \in \mathfrak{h}_\mathbb{R}^*$ such that:

(1) $\langle \alpha_i^\vee, \alpha_j \rangle = c_{ij}$.
(2) α_i are linearly independent.
(3) $\alpha_i^\vee$ are linearly independent.
(4) $\dim \mathfrak{h}_\mathbb{R} = |I| + \operatorname{corank}(C)$, where $\operatorname{corank}(C) = \dim \operatorname{Ker} C$.

Elements α_i are called *simple roots*, and $\alpha_i^\vee$ are called *simple coroots*.

It is easy to show that a realization exists: for example, one can take $\mathfrak{h}_\mathbb{R}^* = L_\mathbb{R} \oplus (\operatorname{Ker} C)^*$. Moreover, a realization is unique up to an isomorphism.

From now on, we assume that we have chosen a realization $\mathfrak{h}_\mathbb{R}$. We will denote by $\mathfrak{h}$ the complexification of $\mathfrak{h}_\mathbb{R}$:

$$\mathfrak{h} = \mathfrak{h}_\mathbb{R} \otimes_\mathbb{R} \mathbb{C}.$$

A.2. Weight lattice

Choose elements $\Lambda_i \in \mathfrak{h}_\mathbb{R}^*$, $i \in I$, so that

(A.3) $$\langle \Lambda_i, \alpha_j^\vee \rangle = \delta_{ij}.$$

Elements Λ_i will be called *fundamental weights*.

Definition A.4. The weight lattice $P \subset \mathfrak{h}_\mathbb{R}^*$ is the lattice generated by fundamental weights Λ_i and simple roots α_j.

Elements $\lambda \in P$ will be called *integral weights*.

If matrix C is nondegenerate, then (A.3) uniquely determines Λ_i; in this case, P is the lattice generated by Λ_i. If C is degenerate, both of these statements fail. However, all of the constructions in this book are independent of the choice of fundamental weights Λ_i.

The following lemma is immediate from the definitions.

Lemma A.5.

(1) P is a lattice of maximal rank in $\mathfrak{h}^*_\mathbb{R}$.

(2) For any weight $\lambda \in P$ and a coroot $\alpha_i^\vee$, we have
$$\langle \alpha_i^\vee, \lambda \rangle \in \mathbb{Z}.$$

A.3. Bilinear form and classification of Cartan matrices

Definition A.6. A generalized Cartan matrix C is called symmetrizable if there exists a diagonal matrix D with entries $d_i \in \mathbb{Q}$, $d_i > 0$, such that
$$C = DB$$
for some symmetric matrix B.

In particular, this is automatically so for Cartan matrices C_Q defined by graphs: each such Cartan matrix is symmetric, so we can take $d_i = 1$.

Lemma A.7. *Given a symmetrizable generalized Cartan matrix $C = DB$ and a realization $\mathfrak{h}_\mathbb{R}$, there exists a symmetric bilinear form $(\,,\,)$ on $\mathfrak{h}^*_\mathbb{R}$ such that:*

(1) $(\alpha_i, x) = \langle \alpha_i^\vee, x \rangle / d_i$ *for any $x \in \mathfrak{h}^*_\mathbb{R}$.*

(2) $(\,,\,)$ *is nondegenerate.*

Moreover, this form is unique up to an isomorphism: if $(\,,\,)$, $(\,,\,)'$ are two forms satisfying the conditions above, then $(x, y)' = (\varphi(x), \varphi(y))$ for some isomorphism $\varphi \colon \mathfrak{h}^*_\mathbb{R} \to \mathfrak{h}^*_\mathbb{R}$ such that $\varphi(\alpha_i) = \alpha_i$.

It is immediate from the definition that we have
$$(\alpha_i, \alpha_j) = c_{ij}/d_i = b_{ij}.$$
In particular, $(\alpha_i, \alpha_i) = 2/d_i$, so we have

(A.4) $$c_{ij} = d_i(\alpha_i, \alpha_j) = \frac{2(\alpha_i, \alpha_j)}{(\alpha_i, \alpha_i)}.$$

For example, if $C = C_Q$ is the generalized Cartan matrix C_Q associated with the graph Q, then it is symmetric, so we can take $D = 1$. In this case, the restriction of the bilinear form $(\,,\,)$ to $L_\mathbb{R}$ is exactly the symmetrized Euler form defined in Section 1.5.

Using this, we can give a classification of generalized Cartan matrices. The following result can be found in [**Kac1990**, Theorem 4.3].

Theorem A.8. *Let C be an indecomposable generalized Cartan matrix. Then exactly one of the following three possibilities holds:*

(Fin) *C is symmetrizable and positive-definite: $C = DB$ for some symmetric positive definite matrix B. In this case $\operatorname{corank}(C) = 0$ and the bilinear form $(\,,\,)$ on $\mathfrak{h}^*_{\mathbb{R}} = L_{\mathbb{R}}$ is positive definite. Such matrices will be called Cartan matrices of finite type.*

(Aff) *C is symmetrizable and positive semidefinite: $C = DB$ for some symmetric positive semidefinite matrix B, with $\operatorname{corank}(B) = 1$. In this case, the restriction of the bilinear form $(\,,\,)$ to $L_{\mathbb{R}}$ is positive semidefinite. Such matrices will be called generalized Cartan matrices of affine type.*

(Ind) *There exists a vector $\mathbf{v} \in L$, $\mathbf{v} > 0$ such that $C\mathbf{v} < 0$, and for a vector $\mathbf{v} \in L_+$, $C\mathbf{v} \in L_+$ iff $\mathbf{v} = 0$. Such matrices will be called generalized Cartan matrices of indefinite type; they are not necessarily symmetrizable.*

It is easy to see that if $C = C_Q$ is symmetric, then it is of finite (respectively, affine) type if and only if the graph Q is Dynkin (respectively, Euclidean). More generally, generalized Cartan matrices of finite and affine type (not necessarily symmetric) can be classified using so-called Dynkin diagrams. For finite type, this classification goes back to Cartan; for affine type, it is due to Kac. For symmetric generalized Cartan matrices, this classification coincides with the classification of Dynkin and Euclidean graphs which was given in Section 1.6.

A.4. Weyl group

Throughout this section, we assume that C is an indecomposable generalized Cartan matrix. We also assume that C is symmetrizable and fix a choice of symmetrization $C = DB$. We denote by $(\,,\,)$ the bilinear form defined in Lemma A.7.

Definition A.9. The Weyl group W of C is the subgroup in $\operatorname{GL}(\mathfrak{h}^*_{\mathbb{R}})$ generated by operators $s_i, i \in I$, which are defined by

(A.5)
$$s_i \colon \mathfrak{h}^*_{\mathbb{R}} \to \mathfrak{h}^*_{\mathbb{R}}$$
$$\alpha \mapsto \alpha - \langle \alpha_i^\vee, \alpha \rangle \alpha_i.$$

Operators $s_i, i \in I$, will be called *simple reflections*.

A.5. Kac–Moody algebra

Example A.10. Let C be the Cartan matrix associated with the Dynkin graph Q of type A_n (see Theorem 1.28). Then $\mathfrak{h}_{\mathbb{R}}^* = L_{\mathbb{R}} = \{\mathbf{x} \in \mathbb{R}^{n+1} \mid \sum x_i = 0\}$, the simple roots are given by $\alpha_i = e_i - e_{i+1}$, $i = 1, \ldots, n$, and the simple reflections are given by $s_i(x_1, \ldots, x_{n+1}) = (x_1, \ldots, x_{i+1}, x_i, \ldots, x_{n+1})$. Thus, in this case $W = S_{n+1}$ is the symmetric group.

The following properties of s_i are immediate from the definition:

(1) Each s_i is an involution: $s_i^2 = \mathrm{id}$.

(2) Action of W preserves the root lattice L and the weight lattice P.

(3) Action of W preserves the bilinear form $(\,,\,)$.

The following theorem can be found in [**Kac1990**, Proposition 3.13, Proposition 4.9].

Theorem A.11. *Let C be a symmetric generalized Cartan matrix and let W be the corresponding Weyl group. Then*

(1) *The group W is a Coxeter group, i.e. it can be described as the group with generators s_i and relations*

(A.6)
$$s_i^2 = 1,$$
$$(s_i s_j)^{m_{ij}} = 1, \qquad i \neq j,$$

where the number m_{ij} depends on the product of entries $c_{ij}c_{ji}$ of the Cartan matrix as follows:

$c_{ij}c_{ji}$	0	1	2	3	≥ 4
m_{ij}	2	3	4	6	∞

(2) *The group W is finite if and only if C is of finite type.*

A.5. Kac–Moody algebra

As before, we assume that C is a generalized Cartan matrix and $\mathfrak{h}$ is a realization of it as defined in Section A.1.

Definition A.12. The Kac–Moody Lie algebra $\mathfrak{g}(C)$ associated with C is the complex Lie algebra with generators

$$e_i, i \in I,$$
$$f_i, i \in I,$$
$$h, h \in \mathfrak{h},$$

and relations

(A.7) $\quad [h_1, h_2] = 0, \quad h_1, h_2 \in \mathfrak{h},$

(A.8) $\quad [h, e_i] = \langle h, \alpha_i \rangle e_i,$

(A.9) $\quad [h, f_i] = -\langle h, \alpha_i \rangle f_i,$

(A.10) $\quad [e_i, f_j] = \delta_{ij} \alpha_i^\vee,$

(A.11) $\quad (\operatorname{ad} e_i)^{1-c_{ij}} e_j = 0,$

(A.12) $\quad (\operatorname{ad} f_i)^{1-c_{ij}} f_j = 0.$

Relations (A.11), (A.12) are known as *Serre relations*. In particular, if $c_{ij} = 0$ they imply that $[e_i, e_j] = 0$; if $c_{ij} = -1$, then Serre relations can be rewritten as

(A.13) $\quad [e_i, [e_i, e_j]] = 0,$

(A.14) $\quad [f_i, [f_i, f_j]] = 0.$

We will frequently use the following decomposition, sometimes referred to as *polarization* of $\mathfrak{g}$.

Lemma A.13. *One has a decomposition*

$$\mathfrak{g} = \mathfrak{n}_- \oplus \mathfrak{h} \oplus \mathfrak{n}_+,$$

where $\mathfrak{n}_+$ is the subalgebra generated by e_i; it can be described as the algebra with generators e_i and relations (A.11). *Similarly, $\mathfrak{n}_-$ is the algebra generated by f_i with relations* (A.12)

Theorem A.14. *The Kac–Moody algebra $\mathfrak{g}(C)$ is finite-dimensional if and only if C is of finite type; in this case, $\mathfrak{g}(C)$ is semisimple. Conversely, every finite-dimensional semisimple Lie algebra over $\mathbb{C}$ can be written in the form $\mathfrak{g}(C)$ for some Cartan matrix C of finite type.*

Example A.15. Let C be the Cartan matrix of the Dynkin graph A_n. Then $\mathfrak{g}(C) \simeq \mathfrak{sl}(n+1, \mathbb{C})$ and the subalgebras $\mathfrak{n}_\pm$ are the subalgebras of upper- (respectively, lower-) triangular matrices.

A.6. Root system

Let C be a generalized Cartan matrix and let $\mathfrak{g}(C)$ be the corresponding Kac–Moody algebra.

Theorem A.16. *One has a root decomposition*

$$\mathfrak{g} = \mathfrak{h} \oplus \Big(\bigoplus_{\alpha \in L - \{0\}} \mathfrak{g}_\alpha \Big),$$

$$\mathfrak{g}_\alpha = \{x \in \mathfrak{g} \mid [h, x] = \langle \alpha, h \rangle x \text{ for all } h \in \mathfrak{h}\}.$$

A.6. Root system

Definition A.17. The set
$$R = \{\alpha \in L, \alpha \neq 0 \mid \mathfrak{g}_\alpha \neq 0\}$$
is called the *root system* of $\mathfrak{g}$. Elements $\alpha \in R$ are called roots; for every root, the number $m_\alpha = \dim \mathfrak{g}_\alpha$ is called the root multiplicity.

It follows from Lemma A.13 that we have a decomposition
$$R = R_+ \sqcup R_-, \qquad R_\pm = R \cap L_\pm.$$
Roots $\alpha \in R_+$ are called *positive roots* and $\alpha \in R_-$ are called *negative*.

Describing explicitly the root system and root multiplicities is a difficult problem. There is one part, however, which is easier to describe. Namely, it can be shown that the root system (and root multiplicities) are invariant under the action of the Weyl group W. Since each simple root α_i is a root with multiplicity one (the corresponding root space is spanned by e_i), each element of the form $w\alpha_i$, $w \in W$, is also a root. Such roots are called *real roots*:
$$R^{re} = W\Pi, \qquad \Pi = \{\alpha_1, \ldots, \alpha_n\}.$$
Note that it is immediate from the definition that if $\alpha = w\alpha_i$ is a real root, then $(\alpha, \alpha) = (\alpha_i, \alpha_i) > 0$.

Roots which are not real are called *imaginary*. We denote the set of imaginary roots by R^{im}; thus,
$$R = R^{re} \sqcup R^{im}.$$
It is immediate from the definition that we have polarizations
$$R^{re} = R^{re}_+ \sqcup R^{re}_-, \qquad R^{re}_\pm = R^{re} \cap L_\pm,$$
and similarly for imaginary roots.

The following theorem gives an explicit description of all the roots in the finite and affine cases. For simplicity, we only formulate it for symmetric Cartan matrices.

Theorem A.18. *Let C be a symmetric generalized Cartan matrix.*

(1) *If C is of finite type, then there are no imaginary roots, and*
$$R = R^{re} = \{\alpha \in L \mid (\alpha, \alpha) = 2\}.$$

(2) *If C is of affine type, then*
$$R^{im} = \{n\delta, n \in \mathbb{Z} - \{0\}\},$$
$$R^{re} = \{\alpha \in L \mid (\alpha, \alpha) = 2\},$$
where $\delta \in L$ is the generator of the kernel of C (see Theorem 1.28).

(3) *If C is of indefinite type, then there exists a root $\alpha \in R$ such that $(\alpha, \alpha) < 0$.*

Corollary A.19. *If C is a symmetric generalized Cartan matrix of finite or affine type, then*

$$R = \{\alpha \in L - \{0\} \mid (\alpha, \alpha) \leq 2\}.$$

Note that this fails for root systems of indefinite type.

Finally, we can give a description of the set of all imaginary roots which works for any generalized Cartan matrix. Namely, consider the set
(A.15)
$$K = \{\alpha \in L_+ - \{0\} \mid \langle \alpha, \alpha_i^\vee \rangle \leq 0 \text{ for all } i \in I, \operatorname{supp}(\alpha) \text{ is connected}\},$$

where for $\alpha = \sum n_i \alpha_i$, we define $\operatorname{supp}(\alpha) = \{i \mid n_i \neq 0\}$.

The following theorem can be found in [**Kac1990**, Theorem 5.4].

Theorem A.20.
$$R_+^{im} = \bigcup_{w \in W} w(K).$$

A.7. Reduced expressions

Recall that, by definition, the Weyl group W is generated by simple reflections s_i.

Definition A.21. An expression

$$w = s_{i_l} \ldots s_{i_1}$$

is called *reduced* if it has the minimal possible length. This length l is called the *length* of w and denoted $l(w)$.

A reduced expression is not unique: for example, in the Weyl group $W = S_3$ of Dynkin graph A_2, we have $s_1 s_2 s_1 = s_2 s_1 s_2$.

Theorem A.22. *Any two reduced expressions for the same element of the Weyl group can be obtained from each other by using Coxeter relations*

$$s_i s_j \cdots = s_j s_i \cdots, \qquad m_{ij} \text{ factors on each side,}$$

where m_{ij} is as in Theorem A.11.

One can also give an equivalent definition of length based on the action of W on the root system R.

Theorem A.23. *Let*
$$w = s_{i_l} \ldots s_{i_1}$$
be a reduced expression for $w \in W$. Define elements $\gamma_1, \ldots, \gamma_l \in L$ by

(A.16)
$$\begin{aligned}
\gamma_1 &= \alpha_{i_1}, \\
\gamma_2 &= s_{i_1}(\alpha_{i_2}), \\
&\vdots \\
\gamma_l &= s_{i_1} s_{i_2} \ldots s_{i_{l-1}}(\alpha_{i_l}).
\end{aligned}$$

Then for each i, γ_i is a real positive root and $w(\gamma_i) \in R_-$. Moreover,
$$\{\alpha \in R_+ \mid w(\alpha) \in R_-\} = \{\gamma_i, \ldots, \gamma_l\}.$$

In particular, this implies that
$$l(w) = |R_+ \cap w^{-1}(R_-)|.$$

A proof of this result can be found in [**Hum1990**, Section 5.6].

A.8. Universal enveloping algebra

Recall that for any Lie algebra $\mathfrak{g}$, its universal enveloping algebra $U\mathfrak{g}$ is the unital associative algebra with generators $x \in \mathfrak{g}$ and relations
$$xy - yx = [x, y], \qquad x, y \in \mathfrak{g}.$$

This algebra is infinite-dimensional; by the Poincaré–Birkhoff–Witt theorem, if $x_1, \ldots, x_d$ is an ordered basis in $\mathfrak{g}$, then monomials
$$X_{\mathbf{n}} = x_1^{n_1} \ldots x_d^{n_d}, \quad \mathbf{n} = (n_1, \ldots, n_d) \in \mathbb{Z}_+^d,$$
form a basis in $U\mathfrak{g}$.

Consider now the case when $\mathfrak{g} = \mathfrak{g}(C)$ is a Kac–Moody algebra associated with a generalized Cartan matrix C. In this case, polarization (A.13) gives the following polarization of $U\mathfrak{g}$:

(A.17)
$$U\mathfrak{g} = U\mathfrak{n}_- \otimes U^0 \otimes U\mathfrak{n}_+,$$

where $U^0 \subset U\mathfrak{g}$ is the subalgebra generated by $h \in \mathfrak{h}$; this subalgebra is commutative. Algebra $U\mathfrak{n}_+$ is generated by e_i, $i \in I$, with Serre relations (A.11); similarly, $U\mathfrak{n}_-$ is generated by f_i, $i \in I$, with Serre relations (A.12). Note that if $c_{ij} = -1$, Serre relations can be rewritten as follows:

(A.18) $$e_i^2 e_j - 2 e_i e_j e_i + e_j e_i^2 = 0,$$
(A.19) $$f_i^2 f_j - 2 f_i f_j f_i + f_j f_i^2 = 0.$$

We define the modified universal enveloping algebra $\tilde{U}\mathfrak{g}(C)$ as the associative algebra given by

(A.20) $$\tilde{U}\mathfrak{g} = U\mathfrak{n}_- \otimes \tilde{U}^0 \otimes U\mathfrak{n}_+,$$

where $\tilde{U}^0 = \bigoplus_{\lambda \in P} \mathbb{C} a_\lambda$, P is the weight lattice. The multiplication in $\tilde{U}$ is defined by

(A.21)
$$a_\lambda a_\mu = \delta_{\lambda\mu} a_\lambda,$$
$$e_i a_\lambda = a_{\lambda+\alpha_i} e_i,$$
$$f_i a_\lambda = a_{\lambda-\alpha_i} f_i,$$
$$(e_i f_k - f_k e_i) a_\lambda = \delta_{ik} \langle \alpha_i^\vee, \lambda \rangle a_\lambda.$$

Note that $\tilde{U}(\mathfrak{g})$ is not unital: the natural candidate for the unit is $\sum_{\lambda \in P} a_\lambda$, which is not in $\tilde{U}$ but in a certain completion of it.

A.9. Representations of Kac–Moody algebras

Let V be a (not necessarily finite-dimensional) representation of a Kac–Moody algebra $\mathfrak{g}$. We say that $v \in V$ has weight $\lambda \in \mathfrak{h}^*$ if for any $h \in \mathfrak{h}$, we have
$$hv = \langle h, \lambda \rangle v.$$

We denote by $V_\lambda \subset V$ the subspace of vectors of weight λ. It immediately follows from the definitions that
$$e_i(V_\lambda) \subset V_{\lambda+\alpha_i},$$
$$f_i(V_\lambda) \subset V_{\lambda-\alpha_i}.$$

Definition A.24. We say that a representation V has a weight decomposition if for every $\lambda \in P$, subspace $V_\lambda \subset V$ is finite-dimensional and
$$V = \bigoplus_{\lambda \in P} V_\lambda.$$

The set
$$P(V) = \{\lambda \in P \mid V_\lambda \neq 0\}$$
is called the set of weights of V.

Every such representation is naturally a module over the modified universal enveloping algebra $\tilde{U}\mathfrak{g}$, with the action of a_λ given by $a_\lambda = \mathrm{id}_{V_\lambda}$, $a_\lambda = 0$ on $V_\mu, \mu \neq \lambda$.

An important class of representations are *highest weight representations*.

Definition A.25. A highest weight representation of $\mathfrak{g}$, with highest weight λ, is a representation generated by one vector $v \in V_\lambda$ such that $e_i v = 0$ for all i.

A.9. Representations of Kac–Moody algebras

It can be shown that any highest weight representation V has a weight decomposition, and
$$P(V) \subset \lambda - L_+ = \{\lambda - \mu, \mu \in L_+\}.$$

Theorem A.26. *For any $\lambda \in P$, there exists a unique irreducible highest weight representation of $\mathfrak{g}$ which has highest weight λ. This representation will be denoted $L(\lambda)$.*

Let us now study the analogs of finite-dimensional representations. Note that a general Kac–Moody Lie algebra might not have any finite-dimensional representations. Instead, we consider so-called integrable representations.

Definition A.27. *A representation with a weight decomposition V of $\mathfrak{g}$ is called integrable if for any i, elements e_i, f_i act locally nilpotently: for any $v \in V$, there exists $N \in \mathbb{Z}_+$ such that $e_i^N v = 0$, $f_i^N v = 0$.*

Theorem A.28. *An irreducible highest weight representation $L(\lambda)$ is integrable iff $\lambda \in P_+$, where*

(A.22) $$P_+ = \{\lambda \in P \mid \langle \lambda, \alpha_i^\vee \rangle \in \mathbb{Z}_+ \text{ for all } i\}.$$

Weights $\lambda \in P_+$ are called dominant.

Moreover, if V is an irreducible integrable representation of $\mathfrak{g}$ such that weights of V are "bounded above":
$$P(V) \subset \bigcup_{\lambda \in M} \left(\lambda - L_+ \right),$$
where $M \subset P$ is some finite set, then V is of the form $L(\lambda)$ for some $\lambda \in P_+$.

If C is of finite type, so that $\mathfrak{g}$ is a semisimple finite-dimensional Lie algebra, then a representation is integrable iff it is finite-dimensional. In this case Theorem A.28 gives a complete classification of irreducible finite-dimensional representations of $\mathfrak{g}$.

Bibliography

[AM1978] R. Abraham and J. E. Marsden, *Foundations of mechanics*, Benjamin/Cummings Publishing Co., Inc., Advanced Book Program, Reading, Mass., 1978. Second edition, revised and enlarged; With the assistance of Tudor Raţiu and Richard Cushman. MR515141 (81e:58025)

[Arn1989] V. I. Arnol'd, *Mathematical methods of classical mechanics*, 2nd ed., Graduate Texts in Mathematics, vol. 60, Springer-Verlag, New York, 1989. Translated from the Russian by K. Vogtmann and A. Weinstein. MR997295 (90c:58046)

[AHDM1978] M. F. Atiyah, N. J. Hitchin, V. G. Drinfel'd, and Yu. I. Manin, *Construction of instantons*, Phys. Lett. A **65** (1978), no. 3, 185–187, DOI 10.1016/0375-9601(78)90141-X. MR598562 (82g:81049)

[ARS1997] M. Auslander, I. Reiten, and S. O. Smalø, *Representation theory of Artin algebras*, Cambridge Studies in Advanced Mathematics, vol. 36, Cambridge University Press, Cambridge, 1997. Corrected reprint of the 1995 original. MR1476671 (98e:16011)

[Bar2000] V. Baranovsky, *Moduli of sheaves on surfaces and action of the oscillator algebra*, J. Differential Geom. **55** (2000), no. 2, 193–227. MR1847311 (2002h:14071)

[Beĭ1978] A. A. Beĭlinson, *Coherent sheaves on $\mathbf{P}^n$ and problems in linear algebra* (Russian), Funktsional. Anal. i Prilozhen. **12** (1978), no. 3, 68–69. MR509388 (80c:14010b)

[Bel2000] G. Belitskii, *Normal forms in matrix spaces*, Integral Equations Operator Theory **38** (2000), no. 3, 251–283, DOI 10.1007/BF01291714. MR1797705 (2002f:15014)

[BGP1973] I. N. Bernstein, I. M. Gelfand, and V. A. Ponomarev, *Coxeter functors, and Gabriel's theorem* (Russian), Uspehi Mat. Nauk **28** (1973), no. 2(170), 19–33. MR0393065 (52 #13876)

[BB1973] A. Białynicki-Birula, *Some theorems on actions of algebraic groups*, Ann. of Math. (2) **98** (1973), 480–497. MR0366940 (51 #3186)

[Boz] T. Bozec, *Quivers with loops and Lagrangian subvarieties*, available at `arXiv:1311.5396`.

[Bra2009] C. Brav, *The projective McKay correspondence*, Int. Math. Res. Not. IMRN **8** (2009), 1355–1387, DOI 10.1093/imrn/rnn160. MR2496767 (2010h:14024)

[Bri2013]   T. Bridgeland, *Quantum groups via Hall algebras of complexes*, Ann. of Math. (2) **177** (2013), no. 2, 739–759, DOI 10.4007/annals.2013.177.2.9. MR3010811

[BKR2001]   T. Bridgeland, A. King, and M. Reid, *The McKay correspondence as an equivalence of derived categories*, J. Amer. Math. Soc. **14** (2001), no. 3, 535–554 (electronic), DOI 10.1090/S0894-0347-01-00368-X. MR1824990

[CG1997]   N. Chriss and V. Ginzburg, *Representation theory and complex geometry*, Birkhäuser Boston Inc., Boston, MA, 1997. MR1433132 (98i:22021)

[Cox1974]   H. S. M. Coxeter, *Regular complex polytopes*, Cambridge University Press, London-New York, 1974. MR0370328 (51 #6555)

[CB1988]   W. W. Crawley-Boevey, *On tame algebras and bocses*, Proc. London Math. Soc. (3) **56** (1988), no. 3, 451–483. MR931510 (89c:16028)

[CB1992]   _____, *Lectures on representations of quivers* (1992), available at http://www1.maths.leeds.ac.uk/~pmtwc.

[CB2001]   _____, *Geometry of the moment map for representations of quivers*, Compositio Math. **126** (2001), no. 3, 257–293, DOI 10.1023/A:1017558904030. MR1834739 (2002g:16021)

[CBVdB2004] W. Crawley-Boevey and M. Van den Bergh, *Absolutely indecomposable representations and Kac-Moody Lie algebras*, Invent. Math. **155** (2004), no. 3, 537–559, DOI 10.1007/s00222-003-0329-0. With an appendix by Hiraku Nakajima. MR2038196 (2004m:17032)

[DR1976]   V. Dlab and C. M. Ringel, *Indecomposable representations of graphs and algebras*, Mem. Amer. Math. Soc. **6** (1976), no. 173, v+57. MR0447344 (56 #5657)

[DR1980]   _____, *The preprojective algebra of a modulated graph*, Representation theory, II (Proc. Second Internat. Conf., Carleton Univ., Ottawa, Ont., 1979), Lecture Notes in Math., vol. 832, Springer, Berlin, 1980, pp. 216–231. MR607155 (83c:16022)

[DK1990]   S. K. Donaldson and P. B. Kronheimer, *The geometry of four-manifolds*, Oxford Mathematical Monographs, The Clarendon Press, Oxford University Press, New York, 1990. Oxford Science Publications. MR1079726 (92a:57036)

[DF1973]   P. Donovan and M. R. Freislich, *The representation theory of finite graphs and associated algebras*, Carleton Mathematical Lecture Notes, vol. 5, Carleton University, Ottawa, Ont., 1973. MR0357233 (50 #9701)

[Drozd1980] Ju. A. Drozd, *Tame and wild matrix problems*, Representation theory, II (Proc. Second Internat. Conf., Carleton Univ., Ottawa, Ont., 1979), Lecture Notes in Math., vol. 832, Springer, Berlin, 1980, pp. 242–258. MR607157 (83b:16024)

[DK2000]   J. J. Duistermaat and J. A. C. Kolk, *Lie groups*, Universitext, Springer-Verlag, Berlin, 2000. MR1738431 (2001j:22008)

[Dur1979]   A. H. Durfee, *Fifteen characterizations of rational double points and simple critical points*, Enseign. Math. (2) **25** (1979), no. 1-2, 131–163. MR543555 (80m:14003)

[DV1934]   P. Du Val, *On isolated singularities of surfaces which do not affect the conditions of adjunction. I, II, III*, Math. Proc. Cambridge Philos. Soc. **30** (1934), 453–459, 460–465, 483–491.

[EG2002]   P. Etingof and V. Ginzburg, *Symplectic reflection algebras, Calogero-Moser space, and deformed Harish-Chandra homomorphism*, Invent. Math. **147** (2002), no. 2, 243–348, DOI 10.1007/s002220100171. MR1881922 (2003b:16021)

[Fog1968]   J. Fogarty, *Algebraic families on an algebraic surface*, Amer. J. Math **90** (1968), 511–521. MR0237496 (38 #5778)

[FK1988]   E. Freitag and R. Kiehl, *Étale cohomology and the Weil conjecture*, Ergebnisse der Mathematik und ihrer Grenzgebiete (3) [Results in Mathematics and Related Areas (3)], vol. 13, Springer-Verlag, Berlin, 1988. Translated from the German by Betty S. Waterhouse and William C. Waterhouse; With an historical introduction by J. A. Dieudonné. MR926276 (89f:14017)

Bibliography

[Fri1983] S. Friedland, *Simultaneous similarity of matrices*, Adv. in Math. **50** (1983), no. 3, 189–265, DOI 10.1016/0001-8708(83)90044-0. MR724475 (86b:14020)

[Ful1993] W. Fulton, *Introduction to toric varieties*, Annals of Mathematics Studies, vol. 131, Princeton University Press, Princeton, NJ, 1993. The William H. Roever Lectures in Geometry. MR1234037

[Gab1972] P. Gabriel, *Unzerlegbare Darstellungen. I* (German, with English summary), Manuscripta Math. **6** (1972), 71–103; correction, ibid. **6** (1972), 309. MR0332887 (48 #11212)

[Gab1973] _____, *Indecomposable representations. II*, Symposia Mathematica, Vol. XI (Convegno di Algebra Commutativa, INDAM, Rome, 1971), Academic Press, London, 1973, pp. 81–104. MR0340377 (49 #5132)

[Gab1980] _____, *Auslander-Reiten sequences and representation-finite algebras*, Representation theory, I (Proc. Workshop, Carleton Univ., Ottawa, Ont., 1979), Lecture Notes in Math., vol. 831, Springer, Berlin, 1980, pp. 1–71. MR607140 (82i:16030)

[Gan1998] F. R. Gantmacher, *The theory of matrices. Vol. 1*, AMS Chelsea Publishing, Providence, RI, 1998. Translated from the Russian by K. A. Hirsch; Reprint of the 1959 translation. MR1657129 (99f:15001)

[GL1987] W. Geigle and H. Lenzing, *A class of weighted projective curves arising in representation theory of finite-dimensional algebras*, (Lambrecht, 1985), Lecture Notes in Math., vol. 1273, Springer, Berlin, 1987, pp. 265–297, DOI 10.1007/BFb0078849. MR915180 (89b:14049)

[GP1969] I. M. Gelfand and V. A. Ponomarev, *Remarks on the classification of a pair of commuting linear transformations in a finite-dimensional space* (Russian), Funkcional. Anal. i Priložen. **3** (1969), no. 4, 81–82. MR0254068 (40 #7279)

[GP1979] _____, *Model algebras and representations of graphs* (Russian), Funktsional. Anal. i Prilozhen. **13** (1979), no. 3, 1–12. MR545362 (82a:16030)

[GM2003] S. I. Gelfand and Y. I. Manin, *Methods of homological algebra*, 2nd ed., Springer Monographs in Mathematics, Springer-Verlag, Berlin, 2003. MR1950475 (2003m:18001)

[Gin1991] V. Ginzburg, *Lagrangian construction of the enveloping algebra $U(\mathrm{sl}_n)$* (English, with French summary), C. R. Acad. Sci. Paris Sér. I Math. **312** (1991), no. 12, 907–912. MR1111326 (92c:17017)

[Gin2012] _____, *Lectures on Nakajima's quiver varieties* (English, with English and French summaries), Geometric methods in representation theory. I, Sémin. Congr., vol. 24, Soc. Math. France, Paris, 2012, pp. 145–219. MR3202703

[GSV1983] G. Gonzalez-Sprinberg and J.-L. Verdier, *Construction géométrique de la correspondance de McKay* (French), Ann. Sci. École Norm. Sup. (4) **16** (1983), no. 3, 409–449 (1984). MR740077 (85k:14019)

[Gre1995] J. A. Green, *Hall algebras, hereditary algebras and quantum groups*, Invent. Math. **120** (1995), no. 2, 361–377. MR1329046 (96c:16016)

[Gro1996] I. Grojnowski, *Instantons and affine algebras. I. The Hilbert scheme and vertex operators*, Math. Res. Lett. **3** (1996), no. 2, 275–291, DOI 10.4310/MRL.1996.v3.n2.a12. MR1386846 (97f:14041)

[Gro1995] A. Grothendieck, *Techniques de construction et théorèmes d'existence en géométrie algébrique. IV. Les schémas de Hilbert* (French), Séminaire Bourbaki, Vol. 6, Soc. Math. France, Paris, 1995, Exp. No. 221, pp. 249–276. MR1611822

[Gur1979] R. M. Guralnick, *A note on pairs of matrices with rank one commutator*, Linear and Multilinear Algebra **8** (1979/80), no. 2, 97–99, DOI 10.1080/03081087908817305. MR552353 (80k:15002)

[Har1977] R. Hartshorne, *Algebraic geometry*, Graduate Texts in Mathematics, vol. 52, Springer-Verlag, New York, 1977. MR0463157 (57 #3116)

[Hau2010] T. Hausel, *Kac's conjecture from Nakajima quiver varieties*, Invent. Math. **181** (2010), no. 1, 21–37, DOI 10.1007/s00222-010-0241-3. MR2651380 (2011d:14033)

[HLRV2013] T. Hausel, E. Letellier, and F. Rodriguez-Villegas, *Positivity for Kac polynomials and DT-invariants of quivers*, Ann. of Math. (2) **177** (2013), no. 3, 1147–1168, DOI 10.4007/annals.2013.177.3.8. MR3034296

[HRV2008] T. Hausel and F. Rodriguez-Villegas, *Mixed Hodge polynomials of character varieties*, Invent. Math. **174** (2008), no. 3, 555–624, DOI 10.1007/s00222-008-0142-x. With an appendix by Nicholas M. Katz. MR2453601 (2010b:14094)

[HKLR1987] N. J. Hitchin, A. Karlhede, U. Lindström, and M. Roček, *Hyper-Kähler metrics and supersymmetry*, Comm. Math. Phys. **108** (1987), no. 4, 535–589. MR877637 (88g:53048)

[Hum1990] J. E. Humphreys, *Reflection groups and Coxeter groups*, Cambridge Studies in Advanced Mathematics, vol. 29, Cambridge University Press, Cambridge, 1990. MR1066460 (92h:20002)

[IN2000] Y. Ito and H. Nakajima, *McKay correspondence and Hilbert schemes in dimension three*, Topology **39** (2000), no. 6, 1155–1191, DOI 10.1016/S0040-9383(99)00003-8. MR1783852 (2001h:14004)

[IN1999] Y. Ito and I. Nakamura, *Hilbert schemes and simple singularities*, New trends in algebraic geometry (Warwick, 1996), London Math. Soc. Lecture Note Ser., vol. 264, Cambridge Univ. Press, Cambridge, 1999, pp. 151–233, DOI 10.1017/CBO9780511721540.008. MR1714824 (2000i:14004)

[Ive1972] B. Iversen, *A fixed point formula for action of tori on algebraic varieties*, Invent. Math. **16** (1972), 229–236. MR0299608 (45 #8656)

[Jan1996] J. C. Jantzen, *Lectures on quantum groups*, Graduate Studies in Mathematics, vol. 6, American Mathematical Society, Providence, RI, 1996. MR1359532 (96m:17029)

[Kac1980] V. G. Kac, *Infinite root systems, representations of graphs and invariant theory*, Invent. Math. **56** (1980), no. 1, 57–92, DOI 10.1007/BF01403155. MR557581 (82j:16050)

[Kac1983] ———, *Root systems, representations of quivers and invariant theory*, Invariant theory (Montecatini, 1982), Lecture Notes in Math., vol. 996, Springer, Berlin, 1983, pp. 74–108, DOI 10.1007/BFb0063236. MR718127 (85j:14088)

[Kac1990] V. G. Kac, *Infinite-dimensional Lie algebras*, 3rd ed., Cambridge University Press, Cambridge, 1990. MR1104219 (92k:17038)

[KV2000] M. Kapranov and E. Vasserot, *Kleinian singularities, derived categories and Hall algebras*, Math. Ann. **316** (2000), no. 3, 565–576, DOI 10.1007/s002080050344. MR1752785 (2001h:14012)

[KS1997] M. Kashiwara and Y. Saito, *Geometric construction of crystal bases*, Duke Math. J. **89** (1997), no. 1, 9–36, DOI 10.1215/S0012-7094-97-08902-X. MR1458969 (99e:17025)

[KS1990] M. Kashiwara and P. Schapira, *Sheaves on manifolds*, Grundlehren der Mathematischen Wissenschaften [Fundamental Principles of Mathematical Sciences], vol. 292, Springer-Verlag, Berlin, 1990. With a chapter in French by Christian Houzel. MR1074006 (92a:58132)

[Kem1978] G. R. Kempf, *Instability in invariant theory*, Ann. of Math. (2) **108** (1978), no. 2, 299–316. MR506989 (80c:20057)

[KN1979] G. Kempf and L. Ness, *The length of vectors in representation spaces*, Algebraic geometry (Proc. Summer Meeting, Univ. Copenhagen, Copenhagen, 1978), Lecture Notes in Math., vol. 732, Springer, Berlin, 1979, pp. 233–243. MR555701 (81i:14032)

[Kin1994]	A. D. King, *Moduli of representations of finite-dimensional algebras*, Quart. J. Math. Oxford Ser. (2) **45** (1994), no. 180, 515–530. MR1315461 (96a:16009)
[Kir2006]	A. Kirillov Jr., *McKay correspondence and equivariant sheaves on $\mathbb{P}^1$* (English, with English and Russian summaries), Mosc. Math. J. **6** (2006), no. 3, 505–529, 587–588. MR2274863 (2008i:14027)
[KT]	A. Kirillov Jr. and J. Thind, *Coxeter elements and periodic Auslander-Reiten quiver*, available at arXiv:math/0703361.
[Kle1884]	F. Klein, *Vorlesungen Über das Ikosaeder und die Auflasung der Gleichungen vom fanften Grade*, Teubner, Leipzig, 1884; English transl., *Lectures on the icosahedron and the solution of equations of the fifth degree*, Second and revised edition, Dover Publications, Inc., New York, N.Y., 1956. Translated into English by George Gavin Morrice. MR0080930 (18,329c).
[Kle2004]	M. Kleiner, *The graded preprojective algebra of a quiver*, Bull. London Math. Soc. **36** (2004), no. 1, 13–22. MR2011973 (2004i:16022)
[KP1985]	H. Kraft and V. L. Popov, *Semisimple group actions on the three-dimensional affine space are linear*, Comment. Math. Helv. **60** (1985), no. 3, 466–479. MR814152 (87a:14039)
[KP1979]	H. Kraft and C. Procesi, *Closures of conjugacy classes of matrices are normal*, Invent. Math. **53** (1979), no. 3, 227–247, DOI 10.1007/BF01389764. MR549399 (80m:14037)
[KR1986]	H. Kraft and Ch. Riedtmann, *Geometry of representations of quivers*, Representations of algebras (Durham, 1985), London Math. Soc. Lecture Note Ser., vol. 116, Cambridge Univ. Press, Cambridge, 1986, pp. 109–145. MR897322 (88k:16028)
[Kro1989]	P. B. Kronheimer, *The construction of ALE spaces as hyper-Kähler quotients*, J. Differential Geom. **29** (1989), no. 3, 665–683. MR992334 (90d:53055)
[KN1990]	P. B. Kronheimer and H. Nakajima, *Yang-Mills instantons on ALE gravitational instantons*, Math. Ann. **288** (1990), no. 2, 263–307, DOI 10.1007/BF01444534. MR1075769 (92e:58038)
[Kuz2007]	A. Kuznetsov, *Quiver varieties and Hilbert schemes* (English, with English and Russian summaries), Mosc. Math. J. **7** (2007), no. 4, 673–697, 767. MR2372209 (2008j:14025)
[Lam1986]	K. Lamotke, *Regular solids and isolated singularities*, Advanced Lectures in Mathematics, Friedr. Vieweg & Sohn, Braunschweig, 1986. MR845275 (88c:32014)
[Lau1981]	G. Laumon, *Comparaison de caractéristiques d'Euler-Poincaré en cohomologie l-adique* (French, with English summary), C. R. Acad. Sci. Paris Sér. I Math. **292** (1981), no. 3, 209–212. MR610321 (82e:14030)
[LBP1990]	L. Le Bruyn and C. Procesi, *Semisimple representations of quivers*, Trans. Amer. Math. Soc. **317** (1990), no. 2, 585–598. MR958897 (90e:16048)
[Leh2004]	M. Lehn, *Lectures on Hilbert schemes*, Algebraic structures and moduli spaces, CRM Proc. Lecture Notes, vol. 38, Amer. Math. Soc., Providence, RI, 2004, pp. 1–30. MR2095898 (2005g:14010)
[LS2010]	A. Licata and A. Savage, *Vertex operators and the geometry of moduli spaces of framed torsion-free sheaves*, Selecta Math. (N.S.) **16** (2010), no. 2, 201–240, DOI 10.1007/s00029-009-0015-1. MR2679481 (2011j:14004)
[Lus1990a]	G. Lusztig, *Canonical bases arising from quantized enveloping algebras*, J. Amer. Math. Soc. **3** (1990), no. 2, 447–498. MR1035415 (90m:17023)
[Lus1990b]	_____, *Canonical bases arising from quantized enveloping algebras. II*, Progr. Theoret. Phys. Suppl. **102** (1990), 175–201 (1991). Common trends in mathematics and quantum field theories (Kyoto, 1990). MR1182165 (93g:17019)
[Lus1991]	_____, *Quivers, perverse sheaves, and quantized enveloping algebras*, J. Amer. Math. Soc. **4** (1991), no. 2, 365–421. MR1088333 (91m:17018)

[Lus1992] _____, *Affine quivers and canonical bases*, Inst. Hautes Études Sci. Publ. Math. **76** (1992), 111–163. MR1215594 (94h:16021)

[Lus1993] _____, *Introduction to quantum groups*, Progress in Mathematics, vol. 110, Birkhäuser Boston Inc., Boston, MA, 1993. MR1227098 (94m:17016)

[Lus1998a] _____, *Canonical bases and Hall algebras*, Representation theories and algebraic geometry (Montreal, PQ, 1997), NATO Adv. Sci. Inst. Ser. C Math. Phys. Sci., vol. 514, Kluwer Acad. Publ., Dordrecht, 1998, pp. 365–399. MR1653038 (2000d:17020)

[Lus1998b] _____, *On quiver varieties*, Adv. Math. **136** (1998), no. 1, 141–182. MR1623674 (2000c:16016)

[Lus2000] _____, *Remarks on quiver varieties*, Duke Math. J. **105** (2000), no. 2, 239–265. MR1793612 (2002b:20072)

[Mac1974] R. D. MacPherson, *Chern classes for singular algebraic varieties*, Ann. of Math. (2) **100** (1974), 423–432. MR0361141 (50 #13587)

[MS1998] D. McDuff and D. Salamon, *Introduction to symplectic topology*, 2nd ed., Oxford Mathematical Monographs, The Clarendon Press Oxford University Press, New York, 1998. MR1698616 (2000g:53098)

[McK1980] J. McKay, *Graphs, singularities, and finite groups*, The Santa Cruz Conference on Finite Groups (Univ. California, Santa Cruz, Calif., 1979), Proc. Sympos. Pure Math., vol. 37, Amer. Math. Soc., Providence, R.I., 1980, pp. 183–186. MR604577 (82e:20014)

[Muk2003] S. Mukai, *An introduction to invariants and moduli*, Cambridge Studies in Advanced Mathematics, vol. 81, Cambridge University Press, Cambridge, 2003. Translated from the 1998 and 2000 Japanese editions by W. M. Oxbury. MR2004218 (2004g:14002)

[MFK1994] D. Mumford, J. Fogarty, and F. Kirwan, *Geometric invariant theory*, 3rd ed., Ergebnisse der Mathematik und ihrer Grenzgebiete (2) [Results in Mathematics and Related Areas (2)], vol. 34, Springer-Verlag, Berlin, 1994. MR1304906 (95m:14012)

[Nak1994] H. Nakajima, *Instantons on ALE spaces, quiver varieties, and Kac-Moody algebras*, Duke Math. J. **76** (1994), no. 2, 365–416. MR1302318 (95i:53051)

[Nak1996] _____, *Varieties associated with quivers*, Representation theory of algebras and related topics (Mexico City, 1994), CMS Conf. Proc., vol. 19, Amer. Math. Soc., Providence, RI, 1996, pp. 139–157. MR1388562 (97m:16022)

[Nak1998] _____, *Quiver varieties and Kac-Moody algebras*, Duke Math. J. **91** (1998), no. 3, 515–560. MR1604167 (99b:17033)

[Nak1999] _____, *Lectures on Hilbert schemes of points on surfaces*, University Lecture Series, vol. 18, American Mathematical Society, Providence, RI, 1999. MR1711344 (2001b:14007)

[Nak2001a] _____, *Quiver varieties and finite-dimensional representations of quantum affine algebras*, J. Amer. Math. Soc. **14** (2001), no. 1, 145–238, DOI 10.1090/S0894-0347-00-00353-2. MR1808477 (2002i:17023)

[Nak2001b] _____, *Quiver varieties and McKay correspondence – Lectures at Hokkaido University, December 2001* (2001), available at http://www.kurims.kyoto-u.ac.jp/~nakajima/TeX/hokkaido.pdf.

[Nak2007] _____, *Sheaves on ALE spaces and quiver varieties* (English, with English and Russian summaries), Mosc. Math. J. **7** (2007), no. 4, 699–722, 767. MR2372210 (2008k:14024)

[Nak2009] _____, *Quiver varieties and branching*, SIGMA Symmetry Integrability Geom. Methods Appl. **5** (2009), Paper 003, 37, DOI 10.3842/SIGMA.2009.003. MR2470410 (2010f:17034)

[Naz1973]   L. A. Nazarova, *Representations of quivers of infinite type* (Russian), Izv. Akad. Nauk SSSR Ser. Mat. **37** (1973), 752–791. MR0338018 (49 #2785)

[New1978]   P. E. Newstead, *Introduction to moduli problems and orbit spaces*, Tata Institute of Fundamental Research Lectures on Mathematics and Physics, vol. 51, Tata Institute of Fundamental Research, Bombay; by the Narosa Publishing House, New Delhi, 1978. MR546290 (81k:14002)

[New2009]   _____, *Geometric invariant theory*, Moduli spaces and vector bundles, London Math. Soc. Lecture Note Ser., vol. 359, Cambridge Univ. Press, Cambridge, 2009, pp. 99–127. MR2537067 (2010m:14059)

[OV1990]   A. L. Onishchik and E. B. Vinberg, *Lie groups and algebraic groups*, Springer Series in Soviet Mathematics, Springer-Verlag, Berlin, 1990. Translated from the Russian and with a preface by D. A. Leites. MR1064110 (91g:22001)

[Rei2008]   M. Reineke, *Moduli of representations of quivers*, Trends in representation theory of algebras and related topics, EMS Ser. Congr. Rep., Eur. Math. Soc., Zürich, 2008, pp. 589–637. MR2484736

[Ric1977]   R. W. Richardson, *Affine coset spaces of reductive algebraic groups*, Bull. London Math. Soc. **9** (1977), no. 1, 38–41. MR0437549 (55 #10473)

[Rie1994]   C. Riedtmann, *Lie algebras generated by indecomposables*, J. Algebra **170** (1994), no. 2, 526–546. MR1302854 (96e:16013)

[Rin1976]   C. M. Ringel, *Representations of K-species and bimodules*, J. Algebra **41** (1976), no. 2, 269–302. MR0422350 (54 #10340)

[Rin1990a]   _____, *Hall algebras*, Topics in algebra, Part 1 (Warsaw, 1988), Banach Center Publ., vol. 26, PWN, Warsaw, 1990, pp. 433–447. MR1171248 (93f:16027)

[Rin1990b]   _____, *Hall algebras and quantum groups*, Invent. Math. **101** (1990), no. 3, 583–591. MR1062796 (91i:16024)

[Rin1990c]   _____, *Hall polynomials for the representation-finite hereditary algebras*, Adv. Math. **84** (1990), no. 2, 137–178. MR1080975 (92e:16010)

[Sai2002]   Y. Saito, *Crystal bases and quiver varieties*, Math. Ann. **324** (2002), no. 4, 675–688, DOI 10.1007/s00208-002-0332-6. MR1942245 (2004a:17023)

[Sch2006]   O. Schiffmann, *Lectures on Hall algebras* (2006), available at `arXiv:math/0611617`.

[Sch1994]   L. Schneps (ed.), *The Grothendieck theory of dessins d'enfants*, London Mathematical Society Lecture Note Series, vol. 200, Cambridge University Press, Cambridge, 1994. Papers from the Conference on Dessins d'Enfant held in Luminy, April 19–24, 1993. MR1305390 (95f:11001)

[Ser2001]   J.-P. Serre, *Complex semisimple Lie algebras*, Springer Monographs in Mathematics, Springer-Verlag, Berlin, 2001. Translated from the French by G. A. Jones; Reprint of the 1987 edition. MR1808366 (2001h:17001)

[Shm2012]   D. A. Shmelkin, *Some remarks on Nakajima's quiver varieties of type A* (English, with English and French summaries), Geometric methods in representation theory. II, Sémin. Congr., vol. 24, Soc. Math. France, Paris, 2012, pp. 419–427. MR3203036

[Slo1980]   P. Slodowy, *Four lectures on simple groups and singularities*, Communications of the Mathematical Institute, Rijksuniversiteit Utrecht, vol. 11, Rijksuniversiteit Utrecht Mathematical Institute, Utrecht, 1980. MR563725 (82b:14002)

[Slo1983]   _____, *Platonic solids, Kleinian singularities, and Lie groups*, Algebraic geometry (Ann Arbor, Mich., 1981), Lecture Notes in Math., vol. 1008, Springer, Berlin, 1983, pp. 102–138, DOI 10.1007/BFb0065703. MR723712 (85f:14037)

[Spr1998]   T. A. Springer, *Linear algebraic groups*, 2nd ed., Progress in Mathematics, vol. 9, Birkhäuser Boston Inc., Boston, MA, 1998. MR1642713 (99h:20075)

[Ste1959]   R. Steinberg, *Finite reflection groups*, Trans. Amer. Math. Soc. **91** (1959), 493–504. MR0106428 (21 #5160)

[Ste1985]   _____, *Finite subgroups of* SU_2, *Dynkin diagrams and affine Coxeter elements*, Pacific J. Math. **118** (1985), no. 2, 587–598. MR789195 (86g:20016)

[VV1999]   M. Varagnolo and E. Vasserot, *On the K-theory of the cyclic quiver variety*, Internat. Math. Res. Notices **18** (1999), 1005–1028, DOI 10.1155/S1073792899000525. MR1722361 (2000m:14011)

Index

Adapted sequence 40
ADHM construction 235
Asymptotically locally Euclidean (ALE)
    spaces 255
Anti-self-dual connection 236
    finite energy 238, 255
Auslander–Reiten quiver 94

Barycentric subdivision 136
Binary polyhedral group 134
Bipartite graph 88
Borel–Moore homology 260

Cartan matrix 20, 273
    decomposable 273
    finite type 276
    affine type 276
    indefinite type 276
    symmetrizable 275
Complexification of a real group 183
Composition algebra 78
Constructible
    subset 61
    function 62
Convolution algebra 262
Coroots 274
Coxeter element 42
    adapted to an orientation 43
    affine 108
Coxeter functors 85
Coxeter group 277
Coxeter number 42

Defect 109

Euler form 15
    symmetrized 16, 21 275
Exceptional fiber 220, 242,
Extending vertex, 107

Flag variety 181, 213
Framing
    of a quiver representation 200
    of a sheaf 232
    of a vector bundle with connection
        239, 256

Gabriel's theorem 32
Generic element 208
GIT quotient 161
    twisted 165
Graded dimension 11
Grothendieck group 11

Hall algebra 48
    of constructible functions 62
    universal 51
Hamiltonian
    action 175
    reduction 177
Hecke correspondence 266
Hereditary category 14
Hilbert scheme 225
Hilbert–Chow morphism 226
Hyperkähler
    manifold 186
    moment map 187
    quotient 187

Instanton 238, 255
   framed 239, 256
Isotropic submanifold 174

Jordan quiver 4

K-group 11
   equivariant 250, 151
Kac conjecture 131
Kac theorem 130
Kac–Moody algebra 277
Kähler manifold 183
Kronecker quiver 4, 111
Kleinian singularity 241

Lagrangian submanifold 174
Longest element in the Weyl group 44

McKay correspondence 141
Moment map 175
   hyperkähler 187
Morphism
   étale 162
Mumford criterion 168

Nilpotent
   cone 181
   element 73

Path in a quiver 7
Path algebra 7
Poincare–Birkhoff–Witt (PBW) basis 60, 281
Poisson structure 172
Preprojective algebra 70, 146

Quiver 3
   cyclic 106
   double 69
   double loop 104
   Dynkin 16, 32
   Euclidean 16
   of finite type 31
   Jordan 4, 227
   Kronecker 4, 118
   tame 103
   wild 104
Quiver variety 205

Real form of an algebraic group 183
Regular polyhedron 134
   spherical 134
Regular point 168, 210

Reductive group 161
Reduced expression, 280
   adapted to $\vec{Q}$ 40
Reflection functors 33
Relative invariants 164
Representation of Kac–Moody Lie algebra
   highest weight 282
   integrable 283
   with weight decomposition 282
Representation of a quiver 3
   framed 200
   indecomposable 8
   injective 15
   projective 11
   preinjective 86, 112
   preprojective 86, 112
   regular 86, 112
   semisimple 8
   simple 8
Representation space 23
Resolution of singularities 169, 181, 227
   crepant 245
   minimal 242, 245
   symplectic 180
Ringel theorem 55
Root 279
   lattice 20, 274
   positive, negative 21, 279
   preinjective 87
   preprojective 87
   real, imaginary 21, 279
   simple 21, 274
   system 21

Schläfli symbol 134
Semisimplification 28
Semistable
   point 165
   representation 193, 198, 207
Serre relations 54, 65 278
Sheaves
   equivariant 146
   torsion 151
   torsion free 230
Simple reflection 21, 276
Sink 32
Slice 89
Source 32
Springer resolution 181
Stability parameter 192

Index

Stable
  point 167
  representation 193, 198, 207
Standard resolution 13
Steinberg variety 264
Symplectic manifold 172

Tits form 16
Translation quiver 88
Tube 116

Universal enveloping algebra 281

Weight
  fundamental 274
  lattice 274
Weyl group 21, 276
  affine 107

Selected Published Titles in This Series

174 Alexander Kirillov Jr., Quiver Representations and Quiver Varieties, 2016
172 Jinho Baik, Percy Deift, and Toufic Suidan, Combinatorics and Random Matrix Theory, 2016
171 Qing Han, Nonlinear Elliptic Equations of the Second Order, 2016
170 Donald Yau, Colored Operads, 2016
169 András Vasy, Partial Differential Equations, 2015
168 Michael Aizenman and Simone Warzel, Random Operators, 2015
167 John C. Neu, Singular Perturbation in the Physical Sciences, 2015
166 Alberto Torchinsky, Problems in Real and Functional Analysis, 2015
165 Joseph J. Rotman, Advanced Modern Algebra: Third Edition, Part 1, 2015
164 Terence Tao, Expansion in Finite Simple Groups of Lie Type, 2015
163 Gérald Tenenbaum, Introduction to Analytic and Probabilistic Number Theory, Third Edition, 2015
162 Firas Rassoul-Agha and Timo Seppäläinen, A Course on Large Deviations with an Introduction to Gibbs Measures, 2015
161 Diane Maclagan and Bernd Sturmfels, Introduction to Tropical Geometry, 2015
160 Marius Overholt, A Course in Analytic Number Theory, 2014
159 John R. Faulkner, The Role of Nonassociative Algebra in Projective Geometry, 2014
158 Fritz Colonius and Wolfgang Kliemann, Dynamical Systems and Linear Algebra, 2014
157 Gerald Teschl, Mathematical Methods in Quantum Mechanics: With Applications to Schrödinger Operators, Second Edition, 2014
156 Markus Haase, Functional Analysis, 2014
155 Emmanuel Kowalski, An Introduction to the Representation Theory of Groups, 2014
154 Wilhelm Schlag, A Course in Complex Analysis and Riemann Surfaces, 2014
153 Terence Tao, Hilbert's Fifth Problem and Related Topics, 2014
152 Gábor Székelyhidi, An Introduction to Extremal Kähler Metrics, 2014
151 Jennifer Schultens, Introduction to 3-Manifolds, 2014
150 Joe Diestel and Angela Spalsbury, The Joys of Haar Measure, 2013
149 Daniel W. Stroock, Mathematics of Probability, 2013
148 Luis Barreira and Yakov Pesin, Introduction to Smooth Ergodic Theory, 2013
147 Xingzhi Zhan, Matrix Theory, 2013
146 Aaron N. Siegel, Combinatorial Game Theory, 2013
145 Charles A. Weibel, The K-book, 2013
144 Shun-Jen Cheng and Weiqiang Wang, Dualities and Representations of Lie Superalgebras, 2012
143 Alberto Bressan, Lecture Notes on Functional Analysis, 2013
142 Terence Tao, Higher Order Fourier Analysis, 2012
141 John B. Conway, A Course in Abstract Analysis, 2012
140 Gerald Teschl, Ordinary Differential Equations and Dynamical Systems, 2012
139 John B. Walsh, Knowing the Odds, 2012
138 Maciej Zworski, Semiclassical Analysis, 2012
137 Luis Barreira and Claudia Valls, Ordinary Differential Equations, 2012
136 Arshak Petrosyan, Henrik Shahgholian, and Nina Uraltseva, Regularity of Free Boundaries in Obstacle-Type Problems, 2012

For a complete list of titles in this series, visit the
AMS Bookstore at www.ams.org/bookstore/gsmseries/.